T0296354

The tiger and the shark

Empirical roots of wave–particle dualism

It is like a struggle between a tiger and a shark,
each is supreme in his own element,
but helpless in that of the other.
J. J. Thomson, 1925

The tiger and the shark

Empirical roots of wave–particle dualism

Bruce R. Wheaton

with a foreword by Thomas S. Kuhn

CAMBRIDGE UNIVERSITY PRESS

CAMBRIDGE UNIVERSITY PRESS
Cambridge, New York, Melbourne, Madrid, Cape Town, Singapore, São Paulo

Cambridge University Press
The Edinburgh Building, Cambridge CB2 8RU, UK

Published in the United States of America by Cambridge University Press, New York

www.cambridge.org
Information on this title: www.cambridge.org/9780521250986

First published 1983
First paperback edition 1991
Reprinted 1992

A catalogue record for this publication is available from the British Library

Library of Congress Cataloguing in Publication data

Wheaton, Bruce R.
The tiger and the shark.
Bibliography: p.
Includes index.
1. Wave-particle duality – History.   2. Radiation – His-
tory.   3. X-rays – History.   4. Gamma rays – History.
I. Title.
QC476.W38W45   1983   539   82-22069

ISBN 978-0-521-25098-6 hardback
ISBN 978-0-521-35892-7 paperback

Transferred to digital printing 2008

For Ariana and Geoffrey;
may they transcend the
indeterminism of their elders

CONTENTS

FOREWORD

The first three decades of the twentieth century embraced two great changes in physical theory: relativity, both general and special, and quantum theory, both the old and the new. Among the innovations that constituted these theories, none was more difficult to accept and assimilate than Einstein's suggestion that light displays particulate properties, especially at high frequencies. Together with the related recognition of the wavelike properties of material particles, Einstein's light-particle theory has proved a fundamental constituent of modern physics, perhaps the single feature that most sharply distinguishes it from the generally Newtonian physics of the preceding three hundred years. But for nearly twenty years after Einstein's proposal in 1905, the concept of light-particles was almost everywhere rejected. Even R. A. Millikan, who in 1914–16 provided the first unequivocal evidence for Einstein's surprising law of photoelectric emission, continued, equally unequivocally, to disdain the light-particle hypothesis from which that law had been derived. Only after 1923, when Compton and Debye independently used the light-particle hypothesis to explain the shift in frequency of scattered x-rays, did more than a very few isolated physicists begin to take seriously the idea that electromagnetic radiation often behaves like particles.

That, in outline, is the way that historians of physics have recently been telling the story of the light-particle hypothesis, and with respect to Einstein it remains very nearly the way the story should be told. But, as Bruce R. Wheaton amply demonstrates in the pages that follow, Einstein's was only one approach to conceiving radiation as particulate. A second, far less well known, was associated with observations on x-rays and γ-rays, both discovered during the decade before Einstein's hypothesis was enunciated and neither unequivocally identified with light for another decade. By

1900, five years after their discovery, x-rays were almost everywhere assumed to be impulses, formed when fast-moving electrons are decelerated at the target electrode of an x-ray tube and thereafter propagated radially from the target through the electromagnetic ether. This in itself, as Dr. Wheaton shows, had a significant and heretofore unappreciated effect on the early accommodation of classical ideas of radiant energy with the new quantum theory. That x-ray energy is transported by impulses required physicists to take explicitly into account the number of pulses that pass as well as their individual energies. This set the stage for a reinterpretation of radiant intensity, one that increasingly diverged from the classical dependence of electromagnetic effect on amplitude toward one in which the temporal impulse width, or later the frequency, was the determining factor.

A similar impulse explanation was widely applied by 1905 to γ-rays as well, with acceleration of an electron ejected from an atom replacing deceleration at the target of an x-ray tube. Supported by existing theories and suggested by a variety of experiences, these explanations were nevertheless difficult to reconcile with two recognized sets of observations on the ionization of gases by the new rays. Those observations were in turn the source of a continuing series of suggestions, often independent both of each other and of Einstein, that x-rays and γ-rays be viewed as particulate.

The first set of observations, readily made but not of great force by themselves, gave rise to what Dr. Wheaton calls the *paradox of quantity.* Passing through a container filled with gas, an advancing impulse should affect all gas atoms equally, but in practice it ionizes only a minuscule portion of them. How can the pulse ionize any atoms if it does not ionize all or most of them? A second set of observations, far harder to establish but also far more difficult to explain away, constitute Dr. Wheaton's *paradox of quality.* An electron torn from an atom in the process of ionization by x-rays carries with it energy of the same order of magnitude as that of the electron decelerated in forming the original impulse. It is as though, at the moment of ionization, all the energy in an isotropically expanding x-ray pulse were suddenly concentrated at the point on the pulse's spherical surface where ionization was to take place. Experimental data from which these paradoxes might have been identified were available in 1900, and more accumulated steadily thereafter. By 1908, at least two well-known experimental physicists, William Henry Bragg and Johannes Stark, had based corpuscular theories of radiation upon them. But such evidence of corpuscular properties impressed few physicists until two additional conditions were fulfilled. Less paradoxical explanations first had to be set aside, for example a variety of hypotheses that derived the energy of an ejected electron not from radiation but from the interior of the atom, radiation serving only to "trigger" emission. In addition, evidence for the paradoxes had to be refined, primarily, Dr. Wheaton shows, by the detailed study of x-ray absorption spectra and the development of techniques for β-ray spectroscopy.

With these new data in hand, it became possible to strike precise balances between the energy lost at the target by the cathrode-ray electron in an x-ray tube, the orbital energies of the interior electrons of the target atom, and the energy hv (and thus the frequency v) of the resulting x-ray. The last of those energies could, in turn, be shown to be equal to the sum of the kinetic energy of an electron ejected during ionization and its orbital energy prior to ejection. In the laboratories where these techniques were developed and these energy balances observed, it was impossible not to notice that, with respect to spatial localization and energetic relations, radiation behaved precisely like particles.

Those conditions were not, however, fulfilled before 1920, and then in only a very few places. One of these few was the private laboratory set up in his Paris home by the French nobleman, Maurice de Broglie. By 1921 de Broglie's laboratory had mapped the x-ray energy levels of many atoms, and during that year an improved β-ray spectrometer was used to study the velocities of the photoelectrons ejected by x-rays of known frequency. Reporting the results to the Third Solvay Congress in April of that year, de Broglie insisted that the radiation "must be corpuscular, or, if it is undulatory, its energy must be concentrated in points on the surface of the wave" (p. 270). The following year, with additional results in hand, de Broglie wrote that the phenomena "present facts in such a way that their qualities are sometimes described in terms of the wave theory, sometimes in terms of the [particulate] emission theory" (p. 277).

Remarks of that sort, both from de Broglie and others, prepared the way for a reevaluation of Einstein's work even before the

particulate interpretation of the Compton effect in 1923. Furthermore, the first major contribution to that reevaluation was made
starting in 1922 by Maurice's younger brother, Louis de Broglie.
Louis had been trained by France's leading theoretical physicist,
Paul Langevin, a man who knew and greatly admired Einstein; he
had also worked closely with Maurice on the interpretation of
laboratory results. Louis de Broglie was strategically placed to
bring together the long-separate theoretical and experimental approaches to the concept of light-particles. In the process he took
major steps toward the emergence in 1926 of Schrödinger's wave
mechanics.

From Wheaton's perspective, the independence of Schrödinger's wave mechanics from the slightly earlier matrix mechanics of Heisenberg can be explained by the different emphasis
each investigator placed on the paradoxes of free radiation. To
Heisenberg, a product of Bohr's school, atomic structure held the
center of interest. To Schrödinger, as before for Einstein and de
Broglie, the nature of light and its behavior in traversing matter
were the most pressing problems. Although theoretical research on
both atoms and light utilized some of the same empirical data –
most notably from x-ray spectroscopy – Einstein and de Broglie
were virtually alone before 1923 in directing it toward a resolution
of the nature of radiation. The development of two independent
but ultimately equivalent forms of the new quantum mechanics
thus arose in part from the lack of interest in the nature of light
shown by most theoretical physicists in the decade before 1923.

The preceding remarks suggest only some main themes of Dr.
Wheaton's narrative. The following pages supply others together
with the rich circumstantial detail, the accounts of false starts and
local triumphs that make history of science a story about people.
The fascination of that detail, much of it here recorded for the first
time, is for me the greatest of the rewards that Wheaton's text
offers readers, but it also qualitatively alters a standard picture of
the development of quantum theory. By emphasizing experiments
and the people who perform them, it redresses the emphasis on
abstract theory characteristic of most previous historical literature
on quantum theory. Simultaneously, by emphasizing x- and γ-ra-
diation, it redresses that literature's special emphasis on optical
spectra. These two changes may, in turn, permit a third. Dr.
Wheaton's is the first account of quantum theory known to me in

which Louis de Broglie appears less as a surprising intruder than as a person with just the background required to play the role for which he is known.

THOMAS S. KUHN

PREFACE

Leopards break into the temple and drink dry the
sacrificial pitchers; this occurs again and again until it
can be predicted, and it becomes part of the ceremony.
F. Kafka, 1935

The first years of this century witnessed the final rejection of
determinism in physical theory; there is no more compelling
example of this than the synthesis forged in the early 1920s between theories of matter and theories of light. The insight of Louis
de Broglie that led to the most complete formulation of wave-
particle dualism was the last act in a series of preliminary attempts
by physicists to resolve paradoxes that had arisen in theories of
radiation following the discovery of x-rays. Historians have not
directed sufficient attention either to radiation theory or to experimental studies of recent physics. In the case at hand, significant
empirical data were recognized to challenge classical radiation
theory long before the theory was successfully modified to agree
with them. The gradual recognition, based on these experimental
results, that internal consistency is unattainable by electromechanical interpretations of radiation forms the subject of this book.

This study grew out of concerns first raised while I was a student
of physics. The inadequacy of most textbook discussions of historical and epistemological issues led me back to the original papers
and then to the literature on history of science. I was fortunate
to be introduced to the latter by John Heilbron, whose critical
approach and demand for clarity in expression showed how insight
can be won in historical analysis of scientific thought. When I
undertook a study of the photoelectric effect – the liberation of
electrons from metals by ultraviolet light[1] – I found that many of
the textbook accounts that had confused me were demonstrably
false when viewed historically.[2]

---

[1] Wheaton, *Photoelectric effect* (1971). Abbreviations used in the footnotes and
the bibliography are identified in the notes on sources.
[2] For example, see Eisberg, *Fundamentals* (1961), 76–81; Jammer, *Conceptual
development* (1966), 35.

After an invaluable exposure to the modern descendant of the European culture within which these issues developed, I was again fortunate to have the opportunity to work on the more general historical problem of wave–particle dualism with Thomas Kuhn as guide. When I extended my original concerns to include high-frequency radiations, I found that the experimental evidence amassed by 1911 concerning x-rays played a far more significant role in preparing physicists to accept dualistic theory than did evidence regarding ordinary light. My doctoral dissertation formed the second stage in the development of this book.[3] In it I clarify the extent to which the then standard impulse interpretation of x-rays led to an implicit tension of a kind destined to bring reformulation of classical concepts regarding the distribution of energy in all forms of radiation in the 1920s.

But the story does not end there, and in this book I bring the discussion to its historical conclusion. The results of attempts to force consistency on electromechanical interpretations of radiation rebounded to affect also the theory of matter. While reconsidering Einstein's lightquantum hypothesis in 1921, Louis de Broglie tried to find a theory of light that would combine the macroscopically incompatible representations of wave and particle. He based this work on the solid foundation provided by his elder brother's experimental corroboration of Einstein's photoelectric law for x-rays. The result in 1923 was a hypothetical synthesis of matter and light: Each was to be considered to possess both particle and wave attributes that make their presence felt to a greater or lesser extent according to experimental conditions. The successful elaboration of this remarkable hypothesis signaled the end of strictly deterministic representations of both matter and of light.

The reader may wonder why the name of Albert Einstein does not figure more heavily in this account of evolving understanding that neither a wave nor a particle characterization of radiation is alone sufficient. After all, Einstein introduced the lightquantum hypothesis in 1905 and early on recognized that a synthetic theory was needed. The reasons are straightforward. Einstein's revolutionary hypothesis had almost no followers and very little influence before 1921, and the growth of its acceptance by others marks the limit of our direct concerns. His lightquantum was not initially intended to apply either to x-rays or to γ-rays, but rather only to

[3] Wheaton, *Nature of x- and gamma rays* (1978).

visible and ultraviolet light. Before 1911 it was not clear to most physicists that x-rays and visible light are common species of radiation. Consequently the experimental evidence that had accumulated by that time that x-rays and γ-rays *do* transfer energy in individual units did not justify Einstein's lightquantum in the eyes of most physicists. The difficulties of any corpuscular theory in explaining interference properties of radiation were too formidable. And for almost a decade following Niels Bohr's quantum theory of the atom in 1913, most mathematical physicists showed little interest in the paradoxes that complicated a consistent electromechanical interpretation of free radiation.

Thus, our story developed quite independently of the imaginative and prescient statistical treatment of light of which Einstein was the chief architect. Einstein's own developing realization of the need for duality in radiation theory has been discussed historically by Martin Klein.[4] Here we analyze other physicists' efforts to wrestle with closely related issues, discuss their rejection of and eventual accommodation to dualistic ideas in physical theory, and trace the way in which this development came to alter ideas not simply about radiation and matter but about humans' ability to construct consistent models of physical phenomena based on their experiences in the macroscopic world of human senses.

The present narrative evolved out of an extensive search in the physics literature of the period, all significant results of which appear in the notes. Documents in several manuscript collections, some in archives and others privately held, were invaluable for reconstructing contemporary insights and opinions, for supporting evidence, and for opening new areas for research. In particular, I benefited from examination of the Archive for History of Quantum Physics,[5] the Bohr archive, the Cherwell papers, the Einstein archive, Paul Langevin's papers, the Lorentz papers, holdings of the Nobel archives, the archives of the Paris Academy of Sciences, the Rutherford papers, the Schwarzschild papers, and the archives of the Solvay Institute. Full identifications and locations of these and other collections are given in the notes on sources. I am grateful to these and other repositories for the kind access granted me, and to the holders of literary rights for permission to quote documents. I was greatly assisted in my research by

---

[4] Klein, *Natural philosopher, 3* (1964), 3–49.
[5] Described in Kuhn, Heilbron, Forman, and Allen, *Sources* (1967).

concurrent work on the *Inventory of sources for history of twentieth-century physics,* being led to several important and formerly untapped collections.[6] The immense quantity of data the ISHTCP has compiled and their as yet incomplete description encourage further mining of their riches.

An undertaking of this magnitude is possible only with the help of colleagues and of prior work in the field.[7] I am pleased here to acknowledge debts of both sorts. I am grateful for the comments of several anonymous reviewers of the typescript before publication. I have already mentioned the importance of Martin Klein's work on the early work of Einstein; other studies by him are mentioned in the Notes. The quantum interpretation of the Compton effect in 1922 was an event of great significance; Roger Stuewer's detailed review of the evolution of Compton's approach to it made it unnecessary to recount that episode in depth here.[8] Although Stuewer's and my interpretations differ on the importance of the electromagnetic impulse hypothesis of x-rays and about the renaissance of interest in the lightquantum, his study is a significant addition to the meager literature on the development of experimental physics in this century.[9]

I have been fortunate, professionally and personally, in my own introduction to history of science to have had as guides both Thomas Kuhn and John Heilbron. I have benefited greatly from discussions with both on many aspects of physical theory. Professor Kuhn's reassessment of the conceptual origins of quantum theory has influenced this study in many ways.[10] Only one who has had the benefit of his insightful criticism and sympathetic ear can understand how much of a debt I owe him. His conceptualization of scientific advance encourages historical analysis within which the humanistic aspects of science are no longer submerged in mythic objectivity. We are all in his debt for that insight.

---

[6] This inventory identifies the location, author, recipient, and approximate date of half a million letters from or to physicists active between 1896 and 1952. It will soon be published in microfiche/book form. Contact Springer Verlag, New York for information on ordering.

[7] For background on the material basis of national style in physics, see Forman, Heilbron, and Weart, *HSPS,* 5 (1975), 1–185.

[8] Stuewer, *Compton effect* (1975).

[9] For the neglect of experimental physics, see Heilbron and Wheaton, *Literature* (1981).

[10] Kuhn, *Black-body theory* (1978).

I would be pleased if this study of radiation theory were taken as complementary to John Heilbron's analysis of concurrent developments in atomic theory.[11] Matter and light are the two perceivable manifestations of nature. Their interaction provided the evidence that showed that electromechanical representations of nature would fail. But although the classical–quantum compromise inherent in the Bohr theory of the atom acted to delay recognition of failure for the electromechanical atom until the early 1920s, the paradoxical behavior of radiation was evident to some as early as 1908. The problems that led to the rejection of deterministic physical theory presented themselves early, and most clearly, in attempts to bring consistency to the theory of radiation.

BRUCE R. WHEATON

[11] Heilbron, *Atomic structure* (1964); contribution in *Twentieth century physics* (1977), 40–108.

NOTES ON SOURCES

PUBLISHED SOURCES

All works cited in short form in the footnotes will be found in the bibliography under the author's name, but the footnote references are intended to be sufficient to lay hands on the work itself. In general, and following Library of Congress practice, names of journals are listed under the name of the issuing organization. Because many are cited frequently I have used the following abbreviations:

AHES	*Archive for History of Exact Sciences*
AJP	*American Journal of Physics*
AJS	*American Journal of Science*
Amsterdam	Koninklijke Akademie van Wetenschappen te Amsterdam. Wis- en Natuurkundige Afdeeling
AP	*Annalen der Physik*
ASPN Genève	*Bibliothèque universelle. Archives des sciences physique et naturelle, Genève*
AusAAS	Australasian Association for the Advancement of Science
BAAS	British Association for the Advancement of Science
Berlin *Mb, Sb*	Königliche preussische Akademie der Wissenschaften, Berlin. Physikalisch–mathematische Klasse, *Monatsberichte, Sitzungsberichte*
BJHS	*British Journal for the History of Science*
Bologna	Reale Accademia della Scienze dell'Istituto di Bologna
CR	Academie des Sciences, Paris. *Comptes rendus hebdomadaires des séances*
DSB	*Dictionary of Scientific Biography*, 14 vols., C. C. Gillispie, ed. (New York, 1970–6)

GDNA	Gesellschaft Deutscher Naturforscher und Ärzte
Göttingen	Königliche Gesellschaft der Wissenschaften zu Göttingen
Halle	Naturforschende Gesellschaft zu Halle
Heidelberg *Sb*	Heidelberger Akademie der Wissenschaften, *Sitzungsberichte* (listed by Abhandlung number)
HSPS	*Historical Studies in the Physical Sciences*
JFI	Franklin Institute, *Journal*
JHI	*Journal of the History of Ideas*
JP	*Journal de physique et le radium*
JRE	*Jahrbuch der Radioaktivität und Elektronik*
Königsberg	Physikalisch- ökonomisch Gesellschaft zu Königsberg
Lincei	Reale Accademia dei Lincei, Rome
Manchester *MP*	Manchester Literary and Philosophical Society, *Memoires and Proceedings*
München *Sb*	Königliche Bayerische Akademie der Wissenschaften zu München, *Sitzungsberichte*
NAS	National Academy of Sciences, Washington, D.C.
NNGC	Nederlandsch natuur- en geneeskundig congres
NRC	National Research Council, Washington, D.C.
PCPS	Cambridge Philosophical Society, *Proceedings*
PM	*Philosophical Magazine*
PPSL	Physical Society of London, *Proceedings*
PR	*Physical Review*
PRI	Royal Institution, *Proceedings*
PRS	RSL, *Proceedings*
PTRS	RSL, *Philosophical Transactions*
PZ	*Physikalische Zeitschrift*
RGS	*Revue général des sciences pures et appliquées*
RDS *Sci trans*	Royal Dublin Society, *Scientific Transactions*
RSL	Royal Society of London
SFP *PVRC*	Société Française de Physique, *Procès-verbaux et résumé des communications*
SHPS	*Studies in History and Philosophy of Science*
TCPS	Cambridge Philosophical Society, *Transactions*
TPRRSSA	Royal Society of South Australia, *Transactions and Proceedings and Report*
VDNA	GDNA, *Verhandlungen*
VDpG	Deutsche physikalische Gesellschaft, *Verhandlungen* (usually bound with *Berichte*)
VI	Victoria Institute
WAS	Washington Academy of Science
Wien *Sb*	Kaiserlich-Königliche Akademie der Wissenschaften,

	Wien. Mathematisch-naturwissenschaftlichen Klasse, *Sitzungsberichte*
Würzburg *Sb*	Physikalisch-medizinische Gesellschaft zu Würzburg, *Sitzungsberichte*
ZP	*Zeitschrift für Physik*
Zürich	Physikalische Gesellschaft zu Zürich

UNPUBLISHED SOURCES

A work of this kind depends on both published and unpublished records of contemporary opinion and results. We are fortunate that the subject at hand developed after the typewriter was invented and before the telephone was widely used. Carbon copies of typed letters double the chance that they will be retained, and the lack of telephone use encourages creation of a written record.

In my research I used many sources of unpublished documents, and I am indebted to several people for permission to quote from them: Victor de Pange for permission to quote Maurice de Broglie, the American Friends of the Hebrew University to quote Albert Einstein, Mme. Luce Langevin to quote Paul Langevin, Dr. Bent Nagel to quote transactions of the Nobel Prize committee, and Ruth Braunizer for permission to quote Erwin Schrödinger. I am grateful to Professor Louis de Broglie, who graciously set aside several hours for a discussion with me about this research. Of course I, not he, am answerable for the interpretation given here.

I am no less grateful to many other libraries and archives in Europe and America for making their collections of correspondence available to me. Consulting them was essential for background information, for leads to new sources, and to ensure that the general picture I painted was true to life. Many of the letters cited here are quoted in part or whole in other published works. References appear in the microfiche index *An inventory of published letters to and from physicists,* 1900–1950, listed in the bibliography under Wheaton and Heilbron (1982). Many letter collections related to modern physics are now described in detail in the *Inventory of sources for history of twentieth-century physics* (a database soon to be published in microfiche/book form by Springer Verlag) including all that are noted below, save the Nobel archives.

AHQP. Archive for history of quantum physics. Berkeley, Copenhagen, Minneapolis, New York, Philadelphia, and Rome. Microfilm references follow the form "*x, y*" where *x* is reel number, *y* is section number. For a description, see Kuhn, Heilbron, Forman, and Allen, *Sources* (1967). Sources marked with a *are available on microfilm at several of these locations

Bohr (Niels) archive.* Niels Bohr Institutet, Copenhagen, Denmark

L. de Broglie dossier. Archives of the Academie des Sciences, Paris

Cherwell (F. A. Lindemann) papers. Nuffield College Library, Oxford, England

Ehrenfest (Paul) archive.* Museum Boerhaave, Leiden, Netherlands

Einstein (Albert) papers. Institute for Advanced Study, Princeton, New Jersey

Langevin (Paul) papers. Formerly privately held in Paris, now at "Fonds des ressources historique Langevin," Ecole superieur de physique et chemie de la ville de Paris, France.

Lorentz (Hendrik) papers.* Algemeen Rijksarchief, the Hague, Netherlands

Nobel archives. Nobelstifting, K. Vetenskapsakademiens Nobelkommitteer, Stockholm, Sweden

Planck (Max) papers.* Staatsbibliothek Preussischer Kulturbesitz, Berlin

Richardson (Owen) papers.* Humanities Research Center, University of Texas at Austin

Rutherford (Ernest) papers. Cambridge University Library, Cambridge, England

Schrödinger (Erwin) papers.* Privately held in Alpbach, Austria

Schwarzschild (Karl) papers.* Niedersächsische Staats- und Universitätsbibliothek, Göttingen

Solvay (Institute) archives. Archives de l'Université Libre, Brussels, Belgium

Sommerfeld (Arnold) papers. Sondersammlung, Deutsches Museum, Munich

1

Introduction

For all practical purposes the wave theory of light is a
certainty.[1]

This is the story of a radical change in humans' concept of light.In
1896 most physicists were convinced that light consisted of wave
disturbances in a medium, the electromagnetic aether. The energy
transported in radiation was thought to propagate spherically
outward from its source and to spread out over successively larger
volumes in space. Thirty years later, a remarkably different concept of light prevailed. Physicists then took seriously the evidence
collected over the preceding two decades showing that the energy
of radiation does not spread in space. Under certain conditions,
light behaves like a stream of particles.

Our subject is more than the story of a shift from one theoretical
explanation of radiation to another. This reconsideration of the
nature of light was a significant event in our scientific understanding of the world. To resolve the paradox that faced them, physicists
rejected the venerable Platonic dictum that the microscopic realm
recapitulates the macroscopic; that laws generalized from the
behavior of objects in the perceivable world may be applied to the
imperceivable one. By 1927 physicists had assigned to all forms of
radiation a curious amalgam of wave and particle behavior. Waves
spread energy over larger and larger volumes of space; particles do
not. Reconciliation of these conflicting properties was possible
only through appeal to an ontology that transcended mechanical
incompatibility.

Nature was declared to be only imperfectly rationalizable in
terms of human experience with macroscopic interactions. The
programmatic goal formulated in the seventeenth century to re-

[1] Hertz, GDNA *Tageblatt* (1890), 144.

duce all physical phenomena to consistent mechanical representations was here recognized to be unattainable. This was realized in part because of paradoxes found in the behavior of radiation, particularly x-rays. This study is therefore intended to clarify the developing understanding of x-rays and related phenomena early in the twentieth century. But it is physicists' growing awareness that electromechanical explanations of radiation would not work in principle that forms the real subject of this book.

STRUCTURE OF THE ARGUMENT

In the earliest period of our concern, analogies were drawn between known properties of light and the growing empirical evidence about new forms of radiation. Shortly after the discovery of x-rays, it was proposed that the new rays were impulses, not periodic waves, propagating in the aether. Empirical corroboration followed, matched by increasing acceptance; 1907 marks the high-water mark of the pulse hypothesis. Contemporary studies on radioactivity contributed a new example of an electromagnetic impulse: When the properties of the γ-rays were closely investigated, they came increasingly to be explained in the same way as were x-rays. X-rays were the spherically propagating impulses due to the deceleration of cathode-ray electrons; γ-rays were the spherically propagating impulses caused by β-ray electrons accelerating from a disintegrating atom.

The second period brought the first clear attempts to formulate a consistent theoretical picture of the process of ionization by x-rays and γ-rays. Problems were soon recognized when seemingly contradictory aspects of x-ray behavior were discovered. Increasing discussion threw doubt on the validity of *any* electromagnetic explanation of x-rays. J. J. Thomson and W. H. Bragg in England, and Wilhelm Wien and Johannes Stark in Germany, each abandoned critical aspects of the impulse hypothesis. The energy that x-rays pass on to atoms was more than could be expected from any spreading impulse. Moreover, only a few atoms passed over by x-rays emitted an electron; and some physicists asked whether the energy of x-radiation was not propagated only into narrow solid angles centered on the source.

The paradoxical behavior of x-rays provoked two different kinds of investigations in the third period. There were experimental studies of x-rays, light, and γ-rays to determine the angular extent of the effective region that acts on individual atoms. The results indicated that the radiations do not propagate their energy isotropically, but instead concentrate it in a cone of limited extent. At the same time, theoretical attempts were made to explain the restricted solid angle in a way that was compatible with electromagnetic theory. The strongest defender of impulse x-rays was Arnold Sommerfeld, who found explanations for some of the difficulties. But Sommerfeld managed this only by stretching the impulse model to extremes, and by 1912 other problems had arisen. It had become clear that x-rays and light are common forms of radiation; consequently, the paradoxes that plagued x-rays applied just as forcefully to ordinary light. The photoelectric effect in particular became a serious problem after 1911, and several unsuccessful attempts were made to find a classical explanation for it. The basic question was: How can periodic light waves apparently concentrate their entire quantity, or quantum, of energy on an individual electron? Sommerfeld had no answer.

The fourth period began with a new discovery that seemed to corroborate both sides in the debates of the preceding years. In 1912 it was found that x-rays can interfere. On the one hand, x-ray spectroscopy provided unquestionable evidence that some x-rays possess a periodic structure and are therefore no different from ordinary light. X-rays from a given metal produce a characteristic pattern of lines in the spectroscope that is entirely equivalent to optical emission spectra. On the other hand, the new spectroscopy supplied precisely the tool needed to confirm the suspicion that x-rays transfer energy to matter only in discrete units. The quantity of energy that they pass to an electron was found to be roughly equivalent to the total quantum of energy in the radiation. This seemed to leave no alternative but to accept the fact that radiation concentrates its energy in specific directions and does not spread that energy to any extent in space. The same techniques that by 1914 showed x-rays and γ-rays to be bona fide waves soon provided the strongest evidence that they, and ordinary light as well, consist of streams of hypothetical *lightquanta,* localized particle-like concentrations of energy in proportion to frequency or impulse width.

The evidence corroborating the failure in principle of a classical representation of radiation had only begun to appear when Niels

Bohr proposed his quantum theory of the atom. The lightquantum hypothesis then won almost no support among those whose major goal was the construction of an internally consistent mathematical representation of nature. Throughout most of the decade 1911–21, mathematical physicists simply avoided the issues that complicated understanding of free radiation. The Bohr theory of the atom provided a promising field of investigation that diverted attention from the paradoxes of radiation. During World War I, interest in these problems was maintained almost exclusively by empiricists, those whose views were derived from experiments on the absorption of radiation. Some of these physicists accepted the lightquantum as a valuable heuristic concept before 1920, but none of them inquired deeply into the structure of radiation itself.

The fifth period began in 1921 when influential members of this small group of empirical physicists began to call for a serious reconsideration of the problematic status of light. They based their concern on the successful corroboration of the quantum relation for the x-ray and γ-ray photoelectric effects. In an environment of increasing acceptance of Einstein's lightquantum, two compelling results were achieved. First, in 1922 Arthur Compton showed that x-rays conserve linear momentum in their interaction with electrons. Were the quantum nature of x-rays assumed, remarkable harmony could be brought to the assembled experimental evidence about such interactions. Second, in 1923 Louis de Broglie forged a connection between the properties of light and those of atoms. Each was to have both wave and particle characteristics; only through the interaction of both properties could the true behavior of either matter or radiation be demonstrated. The final reconciliation of the problems of radiation occurred through appeal to indeterministic theories. Soon thereafter, powerful mathematical techniques were applied by Schrödinger and others to provide a full quantum-mechanical solution to problems with both radiation theory and atomic theory.

The polarization between mathematical and empirical approaches to problems about the nature of free radiation ended at just the time that the quantum theory of the atom confronted equally serious problems. Two separate strains of thought, one concerned with atoms and the other with light, were reunited after a decade of unprofitable isolation. It is no coincidence that this recombination of matter theory and radiation theory produced two essentially independent formulations of the new quantum mechanics. One was addressed to the problems of the atom; the other was a development from ideas already formulated to interpret the absorption of light.

BRITISH MECHANISTIC PHYSICS

Most British physicists in the late nineteenth century interpreted nature as if the laws of ordinary mechanics were valid on the microscopic level. Many thought that the electromagnetic aether could be represented in terms of mechanical models. Their approach was based on the conviction that the laws governing the interaction of perceived objects could be transferred bodily to the actions of molecules and atoms. For some, like Maxwell, this was no slavish dependence on physical models, but rather a conviction that, however accurate the mathematical representation of nature might be, the creative impulses of physicists are best served by an appeal to mechanical representations.[2] The British proclivity to seek mechanical models in physics played an essential role in the early recognition that an electromechanical representation is inherently impossible for light.

For our purposes, the most significant example of this attitude in British physics concerns the study of electric discharges through vacuum. It had been known since midcentury that reproducible displays of light and shadow accompany the discharge of electricity through a tube partially evacuated of air. After the improvement of vacuum pumps in the 1860s, it was possible to study the effects of the discharge after most of the gas had been removed and the normal discharge pattern had entirely disappeared. Under these conditions, some sort of disturbance still traverses the tube. Although the responsible "rays" themselves are invisible, they create a bright glow where they strike the glass walls of the discharge tube. Because they seem to come exclusively from the negative electrode, or cathode, they were called *cathode rays*.[3]

One should not be misled by the denotation "rays." To most British physicists, cathode rays were a stream of "charged material molecules." This interpretation had been offered first by William

[2] Kargon, *JHI, 30* (1969), 423–36.
[3] Cathode rays were discovered by Plücker, *AP, 103* (1858), 88–106; named by Goldstein, Berlin *Mb* (1888), 82–124.

Crookes in 1879 after he showed that the beam is deflected by a magnet, can exert force on an object placed within the tube, and casts sharply defined shadows even when the emitting cathode surface is quite large.[4] The last observation argued against the contention that the rays are emitted isotropically from the cathode, as light would be. There were variations on this theme: Some thought that the particles were complete molecules that had accumulated a net charge; others thought they were the negatively charged products of molecular dissociation. But the majority of British physicists agreed that the cathode rays were material.

Significant research was based on this interpretation. Crookes claimed that he had discovered a "fourth state of matter" that was neither solid, liquid, nor gas because of the extremely low density of the particles. In 1884 Arthur Schuster used the magnetic deviation of the cathode beam to estimate the ratio of charge to mass (e/m) of the presumed particles.[5] As is well known, Joseph John Thomson soon found this ratio to be the same for the beam regardless of the nature of the emitting cathode and the residual gas in the tube. The value of e/m that he found was larger by a factor of a thousand than that of the smallest known atom, and Thomson claimed in 1897 that the cathode beam particles are parts of atoms previously undetected.[6] Thomson's particle is today called the *electron.*

Of most significance in this investigation is Thomson's and Schuster's conviction that the laws of electrostatic deflection, derived from the behavior of macroscopic objects, should necessarily apply to electrons. They were rewarded in this belief by the discovery of new phenomena. Very soon thereafter, Thomson, in particular, encountered perplexing inconsistencies when he tried to apply the same representational standards to the behavior of radiation.

THE ELECTROMAGNETIC WORLD VIEW

Many German physicists regarded the mechanistic bias of British physicists as superficial and potentially misleading. They considered the dependence on mechanical models to be a limitation on

[4] Crookes, *PTRS, 170* (1879), 135–64. Crookes' lecture "On radiant matter" to the BAAS does not appear in the 1879 *Report* but was reprinted privately the same year.

[5] Schuster, *PRS, 57* (1890), 526–59.

[6] Thomson, *PM, 44* (1897), 293–316. See also Topper, *AHES, 7* (1971) 393–410.

the mind rather than an aid to creative thought. The origin of this deeply ingrained cultural tendency in German physics has not been fully explained. One source of influence is assuredly the prevalent atmosphere of idealistic philosophy to which most turn-of-the-century German-speaking physicists were introduced in their formative or university education. Also, a more abstract interpretation of natural law was common in German-speaking countries, one that placed less emphasis on material interpretation of phenomena than was true in Britain.

In part because of this predisposition, German physicists did not accept British theories about cathode rays. Most of them described the rays as a form of light that traverses the tube.[7] The earliest justification for this view was the belief that only light could stimulate fluorescence in matter like that of the discharge tube walls. The magnetic deflection of the beam could be attributed to an abrupt alteration of the refractive index in the residual gas near the magnet. Heinrich Hertz provided the strongest evidence for the German view in 1883 when he was unable to deflect the cathode-ray beam in an electrostatic potential. He also demonstrated that the beam travels straight ahead even if the course of the electrical discharge curves to the side.[8]

There were therefore very good experimental reasons for German physicists to accept the radiant interpretation of cathode rays. Philipp Lenard, for example, based a Nobel-prize-winning series of researches on the aetherial view of cathode rays. He exploited the fact that the rays could pass right through thin metal foils.[9] This too supported the aetherial explanation; solid matter stops streams of molecules, but some forms of matter are transparent to visible or electric waves. Even after Thomson identified the electron by measuring its e/m ratio, some German physicists – both experimentalists and mathematical theorists – persisted in attempts to explain its behavior entirely in terms of the aether.

German mathematical physicists were not deterred by an inability to represent physical concepts mechanically. A school of thought at this time in German-speaking countries sought to reformulate ideas of matter solely in terms of forces in the aether.[10]

---

[7] Lehmann, *Lichterscheinungen* (1898), 518–47.

[8] Hertz, *AP, 19* (1883), 782–816.

[9] Lenard, *AP, 51* (1894), 225–68.

[10] On continental rejection of mechanical models, see Klein, *Centaurus, 17* (1972), 58–82. Joseph Larmor in Britain held views similar to those of many Germans; see his *Aether and matter* (1900).

It was a form of electromagnetic reductionism and was heavily influenced by Kantian principles of epistemology. In a famous talk "On the relations between light and electricity" in 1889, Heinrich Hertz conjured up a vision of this way of thinking. Like the pre-Socratic monists who attempted to find the single principle underlying all apparent phenomena, Hertz claimed that physicists were confronted with the question of "whether everything that is, is fashioned from the aether." The aether offers understanding not only of the imponderables, he said, "but even of the nature of matter itself and its most intimate properties, weight and mass."[11] The forces that define our perception of mass and solidity were to be given higher ontological significance than was the idea of matter.

Not all German-influenced physicists agreed with the fullest formulation of this viewpoint. By the mid-1890s, some success had attended a modified version of the reductionist program. This was the electron theory of the eminent Dutch physicist H. A. Lorentz.[12] According to Lorentz, all interaction between matter and the aether is mediated solely by then-hypothetical electrons. Electrons act like particles, but their mass and momentum are only reflections of the effect that a center of electric force has on the continuum of force that constitutes the aether. The mechanical properties of electrons are due to the reaction between their inherent charge and the surrounding electric and magnetic field. Lorentz' detailed theory was able to account for several perplexing observations, including the failure of interference experiments to find evidence for the relative motion of the earth and the aether.

But other physicists interpreted the issues Hertz raised in an extreme form. Some felt that they stood "on the threshold of a new epoch in physics," one in which suitably defined discontinuities in the continuous electromagnetic field would entirely replace mechanical concepts.[13] Emil Wiechert and Wilhelm Wien, for example, felt that the very concept of electron could be reduced to the idea of a form or structure in the electromagnetic medium.[14] Those who subscribed to this all-encompassing reductionist program

---

[11] Hertz, GDNA *Tageblatt* (1890), 149.

[12] Hirosige, *HSPS, 1* (1969), 151–209.

[13] Quincke, GDNA *Tageblatt* (1890), 149.

[14] Wiechert, Königsberg *Schriften, 35* (1894), 4–11. Wien, *Archives Néerlandaises (2), 5* (1900), 96–107. Abraham, *Elektrizität, 2* (1905), 147ff.

described their goal as the "electromagnetic world picture." To them, only the electromagnetic aether existed. It was all that was necessary to explain the phenomena of the world.[15]

THE WAVE THEORY OF LIGHT

Throughout the second half of the nineteenth century, light was thought to be a wave propagating in an all-pervading medium. Its properties – diffraction, interference, and polarization – convinced its students that visible monochromatic light is a periodic transverse oscillation with a wavelength between about 3×10^{-5} and 7×10^{-5} cm. The origin and development of optical spectroscopy were based on this interpretation; the assignment of a numerical wavelength to each pure color of light brought great advantages in the analysis of terrestrial and astronomical matter. During the nineteenth century, radiations on both sides of the visible spectrum were discovered: the *infrared* and the *ultraviolet* or *chemical* rays, so-called because of their ability to provoke chemical reactions. Like visible light, infrared and ultraviolet light were understood to be periodic transverse waves.

Yet lacking was a physical understanding of the medium of which the undulations constitute light. This omnipresent medium – the aether – had to have seemingly contradictory properties. It had to be extremely rigid to sustain the high-frequency vibrations of light, and yet be sufficiently yielding to allow matter, such as the planets, to pass through it freely. Physicists had attempted to formulate a consistent representation of an elastic solid with such properties. Even if they had succeeded, they would have faced grave problems. The aberration of starlight indicated that the earth moves relative to the aether during its yearly revolution about the sun, but interferometric experiments failed to corroborate this effect. Moreover, if the aether were strictly analogous to an elastic solid, then a longitudinal wave traveling at a greater velocity than the transverse should have been observed. It had not been.

---

[15] McCormmach, *Isis, 61* (1970), 459–97. Hertz himself rejected the electromagnetic force ontology. His *Mechanik* (1894) attributes the forces to insensible particles. McCormmach, *DSB, 6,* 340–50. For representative statements about the electromagnetic view during our general period of concern, see the proposition by Marx, *Grenzen in der Natur* (1908), and the antithesis by Kunz, *Theoretische Physik* (1907).

Maxwell's theoretical formulation of electricity and magnetism in the 1860s offered partial answers to some of the puzzles about the luminiferous medium.[16] He proposed that light is a transverse wave traveling in a continuous medium whose stresses and strains constitute electric and magnetic forces. Like jiggles moving along a rope, oscillations perpendicular to the direction of motion of the wave should occur along the force vectors established by electrified bodies. Maxwell showed that the velocity of these waves would be close to that found for electrical signals in a wire. Even before it received experimental corroboration, the theoretical interpretation that light is a periodic electromagnetic wave provided support for the wave theory of light.

George Gabriel Stokes described in 1884 the "views respecting the nature of light which are at present held, I might say almost universally, in the scientific world."[17] He meant the wave theory of light. He regarded the idea that light is a stream of particles shot from the luminous source as "altogether exploded." "No one who has studied the subject can doubt," he said, that "light really consists of a change of state propagated from point to point in a medium."[18] The colors of thin films and diffraction phenomena show, he said, that periodicity is "naturally, almost inevitably, involved in the fundamental conception."[19]

In Germany, Hermann von Helmholtz began his lectures on the electromagnetic theory with an affirmation of the wave theory. "Hardly a doubt remained about the superiority of the wave theory compared to the emission theory" after polarization was explained by the hypothesis of transverse waves, he claimed; the emission theory is "hardly mentioned anymore."[20] Gustav Kirchhoff based his theory of diffraction in 1882, as Stokes had done before, on the assumption that "light is a transverse undulation in the aether."[21] The leading French mathematical physicist of the time, Henri Poincaré, wrote in 1888 that "of all theories in physics, the least imperfect is that of light based on the work of Fresnel and his successors."[22]

In 1888 Heinrich Hertz completed proofs that "electric waves"

---

[16] Maxwell, *PM, 21* (1861), 161–75, 338–48; *23* (1862), 12–24, 85–95.
[17] Stokes, *Nature of light* (1884), viii.
[18] *Ibid.,* 25.
[19] *Ibid.,* 32.
[20] von Helmholtz, *Vorlesungen* (1897), 3.
[21] Kirchhoff, Berlin *Mb* (1882), 641.
[22] Poincaré, *Théorie mathématique de la lumière, 1* (1889), 1.

radiate from sparks and obey the same laws that govern the behavior of light. They can be refracted and reflected; their wavelength can be measured; their polarization indicates that they are transverse waves; and they propagate at the velocity of light.[23] Maxwell's theory, proposed as an explanation of optics, had achieved a "brilliant victory."[24] The inconsistent properties of the mechanical aether, it seemed, might disappear entirely when reinterpreted on the basis of the electromagnetic continuum supposed by Maxwell's theory.

After Hertz' experiments, physicists' conviction about the wave theory of light became virtual certainty. Hertz provided the modern electromagnetic context for the wave hypothesis and spoke forcibly to German physicists who had found the concept of a mechanical aether alien to their sensibilities. By 1890 most physicists thought that the wave theory of light was as close to certainty as physics could make it. This theory had assumed a status equal to that of the theory of gravitation as an example of a fully corroborated theory in classical physics.

The differences between the British and German approaches to electromagnetic theory did not lead directly to a difference in their respective interpretations of light. Both British and German physicists were convinced that monochromatic light obeys all the laws governing periodic waves. But they did differ on how light waves may be analyzed. Many influential German physicists thought that electromagnetic forces, extending over great distances in space, can, through an appropriate twist or discontinuity, give rise to apparently localized properties in the aether. In Germany, more widely than in Britain, the idea was accepted that a properly selected set of sinusoidal oscillations, if added together, will produce a single localized force. The willingness of some German physicists to combine, or their inability to separate, periodic waves and temporally or spatially localized events in the aether was a crucial advantage early in the century when the behavior of radiation was first interpreted on the basis of quantum principles. And national distinctions did play a significant role in the various hypotheses raised after 1895 when word came from Würzburg about x-rays, an entirely "new type of radiation" that could pass right through matter itself.

[23] Hertz, *AP, 34* (1888), 551–69; Berlin *Sb* (1888), 197–210; *AP, 36* (1889), 769–83.
[24] Hertz, GDNA *Tageblatt* (1890), 148.

PART I
The introduction of temporal discontinuity 1896–1905

2

The electromagnetic impulse hypothesis of x-rays

The Röntgen emanation consists of a vast succession of independent pulses . . .[1]

Between 1898 and 1912, a majority of physicists thought that x-rays were impulses propagating through the electromagnetic field. Only the extremely large number of pulses gave the x-ray beam its seeming continuity. Although this hypothesis was compatible with the wave theory of light, it was a special case of that theory. Impulses are not ordinary waves. Although they propagate spherically outward from their source, pulses are not periodic oscillations. The energy in an impulse is temporally but not spatially localized. It is contained within an ever-expanding shell, but the shell's radial thickness remains constant and small. Along the circumference, energy is distributed uniformly. But radially, from front to back so to speak, electromagnetic energy rises quickly from zero and drops back just as rapidly. When it passes a point in space, an impulse exerts only a single push or a single push – pull. A pulse collides, rather than resonates, with an atom.

In their temporal discontinuity, impulses differ decisively from their periodic-wave cousins. Light has an intrinsic oscillatory character that allows it to interfere; the superposition of two beams of coherent monochromatic light produces alternate regions of constructive and destructive interaction, the well-known interference fringes. A pulse has no oscillatory structure. Its interference properties are qualitatively different from those of light. A truly monochromatic light wave must extend infinitely in time; if it does not, an intrinsic ambiguity arises in the definition of its frequency. A pulse is restricted in temporal extent, and the very concept of

[1] Stokes, *PCPS, 9* (1896), 216.

frequency cannot readily be applied. Recognition of these differences varied according to national styles in research and from physicist to physicist, as we shall see in this chapter.

In a larger context, it should be remembered that the hypothesis of x-ray impulses competed for acceptance with many claims for "new rays." In the first decade of x-ray research, physicists were called upon to pass judgment on numerous explanations for radiant interaction. Some of the newly proposed entities – x-rays, cathode rays, canal rays, and α-, β-, and γ-rays, plus a few radioactive emanations and a bewildering variety of decay products – were later recognized to be legitimate. Others – discharge rays, radioactinic rays, black light, N-rays, Moser rays, spathofluoric rays, and the peculiar light of glowworms – were, at best, special cases of the others and, at worst, totally spurious.[2] No clear distinction could be drawn between the phenomena themselves and the many explanations proposed, nor could one be certain about which new observations were valid at all. In this circumstance, it is hardly surprising that there was early confusion over which of the new rays were particle streams and which were waves in an aether. Impulses came to occupy a position midway between these extremes. A pulse is a spherical wave, but it collides with an atom as if it were a particle. Part of the early appeal of the impulse hypothesis was this ability, chameleonlike, to express the characteristics of both particles and waves.

THE NEW RAYS

Within six months of Wilhelm Conrad Röntgen's announcement of his discovery of x-rays late in 1895, several proposals were heard about the physical nature of the "new type of ray."[3] Suggestions included discharged and uncharged particles,[4] vortices in the

[2] Nye, *HSPS, 11* (1980), 125–56.
[3] Röntgen, Würzburg *Sb* (1895), 132–41; Würzburg *Sb* (1896), 11–16; Berlin *Sb* (1897), 576–92. Reprinted in *AP, 64* (1898), 1–11, 12–17, and 18–37. Future references follow the form "Röntgen I, II, or III," with page references to the *AP* reprinting. The most useful bibliography is Klickstein, *Wilhelm Conrad Röntgen* (1966). Gocht, *Röntgen-Literatur* (1911–25), concentrates on medical and technical issues.
[4] The many claims may be found in Phillips, *Bibliography* (1897). The most influential claimant for uncharged electrons was Bernhard Walter, *AP, 66* (1898), 74–81.

aether,[5] and acoustical[6] or gravitational[7] waves of high frequency. Within a short time, however, the choices were narrowed to three, each a form of electromagnetic disturbance. The first theory, popular in Germany before 1900, stated that x-rays were ordinary light of extremely high frequency. This hyperultraviolet light consisted of periodic transverse oscillations in the electromagnetic field. The second idea, suggested by Röntgen himself, raised the hope that x-rays were the long-sought condensational, or longitudinal, aether wave. The third identified x-rays as temporally discontinuous transverse impulses in the aether.

From the start, most physicists took Röntgen's rays to be a form of electromagnetic wave. In support of this claim, Röntgen provided only three pieces of evidence: (1) The rays, like light, propagate in straight lines. (2) The rays are undeflected by magnetic or electric influence. (3) In contrast to the behavior of cathode rays, x-rays are not absorbed by matter in strict proportion to the density of the absorber.[8] On the other hand, there was strong evidence that x-rays do not behave like ordinary light rays: Röntgen showed that they suffer no measurable reflection, refraction, interference, or polarization. How then can one account for the consensus view that x-rays are waves? The answer lay in the form of the evidence. The overarching exhibit, the technological *fait accompli* that irrevocably cemented Röntgen's name to a phenomenon seen by others prior to 1895, was the photographs. Photography by ordinary light was new enough in 1896 to hold the interest of many physicists. But here was evidence of new light, not visible to the eye, that exposed a photographic plate nonetheless. Technical claims were open to interpretation, but the images of a compass needle with graduations seen through its metal case, of nails in the door frame seen in a photograph taken outside a closed room containing the discharge tube, and, eeriest of all, of the bones inside a living hand were accepted at face value. A new form of light had been recorded.[9]

[5] Michelson, *AJS, 1* (1896), 312–14; *Nature, 54* (1896), 66–7.
[6] Edison, *Electrical engineer, 21* (1896), 353–4.
[7] Lodge, *Electrician, 36* (1896), 438–40, 471–3.
[8] Röntgen I, pp. 3, 5, 6, 8, 9, and 10.
[9] "For me the photography was a means to an end but [the press] made it the most important thing." Röntgen to L. Zehnder, February 1896. Zehnder, *W. C. Röntgen* (1935), 39–44. For a description of the photographs that Röntgen sent, see Schuster, *British medical journal* (18 January 1896), 172–3.

To explain the lack of characteristic radiative behavior, Röntgen proposed tentatively that the rays were longitudinal aether waves, long anticipated on the basis of the elastic solid model of the aether.[10] Röntgen's proposal did not spring *de novo* from his imagination. Indeed, his complete lack of supporting evidence suggests that the source of his confidence lay outside the context of his own investigations. One need not seek far to find it. Both Gustav Kirchhoff and Lord Kelvin had published formulations of electromagnetism that predicted the existence of longitudinal as well as transverse electromagnetic waves.[11] Hermann von Helmholtz and Heinrich Hertz had both toyed with the idea that cathode rays were compression waves,[12] and Gustav Jaumann had developed this concept in a series of papers that predate Röntgen's work.[13] The problems that had complicated a compression-wave model of cathode rays seemed to disappear entirely when x-rays became the subject.

So Röntgen's hypothesis of longitudinal-wave x-rays struck sympathetic chords in representatives of his generation. Ludwig Boltzmann, who had recently analyzed the mechanical properties of the electromagnetic medium, supported Röntgen's idea.[14] He thought the x-rays might be longitudinal waves of sufficiently short period to induce fluorescence but of long enough wavelength to pass easily through matter.[15] An acoustical analogy is helpful here: Long-wavelength sound is less readily absorbed by matter than short-wavelength sound. Henri Poincaré agreed that the lack of refraction cast doubt on any transverse-wave model.[16] Gustav

[10] Röntgen I, p. 11.
[11] [W. Thomson], *Baltimore lectures* (1904), 45, 141. Kirchhoff, Berlin *Mb* (1882), 641–69; *Vorlesungen, 2* (1891), 79–97.
[12] Hertz wrote Helmholtz in 1883 suggesting that cathode rays might be longitudinal waves. Helmholtz replied that he had once thought that cathode rays were electromagnetic waves, likely longitudinal, that propagate from *plötzlichen Stosses*, molecular collisions at the cathode. Helmholtz to Hertz, 29 July 1883. Hertz, *Gesammelte Werke, 3* (1895), xxv–vi.
[13] Jaumann, Wien *Sb, 104:IIa* (1895), 747–92.
[14] Boltzmann, *Elektro-Techniker, 14* (1896), 385–9; *Electrician, 36* (1896), 449. The earlier study: *AP, 48* (1893), 78–99.
[15] The velocity of the compression wave was expected to be greater than that of the transverse wave. Thus, longitudinal waves might possess high frequencies and long wavelengths simultaneously. Both properties help to explain the rays' extraordinary ability to pass through matter.
[16] Poincaré, *RGS, 7* (1896), 55.

Jaumann, who had developed a longitudinal hypothesis for the cathode rays, simply added x-rays to the category.[17] Oliver Lodge "hoped" that the rays were longitudinal oscillations, but he did not think they were electromagnetic. He thought that x-rays might be the gravitational analog to electromagnetic waves: As light is the wave whereby electric and magnetic forces are transmitted, so x-rays might be the wave of gravitational force. Their frequency, he thought, might be as much greater than that of visible light as the frequency of light is above that of acoustical vibration.[18]

Lord Kelvin took the compression-wave hypothesis most seriously. He quickly recognized that partial corroboration of his elastic–solid model of the aether might be forthcoming. He had recently expressed concern that the compression wave had not yet been observed, and had suggested other experimental tests.[19] "There are such waves," he had claimed, "and I believe that the unknown condensational velocity is the velocity of propagation of electrostatic force."[20] Röntgen's rays appeared to hold out new hope, and Kelvin actively encouraged experiments on them.

The critical experiment to disprove Röntgen would have been a demonstration of polarization. But throughout the period 1896–1904, polarization of x-rays remained problematic. Many physicists claimed its demonstration. Karnozhitskiy and Golitsyn in St. Petersburg reported in March 1896 that the transmission of x-rays through tourmaline depends on the relative orientation of the sheets.[21] But the discovery of radioactivity cast doubt on the purity of the x-ray beam that had been used, and little weight was given to the result. J. J. Thomson, in an early experiment, failed to find any effect of tourmaline on x-rays.[22] Some took the lack of polarization evidence as support of the longitudinal hypothesis.

Opinion soon turned against Röntgen's suggestion even without positive evidence. Arthur Schuster pointed out that even transverse waves of sufficiently high frequency would suffer little refrac-

[17] Jaumann, Wien *Sb, 106:IIa* (1897), 533–50.
[18] Lodge, *Electrician, 36* (1896), 438–40.
[19] [W. Thomson], *PRS, 59* (1896), 270–3.
[20] [W. Thomson], *Baltimore lectures* (1904), 141.
[21] Reported in *Nature, 53* (1896), 528. Stokes, Manchester *MP, 41* (1897), no. 15, p. 2. Gifford, *Nature, 54* (1896), 172.
[22] J. J. Thomson, *Nature, 53* (1896), 391–2.

tion, diffraction, or reflection.[23] The principal motive of the longitudinal wave thus removed, Ockham's razor, which slices away
multiple hypotheses where one will suffice, more than compen-
sated for the lack of definite evidence about polarization. By the
fall of 1896, transverse waves had become "the surviving hypothesis."[24]

Major support for the transverse wave came from Henri Becquerel's discovery of radioactivity in 1896, for Becquerel had
claimed to find polarization in his uranium rays.[25] At the end of
1896, most physicists agreed with Thomson when he described the
Becquerel rays as resembling "the Röntgen rays more closely than
any kind of light known hitherto."[26] The Becquerel rays seemed to
be intermediate between light and x-rays. This view "greatly
strengthened the position . . . that the x-rays were of the nature
of ultra-ultraviolet light."[27] Because the Becquerel rays reportedly
showed reflection and refraction, the possibility of these properties
was more easily projected onto x-rays. Lodge concluded, late in
1896, that "it has become almost certain that the x-rays are simply
an extraordinary extension of the spectrum," that is, periodic
transverse waves of high frequency.[28]

THE BRITISH VIEW OF ELECTROMAGNETIC IMPULSES

X-rays are produced when a cathode ray beam strikes the metal
anode or the walls inside an evacuated electrical discharge tube. In
Britain, most researchers were convinced that cathode rays consisted of projected charged particles. It was quickly recognized that
each particle, stopped by the collision, should give rise to a distur-

---

[23] Schuster, *Nature, 53* (1896), 268. That longitudinal waves were not required to
explain the penetrability of x-rays was recognized immediately in Britain. See
Thomson to Lodge, 19 January 1896, in [R. J. Strutt], *Life of Thomson* (1942),
65–6; and FitzGerald to Lodge, 24 January 1896, in Lodge, *Electrician, 37*
(1896), 370–3.

[24] Lodge, *ibid.*

[25] Becquerel, *CR, 122* (1896), 762–7; and earlier studies, *ibid.,* 559–64 and
689–94. For a useful discussion of Becquerel's early work, see Malley, *Hyperphosphorescence* (1976).

[26] The Rede Lecture, delivered 10 June 1896, *Nature, 54* (1896), 304. See also J. J.
Thomson, BAAS *Report* (1896), 703ff.

[27] Lodge, *Electrician, 37* (1896), 370.

[28] *Ibid.,* 371.

bance in the equilibrium of the surrounding electromagnetic field. This dislocation would propagate spherically outward from the impact site and affect charged particles as it passed over them. This *impulse* hypothesis of x-rays was discussed informally in Britain very early in 1896.[29] Late that year, as the material interpretation of cathode rays neared experimental corroboration, the electromagnetic impulse hypothesis of x-rays began to exercise greater influence.

In the English-speaking world, the pulse hypothesis was propounded by Sir George Gabriel Stokes, then in his forty-seventh year as Lucasian Professor of Mathematics at Cambridge.[30] Stokes had not believed Röntgen's compression-wave hypothesis; he cited George Green's arguments against the necessary existence of compression waves.[31] At first, he had taken x-rays to be high-frequency transverse waves.[32] But in March 1896 he recognized that even this "supposition is by no means exempt from difficulty."[33] In July he hinted at the impulse hypothesis. Speaking as president to the Victoria Institute, Stokes cited Becquerel's claims for polarization of uranium rays as if they confirmed the transverse nature of x-rays.[34] He ignored the fact that Becquerel had, on three separate occasions, pointed out that his new rays were *not* x-rays. Stokes had other reasons to support a transverse model of x-rays, reasons that he did not yet specify in detail.

Kelvin was dubious. One of the chief difficulties with any periodic wave model was how extremely small the wavelength had to be to explain how x-rays pass through matter. The French physicist L. Georges Gouy had determined an upper limit on the wavelength of 0.005 μm, or 1% that of green light.[35] Stokes and Kelvin both doubted that any mechanical process could generate oscillations at the rate of 70 thousand million million (7×10^{16}) per second. Thinking that Stokes was proposing periodic transverse

[29] Schuster, *Nature, 53* (1896), 268. Lodge, *Electrician, 37* (1896), 370–3.
[30] For general information on Stokes, see Parkinson, *DSB, 13* (1976), 74–9.
[31] Green held that the aether is incompressible and that the longitudinal wave has no physical existence. See *TCPS, 7* (1839), 2–24, 121–40.
[32] Stokes to S. P. Thompson, 29 February 1896. Stokes, *Memoir, 2* (1907), 495–6.
[33] Stokes to S. P. Thompson, 2 March 1896. Stokes, *Memoir, 2* (1907), 496.
[34] Stokes, VI *Journal of transactions, 30* (1898), 13–25; *Nature, 54* (1896), 427–30. The Victoria Institute was a Christian apologist organization, dedicated to countering "oppositiones falsi nominis scientiae," founded in 1866.
[35] Gouy, *CR, 123* (1896), 43–4. Precht, *AP, 61* (1897), 350–62.

waves of these high frequencies, and still believing that x-rays are compression waves, Kelvin simultaneously praised and gently chided the seventy-seven-year-old Stokes:[36]

In physical science my touchstone as to truth is, "what does Sir George Stokes think of it?" When I hear that he declares for transverse vibrations, and when I hear the strong reasons he has put before us for this conclusion, I myself am very strongly fortified indeed in accepting it. I do not, however, forget that it is put before us not as absolutely demonstrated but as his present opinion . . . he leaves our minds open to the possibility of other explanations – indeed he has tantalized us very much with the idea of the possibility of another explanation to which he has alluded.

Stokes committed himself to the impulse hypothesis in November in a talk to the Cambridge Philosophical Society.[37] According to Maxwell's theory, he said, individual impacting cathode-ray particles should give rise to a "vast succession of independent pulses [in the aether] analogous to the 'hedge-fire' of a regiment of soldiers."[38] Stokes thought that the impulse due to a colliding electron might very well be as short as Gouy's estimated wavelength.

Stokes, the leading British authority on optical diffraction, was thus led to a new conception of refraction – one that, he said, "fairly startled" him at first. In his Wilde lecture to the Manchester Literary and Philosophical Society in July 1897, he worked out the details.[39] Molecules of ponderable matter respond by resonance to passing waves. Energy so extracted serves to slow the beam, and the resulting interference of Huygens' wavelets refracts the ray. Aperiodic pulses, on the other hand, can never reinforce the initial push–pull; they lack the requisite oscillation. Therefore, Stokes concluded, x-ray impulses could pass through matter without

[36] Kelvin's remarks following Stokes' address, VI *Journal of transactions, 30* (1898), 25–7.
[37] Stokes, *PCPS, 9* (1896), 215–16. Stokes also gave "a beautifully clear and animated exposition of his theory of Röntgen rays" at the 1896 meeting of the British Association for the Advancement of Science, according to J. J. Thomson, *Recollections* (1936), 409. That Stokes spoke is reported in *Nature, 54* (1896), 565; no details are given in the 1896 BAAS *Report*.
[38] Stokes, *PCPS, 9* (1896), 216.
[39] Stokes, Manchester *MP, 41* (1897), no. 15. Readers familiar with the so-called wave theory of Christiaan Huygens may recognize that his theory was also one of impulses. The drawings in *Traité de la lumiere* (Leiden, 1678), for example, on p. 17, that appear to be snapshots of periodic wave crests are, in fact, multiple exposures of the progress of an individual impulse.

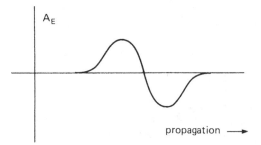

Figure 2.1. G. G. Stokes' doubly directed electromagnetic impulse.

appreciable loss of energy, and consequently suffer little refraction or reflection.

The impulse hypothesis also showed why x-ray diffraction had not been observed: The effect was very small. To show this, Stokes assumed that the pulse was doubly directed; that is, the total positive excursion of impulse amplitude from equilibrium is matched by the negative. Figure 2.1 shows the waveform schematically. The net integral of amplitude taken across the width of the pulse, or over the time of passage, is zero.[40] Thus, argued Stokes, the net transverse impulse radiated toward a point P from that portion of the pulse, shown in Figure 2.2, that lies far from the line joining the source O to P, will be vanishingly small. Seen from an oblique angle greater than ϕ, the positive and negative displacements will arrive at P almost simultaneously. Only a limited portion of the pulse, delimited by angle ϕ in Figure 2.2, can make a sensible contribution at P.

To fix the size of this active region Stokes employed standard diffraction analysis, considering separately the effect at P due to all segments of the impulse that lie within arcs of constant radius r centered on P.[41] (1) If $r > PS$, then either the transverse displace-

---

[40] I have drawn a fairly symmetrical doubly directed impulse, but as long as the area under the positive excursion equals that under the negative one, Stokes' criterion is satisfied. The reasons why another pulse shape was incompatible with the "dynamic theory of diffraction" became clear in 1912, as we shall see in Chapter 8.

[41] Stokes, *TCPS, 9* (1849), 1–62. The final equation of section 22 is the general relation for the transverse displacement at P due to an initial disturbance at O. Several small-angle approximations and symmetry about the azimuthal angle reduce the complex double integral to a simple expression proportional to the transverse displacement in each differential element $dr\,d\theta$.

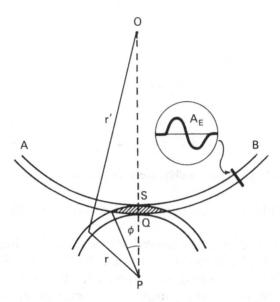

Figure 2.2. Diffraction effect of Stokes' impulse.

ment is everywhere zero (i.e., to the left of, the right of, or fully within the impulse) or its effect arises from the integral across the full impulse. But according to Stokes' assumption, this integral vanishes; viewed from the side, the net contribution at P will then be zero. (2) If $r < PQ$, no part of the pulse can affect P. (3) The only contribution at P comes from the lenticular shaded region for $PS \geq r \geq PQ$. The total transverse displacement at P is just the sum of the contributions from this region, attenuated in the ratio $r'/(r' + r)$. Stokes showed that this region was quite small. For a pulse of width 10^{-6} in., and for reasonable values of r and r' of 4 in., the radius of the active zone is only 0.002 in. The complete decrease in pulse intensity to the side of the shadow occurs in only $1/250$ in., below the threshold of contemporary photographic resolution.

J. J. THOMPSON ON PULSES

Joseph John Thomson, Cavendish Professor of Physics at Cambridge University, carried out the detailed calculation of Stokes' hypothesis. Thomson was an expert on Maxwell's theory. Furthermore, he was actively collecting proofs that cathode rays are subatomic particles, having recently proclaimed them to be such. The impulse hypothesis of x-rays, a corollary to the particulate interpretation of cathode rays, was bound to attract his interest. Thomson had considered the longitudinal wave hypothesis in an early study.[42] But he quickly recognized the merits of Stokes' idea and, in December 1897, put it to a test.[43]

An electron is the source of isotropically diverging electric lines of force. If the electron moves at a velocity that is small compared to that of light, this set of lines remains symmetric as it moves along with and centered on the electron. The accompanying magnetic field is arranged in circles lying in successive planes perpendicular to the direction of electron motion. If the electron stops abruptly, it takes some small time for that fact to be broadcast to the surrounding field. The propagation of the dislocation outward along the field lines constitutes the electromagnetic impulse. Figure 2.3 illustrates the growth of the spherical impulse at successive equal intervals of time after the electron stops.[44] The center marked "x" indicates the position the electron would have occupied had it not stopped. The field outside the impulse remains centered on "x," whereas the field inside diverges from the stationary electron.

Using Figure 2.4, one may easily calculate the ratio of tangential to radial electric force. At time t after the impact, it will be equal to the ratio

$$\frac{\text{NP}'}{\text{NN}'} = \frac{vt \sin \theta}{\delta} \qquad (2.1)$$

where δ is the thickness of the pulse and v is the original velocity of the electron. The normal electric force in the radial direction is $e/4\pi r^2$, where e is the electronic charge. Thus, the total transverse electric force amplitude is given by

$$A_E = \frac{e}{4\pi r \delta} \frac{v \sin \theta}{c} \qquad (2.2)$$

where $c = r/t$ is the velocity of wave transmission along the field

---

[42] J. J. Thomson, *PCPS, 9* (1896), 49–61.
[43] J. J. Thomson, *PM, 45* (1898), 172–83. Thomson mentioned in his Princeton lectures in October 1896 that electromagnetic impulses should arise from cathode ray impacts. *Discharge of electricity* (1898), 192–3.
[44] Heaviside, *PM, 27* (1889), 324–39; *PM, 44* (1897), 503–12.

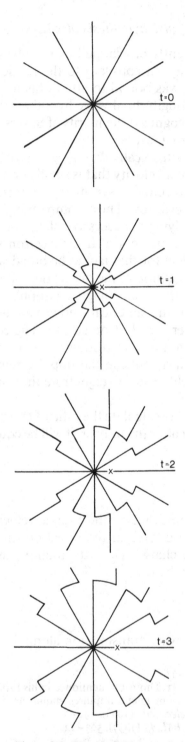

Figure 2.3. Electric field around an electron t units of time after it stops moving to the right. X marks the position the electron would have occupied.

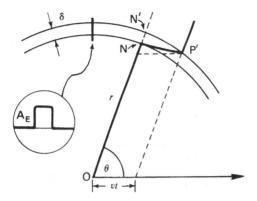

Figure 2.4. J. J. Thomson's derivation of radial to transverse electric force for an impulse. [Based on J. J. Thomson, *Conduction of electricity* (1903), 537.]

lines.[45] The total magnetic force is given by $4\pi c A_E$, or

$$A_H = \frac{e}{r\delta} v \sin \theta \qquad (2.3)$$

The point is that the transverse effect of the impulse drops only as $1/r$, whereas the radial force drops as $1/r^2$. Thus, the electrodynamic effect of the impulse decreases less rapidly than does the normal radial electric force, and will predominate at great distances from the electron.

In 1903, Thomson calculated the energy residing in one impulse. In a given direction θ from the direction of electron motion, the energy density is

$$\frac{\mu}{4\pi} A_H^2 = \frac{\mu}{4\pi} \left(\frac{ev \sin \theta}{r\delta} \right)^2 \qquad (2.4)$$

A_H is the magnitude of the magnetic force, and μ is the magnetic permeability of the medium (in a vacuum, $\mu = 1$). Integrating this expression over the volume of the expanding spherical shell gives the total energy carried by the impulse:

$$E_{\text{total}} = \frac{2\mu \, e^2 v^2}{3\delta} \qquad (2.5)$$

[45] In 1897 Thomson published only his findings on the magnetic force; those on the electric force followed in 1903.

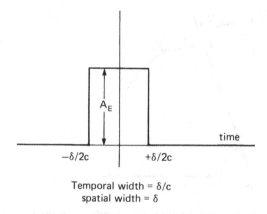

Temporal width = δ/c
spatial width = δ

Figure 2.5. Singly directed square impulse as a function of time.

Note, in particular, that the energy is inversely proportional to the impulse width.[46] The narrower the pulse, the greater the energy carried. Note also that Thomson's impulse is quite different in form from Stokes' (Fig. 2.1). Expressed as a function of time, Thomson's impulse is like that shown in Figure 2.5. The integral taken across this pulse does not vanish. Stokes had stated that the "dynamical theory" of diffraction would not allow a singly directed pulse.[47] Thomson was more concerned with the physical constraint that the deceleration of source electrons occurs in one direction only. He could see no way to justify a doubly directed impulse, and never illustrated it as I have in Figure 2.5.

---

[46] J. J. Thomson, *Conduction of electricity* (1903), 540.

$$E = \frac{\mu e^2 v^2}{4\pi\delta^2} \int \frac{\sin^2 \theta}{r^2} \, dV \text{ and } dV = \delta r d\theta$$

Integrating over a hemisphere:

$$E = \frac{\mu e^2 v^2}{4\pi\delta^2} \int_0^\pi \frac{2\pi r \sin^3 \theta}{r^2} \, r\delta d\theta$$

$$E = \frac{\mu e^2 v^2}{2\delta} \int_0^\pi \sin^3 \theta d\theta$$

Since

$$\int_0^{\pi/2} \sin^3 \theta d\theta = \frac{2}{3}, \qquad E = \frac{2}{3} \frac{\mu e^2 v^2}{\delta}$$

[47] Stokes, *Papers*, 5 (1905), 265–6.

By 1899, Kelvin too had accepted the impulse hypothesis. His original concern had been that an extremely high-frequency longitudinal wave was necessary to explain the low absorption of x-rays. He analyzed the analogous case of oscillations in a mechanical system to chart the relative amplitudes of the transverse and longitudinal waves when the wavelength was about equal to the molecular spacing in matter.[48] He concluded that the amplitude of the condensational wave might drop to negligible values for frequencies above 10^{12} per second.[49] No explicit denial of the longitudinal wave can be found in Kelvin's writings, but it is difficult not to interpret his several attempts in 1899 to calculate the theoretical properties of single, detached impulses as tacit approval of the impulse hypothesis of x-rays.[50]

CONTINENTAL RESPONSE TO X-RAY DIFFRACTION

Outside of Britain, reception of the impulse hypothesis was less favorable and occurred more slowly. This resistance may be partly explained by two views particularly characteristic of contemporary German physics: Cathode rays were widely thought to be periodic waves, not particles, and discontinuities in the aether could be derived from the superposition of periodic waves. Cathode rays induce fluorescence in several materials. This capability was widely thought at the time to be restricted to light.[51] When it was discovered that x-rays originate from the fluorescent spot on the tube, the analogy seemed complete – x-rays are periodic transverse waves, like cathode rays, but of slightly lower frequency to accord with Stokes' law of fluorescence. Some continental physicists saw differences in the properties of x-rays and cathode rays as a matter of degree only.[52]

There were, of course, some German physicists who rejected Röntgen's longitudinal waves. But in the early years, most of these

---

[48] [W. Thomson], *PM, 46* (1898), 494–500; BAAS *Report* (1898), 783–7.

[49] [W. Thomson], *Baltimore lectures* (1904), 159; BAAS *Report* (1898), 784.

[50] [W. Thomson], *PM, 47* (1899), 179–91, 480–93; *48* (1899), 227–36, 388–93.

[51] There was other evidence against the material interpretation of cathode rays. See Chapter 1; Hertz, *AP, 19* (1883), 782–816; Lenard, *AP, 51* (1894), 225–68.

[52] The most outspoken proponent of this interpretation was Philipp Lenard. See Wheaton, *HSPS, 9* (1978), 299–323.

preferred periodic transverse waves to aperiodic impulses. Continental texts often did not mention the impulse hypothesis.[53] Ernst Mach let it be known that he thought very little of the hypothesis of "isolated impulses." He insisted that x-rays were ordinary light of exceedingly short wavelength in order to explain the sharp-edge shadows they cast.[54] Yet partly on the basis of evidence for diffraction, physicists by 1900 would favor impulse x-rays, even on the Continent.

Acceptance of the impulse hypothesis on the Continent was hastened by increasing tolerance for a material interpretation of cathode rays after 1897. Thomson's measurement of e/m for the cathode beam was soon corroborated by Philipp Lenard and Walther Kaufmann.[55] By that time, there were strong theoretical reasons to accept the result. The theory of the electron, in which the charged particle was conceived by Lorentz and others as an intermediary between ponderable matter and the electromagnetic field, had borne fruit. With it Lorentz had given explanations of Fizeau's experiment, of anomalous dispersion, of the Zeeman effect, of Faraday's experiments on the rotation of the plane of polarization in light, and of the Michelson – Morley experiment.[56] The combination of theoretical significance and experimental demonstration was hard to deny. By 1899 the localized concentration of electric charge, if not the material nature of the electron, had become a respectable microreality, and its acceptance was growing. When cathode rays, satisfying all the criteria of charged particles, collide with atoms of the anticathode, sharply defined electromagnetic dislocations should radiate into the surrounding field. The question was only whether the impulses so produced could be identified as x-rays.

It is indicative of the relation between these views that an early electron theorist in Germany, Emil Wiechert, had been the first to

---

[53] Examples include Donath, *Einrichtung zur . . . Roentgen-strahlen* (1899), and Guillaume, *Rayons x* (1896), both of which claimed to discuss current ideas about the nature of x-rays.

[54] Mach, *Zeitschrift für Elektrotechnik, 14* (1896), 259–61.

[55] The most important measurements were those of Kaufmann and Aschkinass, *AP, 62* (1897), 588–95; Lenard, *AP, 64* (1898), 279–89.

[56] Lorentz, Amsterdam *Verslag, 7* (1899), 507–22. On the successes of electron theory, see Hirosige, *HSPS, 1* (1969), 151–209; McCormmach, *Isis, 61* (1970), 459–97.

suggest the impulse hypothesis.[57] In a talk to the Königsberger physikalisch-ökonomische Gesellschaft in May 1896, six months before Stokes' first public suggestion of impulses, Wiechert proposed, but did not develop, his hypothesis of *jähen Lichtstosse,* or irregular impulses of light.[58] He was still equivocal; x-rays might be impulses or high-frequency transverse waves. In either case, he attributed their low absorption by matter to the inability of molecules to react to the extremely rapid changes in the electromagnetic aether.

In February 1899, two Dutch physicists announced that x-rays could be diffracted.[59] By itself the photograph submitted to support this claim would not have been convincing. A rash of false claims of diffraction had hardened most physicists, and the photograph that C. H. Wind and H. Haga had taken was extremely difficult to assess (see Figure 2.6). Working with his institute director at Groningen, Wind exposed a wrapped plate to the x-rays passing through a wedge-shaped slit, narrowing from 14 to 2 μm in width. Exposures of up to 200 hours, extending over 10 days, were required; the *Erschütterungsfreiheit* provided by the institute's free-standing pier was essential. The investigators claimed that they discerned evidence at the 9-μm width (marked ← in the figure) for broadening of the image. This they interpreted as diffraction of periodic waves of approximately 1 Å in length, 10^{-4} times the wavelength of visible light, and a factor 100 times smaller than Gouy's estimate. They provided very little photometric analysis, and the only justification for their claim was that Wind had done this study after a careful analysis of most of the preceding false evidence. He showed that prior claims for diffraction fringes had resulted from optical illusions.[60] Haga and Wind, unlike their fringe-seeking predecessors, based their argument on evidence for simple broadening of the image. They explained away all other

[57] Wiechert, Königsberg *Schriften, 37* (1896), 1–48; *AP, 59* (1896), 283–323, especially 321–2.
[58] "Irregular light blows." There is no question that Wiechert used *jähe* in the temporal sense, and he specified a sense for *Stoss* that made it clear that he meant individual impulses.
[59] Haga and Wind, Amsterdam *Verslag, 7* (1899), 387–8, 500–7.
[60] Wind, Amsterdam *Verslag, 7* (1899), 12–19. Among other results called into question were those of Fomm, München *Sb, 26* (1896), 283–6.

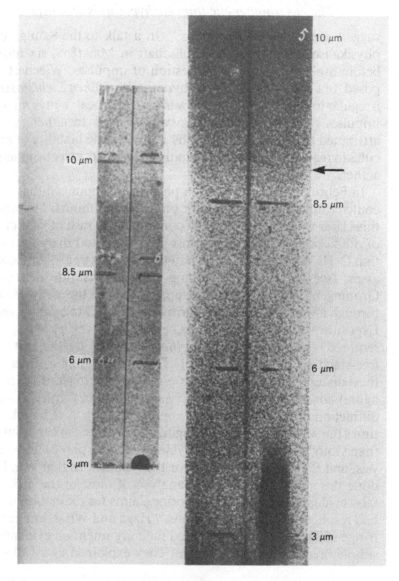

Figure 2.6. Haga and Wind's photograph of x-ray diffraction by a slit. *Left:* Exposure 5, collimation 25 μm, distance from slit to photoplate 1 cm, exposure 29 hr. No discernible diffraction. *Right:* Exposure 2, collimation 14 μm, distance from slit to photoplate 75 cm, exposure 100 hr. Broadening of the image appears at approximately 8 to 9 μm. The dark region at the lower right is the image of a reference hole drilled in one of the 0.5-mm-thick platinum plates that form the slit. It is elongated in the right-hand exposure because of angular extension of the x-ray source. [Haga and Wind, Amsterdam *Verslag, 7* (1899), facing 502.]

possible sources for the broadening and concluded that it must have arisen from diffraction of x-rays.

SPECTRAL DECOMPOSITION OF AN IMPULSE

Wind and Haga's photograph was taken as proof of x-ray diffraction because it provided material for a new treatment of diffraction proposed by Arnold Sommerfeld, then *Privatdozent* at Göttingen. Sommerfeld had already shown himself to be an expert on diffraction; his *Habilitationsschrift* at Göttingen had been an attempt to rectify the faults of Kirchhoff's treatment of the subject, which, in Sommerfeld's view, failed "to satisfy the demands of mathematical precision."[61] Nor was he unfamiliar with the impulse concept of x-rays. As a student in Königsberg, he had studied with Wiechert, whom he characterized as a "profound mathematical–physical thinker."[62] Indeed, Wiechert soon became the model of a physicist for Sommerfeld, and they collaborated on, among other things, a harmonic analyzer for complex waveforms in the years immediately preceding Wiechert's suggestion of impulse x-rays.[63] Only a few years after leaving Königsberg, Sommerfeld claimed that the "Wiechert–Stokes" hypothesis of discontinuous x-ray impulses followed from the "almost certain concept of the nature of cathode rays according to which they consist of ejected electrically charged particles of small mass."[64] But the pulse hypothesis presented a problem for diffraction analysis, one that Sommerfeld thought would provide an ideal opportunity to test the flexibility of his new rigorous theory.

Sommerfeld recognized that diffraction theory would fail for single impulses. He was referring to the square pulse pictured in Figure 2.5. This is the same pulse that Thomson had assumed. Like Thomson, Sommerfeld ignored Stokes' warning that the integral over the impulse must vanish. The result was that when Sommerfeld decomposed the impulse into its Fourier component spectrum, he found that the components of greatest amplitude were

[61] Sommerfeld, Göttingen *Nachrichten* (1894), 341; Deutsche Mathematiker-Vereinigung *Jahresbericht, 4* (1895), 172–4; *Mathematische Annalen, 47* (1896), 317–74.

[62] Sommerfeld, *Geist und Gestalt, 2* (1959), 100–9.

[63] Sommerfeld and Wiechert, Deutsche Mathematiker-Vereinigung *Katalog* (1892), 214–21.

[64] Sommerfeld, *Zeitschrift für Mathematik und Physik, 46* (1901), 11–12.

those for long, actually infinite, wavelengths. "If [however] we regard the impulse as a whole, and not as if decomposed into single sine waves," Sommerfeld argued, "all results become clearer."[65]

A Fourier series assigns a specific amplitude and a definite phase relation to each member of a set of sinusoidal oscillations. The frequency of each wave is a different integral multiple of a *fundamental* frequency. If the various waves are then superimposed, physically added together, the resulting complex wave is fully determined. It repeats its vibration at a rate equal to that of the fundamental frequency. The Fourier series expansion is a straightforward mathematical method used to reduce any repeating complex waveform to its component set of sinusoidal vibrations.

Complex waveforms that are *not* repeated are more difficult to decompose, but conceptually are no different. The required set of component sine waves is no longer limited to harmonic multiples of a fundamental frequency. There is no fundamental frequency. Instead, a Fourier *integral* must be taken over a continuous range of frequency values. What results is, in fact, a spectrum; each value of the frequency is associated with a definite amplitude. If this extremely large number of sine waves, not harmonically related to one another, is superimposed, the initial single impulse is obtained. Everywhere else, the component waves cancel each other out.

The general Fourier integral is given by the expression

$$\int_{-\infty}^{+\infty} f(t)e^{-i\omega t}\, dt \tag{2.6}$$

where $f(t)$ is the known functional dependence of the waveform on time. In the case of the square impulse pictured in Figure. 2.5, the contribution to the integral from the regions to the left and right of the impulse is zero; hence the integral reduces to

$$\int_{-\delta/2c}^{+\delta/2c} A_E e^{-i\omega t}\, dt = \frac{A_E \sin \omega t}{\omega}\bigg|_{-\delta/2c}^{+\delta/2c} = 2A_E\left(\frac{\sin \omega\delta/2c}{\omega}\right) \tag{2.7}$$

If we now plot the absolute value of the integral as a function of the positive values of frequency ω, we obtain Figure 2.7. The height of the function at any frequency gives the amplitude that must be

---

[65] Sommerfeld, comments to Wien following *VDNA* (1900), part 2, first half, p. 24.

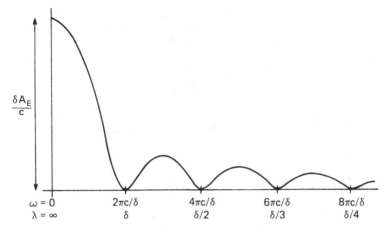

Figure 2.7. Fourier spectrum of square impulse.

assigned to that frequency in order, by superposition of all waves, to re-create the original square impulse. As may be seen, the maximum contribution occurs at zero frequency, and the amplitude for a wavelength corresponding to impulse width δ is zero.[66]

SOMMERFELD'S DIFFRACTION ANALYSIS

Sommerfeld characteristically presented his results in two versions: one for mathematicians interested in rigor, and the other for physicists interested in x-rays.[67] He spoke to the physicists through the pages of the *Physikalische Zeitschrift* in 1899, and in person at the seventy-second *Naturforscherversammlung* the following year.[68] As with any consideration of diffraction, Sommerfeld had to contend with the original disturbance in interaction with its scattered self. The planar impulse is assumed to approach the screen shown in Figure 2.8 from the left, attenuating in amplitude as it traverses space. The impulse scatters from the edge of the

[66] Sommerfeld used the standard symbol for wavelength λ to characterize the impulse width, although he stated quite explicitly that no periodicity is to be expected in the impulses. I will use the neutral symbol δ throughout this study to indicate spatial impulse width.

[67] Sommerfeld's initial statement of the failure of classical diffraction theory begins with two statements of the problem, one *physikalisch ausgedrückt* and the other *mathematisch ausgedrückt*.

[68] Sommerfeld, *PZ, 1* (1899), 105–11; *PZ, 2* (1900), 55–60.

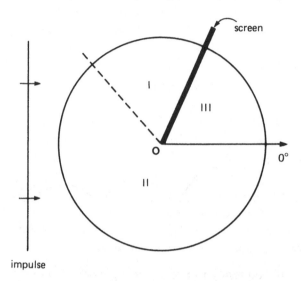

Figure 2.8. Sommerfeld's diagram of x-ray diffraction by an edge.

screen, and by Huygens' principle a second cylindrical impulse results, centered on point O. Sommerfeld split it into the three sections shown. In region I, the net disturbance consists of the scattered impulse and the original impulse reflected from the screen; the boundary between regions I and II is the edge of the reflected impulse. Given the lack of evidence that x-rays are reflected, Sommerfeld took that contribution to be zero. Region III, in the shadow of the screen, contains only the cylindrical scattered impulse, whereas in region II one must add to this the undeviated original planar impulse.

Sommerfeld uncovered one of the more dramatic corollaries of the fact that impulses are not periodic waves: *No diffraction fringes result*. The periodic structure in ordinary light that produces the fringes is not present. In effect, one obtains only the envelope of the classical diffraction pattern: a monotonic decrease in intensity to the side of the shadow. Sommerfeld's three-dimensional represen-tation of the resulting distribution of x-ray intensity is shown in Figure 2.9. The maximum intensity in the forward direction peaks at one-half the amplitude of the unscattered original impulse. The smaller the ratio of pulse width to the distance r from the screen, the more quickly this intensity drops off to the sides. This is shown

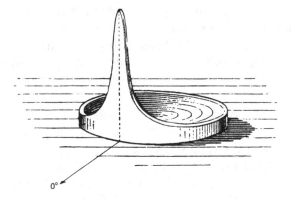

Figure 2.9. Distribution of diffracted impulse x-ray intensity. [Sommerfeld, *PZ, 1* (1899), 108.]

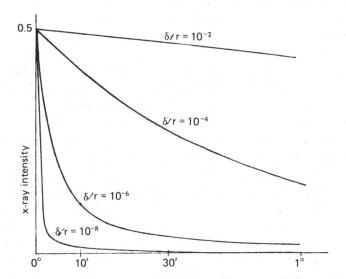

Figure 2.10. Change in diffracted intensity with impulse width. [Based on Sommerfeld, *PZ, 1* (1899), 109.]

in Figure 2.10. The smaller the pulse width, the less the discernible diffraction.

To relate the calculation to the results from Groningen, Sommerfeld had to introduce approximations. He needed to convert electromagnetic amplitude to a measure of photographic blackening. Then again, Wind and Haga's data came from *slit* diffraction;

the introduction of the second edge, a problem Sommerfeld had not solved completely in his 1896 study, introduced error. As an approximation, he took the total broadening of the image as due half to each side. The result was a calculated x-ray pulse width of 33 Å, compared to the Dutch estimate of a 1-Å periodic wavelength.

In the second phase of his study, Sommerfeld the physicist retracted what Sommerfeld the mathematician had done. After some discussion with Wind on sources of error in the photograph, he revised his calculation.[69] This time he reworked the problem explicitly by taking both edges of the slit into account. "Refraction from a slit," he said, "often turns out very different from refraction at an edge."[70] He now found that the "dark core" of the wedge-slit image, his interpretive sketch of which is shown in Figure 2.11, would cease at a slit width of $\sqrt{8\delta r}/10$. Wind's best estimate was that this occurred at 9 μm; Sommerfeld derived a pulse with a spatial width $\delta = 1.3$ Å and a temporal width of 4×10^{-19} sec. He thought this a "much more trustworthy" estimate. To his friend, the astrophysicist Karl Schwarzschild, an expert on photochemical response, Sommerfeld wrote, "with such a small [pulse width] it is easy to understand why x-rays are neither absorbed nor refracted. Molecules simply cannot follow such short impacts."[71]

Sommerfeld went on to propose that a continuum of radiation waveforms exists, one that extends from the purely periodic sine waves of monochromatic light to the temporally localized aperiodic impulses of x-rays. There would generally be a mixture of pulse widths in an x-ray beam. He had proven to his satisfaction that x-ray impulses are aperiodic; the lack of characteristic diffraction fringes was evidence for their intrinsic temporal discontinuity.[72] Thus, Sommerfeld had a physical if not a formal reason to reject the consideration of an impulse as if broken into its Fourier spectrum, as others were suggesting. Wind, for example, was surprised to have his photograph successfully analyzed in terms of aperiodic pulses; recall that he had first claimed the diffraction

[69] Details of Wind's criticism exist in his several letters to Sommerfeld between April and August 1900. Sommerfeld papers; AHQP 34, 14.
[70] Sommerfeld, *PZ*, 2 (1900), 55.
[71] Sommerfeld to Karl Schwarzschild, 24 September 1900. Schwarzschild papers.
[72] Discussion following Wind's presentation to the seventy-second *Naturforscherversammlung*, *PZ*, 2 (1900), 297–8.

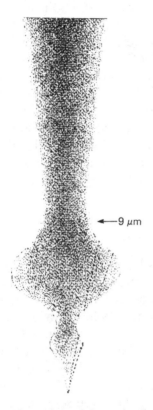

←9 μm

Figure 2.11. Schematic distribution of x-ray impulse intensity after diffraction by a slit. Width exaggerated 100-fold. [Sommerfeld, *PZ, 2* (1900), 293.]

photo as evidence of periodic-wave x-rays. He responded to Sommerfeld's careful analysis that the "wavelength" of 1 Å should be interpreted as the high-amplitude component of the Fourier series that represents the impulse.[73]

Sommerfeld would have none of it. Wind's claim could not be true for the impulses Sommerfeld had treated; their maximum Fourier component fell at much longer wavelengths, and their contribution at the wavelength comparable to the pulse width was zero. At the conclusion of his second study of pulse diffraction, the one for the mathematicians, Sommerfeld contrasted his method to that of Wind. "One cannot speak of the superiority of one or the

[73] Wind, *AP, 68* (1899), 896–901; *PZ, 2* (1900), 189–96.

other method," he remarked; "each has its strengths and weaknesses." Wind's method offered "great generality and adaptability to experimental conditions"; Sommerfeld's provided "conceptual clarity and precision of results."[74] For Sommerfeld, the difference between an impulse and its Fourier representation had assumed a physical as well as a formal significance. There was never any question in his mind that mathematically the two representations were equivalent. But with the idea that a continuum exists extending from periodic light to aperiodic x-rays, Sommerfeld had introduced what would become an influential conceptual framework, one that emphasized the distinction between x-ray pulses and ordinary light.

THE NATURE OF WHITE LIGHT

Whether a discontinuous impulse is better represented as an isolated event or as a superposition of harmonic oscillations had been discussed before; it was clear that differences of opinion cut across national boundaries. The question was, which waveform best represents polychromatic white light? L. Georges Gouy had concluded in 1886 that there is no sense in claiming that white light is a repeating wave.[75] It is the superposition of all colors of light, and as such it cannot have any well-defined wavelength. He thought that the actual waveform consisted of irregular impulses in the aether. It had been claimed that interference experiments give a measure of the degree of periodicity of white light: A beam is split and reunited after one component travels a great number of wavelengths. The resulting interference fringes, it was supposed, prove that the light repeats its oscillation even after these long intervals. Gouy maintained that the regularity is introduced entirely by the spectral analyzer. A diffraction grating separates out of the light only the "different simple movements" that until then had a purely analytical existence.[76] Gouy insisted that we should think only of the total electromagnetic disturbance, the *mouvements lumineux,* which arrive at the eye. The pure frequencies are only analytical tools out of which we fashion the concept of the physical disturbance.

[74] Sommerfeld, *Zeitschrift für Mathematik und Physik, 46* (1901), 97n.
[75] Gouy, *JP, 5* (1886), 354–62. Righi, *JP, 2* (1883), 437–46.
[76] Gouy, *ibid.,* 362.

The light we see from most objects is a mixture of pure frequencies. Green light, for example, is a superposition of many frequencies of light, which are predominantly but not exclusively greenish to the eye. By adding each pure frequency, with its proper amplitude and phase, to all the others, we arrive at the complex disturbance that is presented to the eye. According to Fourier analysis, the complex wave so constructed can have virtually any functional form. If it is a periodically repeating disturbance, it is repeated less rapidly than is any one of the sine waves of which it is composed. Its periodic frequency is the least common factor of its component frequencies.[77]

Polychromatic light contains a virtual continuum of frequencies. As one constructs the complex wave, the least common factor of the frequencies quickly drops to an exceedingly small value. Consequently, the period of the resultant complex wave grows very long; for all practical purposes, it becomes infinite. It is, in a sense, the human attempt to assign quantitative measures that creates the conceptual problem. It is, in fact, just as much of a problem for the representation of green light, but because white light cannot be assigned a perceived frequency, the problem cannot be avoided. If we insist that there is a one-to-one relation between numbers and pure frequencies of light, then the colors added to make white light can no more share a commensurate wavelength than can all points on the number line share a common factor. The "frequency" of the resultant wave is zero, and there is no meaning to the concept of *regularity*. As Lord Rayleigh said, it is truly "nonsense."[78]

But in order for any wave to have a well-defined frequency, it must also have a long temporal extension. A piece of a sine wave does not possess a well-defined frequency; only a pure oscillation that extends to infinity in space can strictly be called monochromatic. It is not simply the rate at which a wave repeats, but also the actual number of oscillations it undergoes, that define a mono-

[77] Counting frequencies limits the apparent discontinuity. Assigning each pure color a number implies that some small commensurate frequency unit exists that relates any two frequencies. In this sense all frequencies are "harmonic," although the fundamental is exceedingly small. In principle, unless frequencies are irrationals, a totally isolated impulse cannot arise, according to Fourier analysis.

[78] [J. W. Strutt], *PM, 27* (1889), 463; *Encyclopaedia Britannica, 24* (1888), 421-59.

chromatic beam. Passing an impulse through an analyzing apparatus sufficiently extends it so as to make interference of its component wavelengths possible. In 1894 Arthur Schuster showed that
both a diffraction grating and a prism will impose their own
periodicity when stimulated by a discontinuous disturbance such
as white light. "There is no distinction between regular and irregular light," he said, "beyond that which is brought out by the
distribution of intensity in the spectrum."[79]

Following Stokes' suggestion of x-ray impulses in 1897, similar
issues were raised in regard to x-rays. Lord Rayleigh, an expert on
acoustics and Fourier analysis, asked in a note to *Nature* why
Stokes maintained that x-ray pulses were physically different from
periodic waves.[80] One need only ascribe to them an extraordinarily
long repetition period, say on the order of days; they can then be
reduced to simple harmonic sums. G. Johnstone Stoney proceeded to do just that, and concluded that this "represents what
really takes place in nature."[81] A year later, Stokes wrote to Rayleigh, asking in disbelief if he had been serious in his assertion that
individual pulses were broken into harmonic waves by the action
of a spectroscope.[82] Rayleigh replied, yes; "*whatever* be the disturbance incident upon a grating, the light sent off in a particular
direction will have been *made* regular."[83]

The repeated objections and continued discussion of this question are puzzling to the modern reader. Mathematically, the two
representations of an impulse are equivalent. What was at stake
was the physical, conceptual vision. Does it in fact mean *anything*
to think of pulses as the resultant of an extremely large number of
waves extending much farther in space than the impulse alone?
Emmanuel Carvallo tried to show in 1900 that impulses cannot
physically be represented by Fourier expansion.[84] Max Planck, still
seeking a complete physical interpretation of the new constant he
had introduced in his thermodynamic interpretation of cavity
radiation, saw a parallel in 1902 to his own work.[85] He argued that a

[79] Schuster, *PM, 37* (1894), 509–45. Larmor, *PM, 10* (1905), 574–84.
[80] [J. W. Strutt], *Nature, 57* (1898), 607.
[81] Stoney, *PM, 45* (1898), 532–6; *46* (1889), 253–4.
[82] Stokes to Rayleigh, 31 July 1899. Stokes, *Memoir, 2* (1907), 123–4.
[83] Rayleigh to Stokes, 10 August 1899 and 12 February 1900. Stokes, *Memoir, 2* (1907), 124–5.
[84] Carvallo, *CR, 130* (1900), 79–82, 130–2. Gouy, *CR, 130* (1900), 241–4, 560–2.
[85] Planck, *AP, 7* (1902), 390–400.

spectroscope could no more separate the energy of a single partial oscillation from white light than could experiments isolate the kinetic energy of a single gas molecule. Joseph Larmor suggested in 1905 that white light differed from x-rays in that the "molecular shocks" of white light "would be regarded as having some sort of statistical order in their distribution," unlike x-ray impulses.[86]

The differences of opinion about the physical nature of pulses symbolized an essential ambiguity that gave electromagnetic impulses their intrinsic adaptability. The impulses that Stokes had proposed, and that Sommerfeld imagined to pass through a wedge-shaped slit, had no periodic properties. They did not occur in regular succession; if they had, then diffraction fringes should have been observed. Impulses are waves. But these waves transport energy in much the same way as do particles. In part due to this flexibility, in part due to electron theory, and in part due to Sommerfeld's successful diffraction analysis, the hypothesis of impulse x-rays was soon as widely accepted on the Continent as it was in Britain. Early discussions, like that in the 1897 edition of Wüllner's *Lehrbuch der Experimentalphysik,* either neglect the pulse hypothesis entirely or make only implicit concession to it. Wüllner described x-rays as "undoubtedly a process in the aether, as are the cathode rays," then ranked them as ordinary light but with a frequency higher than that of ultraviolet light.[87] But by 1902, Lorentz described x-rays as "extremely short violent electromagnetic disturbances of the aether,"[88] and by 1903 the impulse hypothesis had virtually everywhere supplanted the view that x-rays are periodic waves.[89]

C. G. BARKLA AND X-RAY POLARIZATION

Even in the absence of any demonstration of x-ray polarization, there was very little doubt after 1898 that the form of an x-ray aether disturbance is transverse. In 1904 this long-anticipated experimental result was finally demonstrated in Britain, adding indirect support to the pulse hypothesis. There was never evidence

[86] Larmor's editorial comment, Stokes, *Papers,* 5 (1905), 272n.
[87] Wüllner, *Lehrbuch der Experimentalphysik, 3* (1897), 1300–02.
[88] Lorentz, untitled Nobel lecture. *Nobel 1902* (1905), separately paginated, p. 4.
[89] Among examples of the pedagogical acceptance of impulse x-rays is Khvol'son, *Lehrbuch der Physik, 2* (1904), 176.

that x-rays can be reflected. However, early studies had turned up what was called a *diffuse reflection* of x-rays from certain solids.[90] Röntgen had mentioned in his third and final x-ray study that a fluorescent screen might still glow even when shielded from the direct x-ray beam.[91] He concluded that x-rays can stimulate other x-rays in matter; the primary x-ray beam gives rise to secondary x-rays. Sagnac showed in 1898 that the secondary x-rays from irradiated metals differ from the primary by being more easily absorbed.[92] Further, their penetrability seems to depend, for the same primary beam, on the type of material used to produce them.

Sagnac based his research on a modified periodic-wave hypothesis of x-rays.[93] There was some question of whether the impulse hypothesis was also able to explain the result. In the first edition of his book *Conduction of electricity through gases,* J. J. Thomson answered in the affirmative. Secondary impulses should radiate from electrons in the solid. Each primary x-ray pulse gives rise to a large number of secondary pulses, one from each charged particle passed over. Not only might ionic forces in the solid constrain the individual secondary pulses, but the multiplicity of secondaries would be expressed as a semiperiodic wave with a length equal to the spacing of the radiating charged particles. This, Thomson suggested, could cause as much difference as existed "between the sharp crack of lightning and the prolonged roll of thunder."[94] Nevertheless Thomson gave no quantitative argument to show why the penetrability of x-ray impulses would be less in the secondary rays. This problem was especially difficult in the case of irradiated gases, in which ionic bonding would not cause impulse degradation and in which the radiating electrons are spaced relatively far apart.

Enter Charles Glover Barkla. Schooled in Liverpool by Oliver Lodge, in 1899 Barkla went to Cambridge, where he heard lectures by Stokes and learned experimental physics from Thomson. Not surprisingly, he became an early advocate of the pulse hypothesis and, before leaving Cambridge in 1903, had begun studies on secondary x-rays. His first experiments dulled Sagnac's criticism;

---

[90] [W. Thomson], *PRS, 59* (1896), 332 – 3. J. J. Thomson, *PCPS, 9* (1898), 393 – 7.
[91] Röntgen III, p. 18.
[92] Sagnac, *CR, 125* (1897), 942 – 4; *126* (1898), 467 – 70, 521 – 3.
[93] Sagnac, *Thèses* (1900), 163.
[94] J. J. Thomson, *Conduction of electricity* (1903), 270.

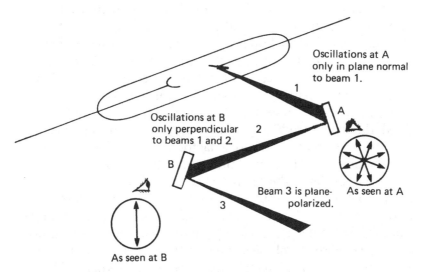

Figure 2.12. Schematic diagram of Barkla's demonstration of x-ray polarization.

the penetrability of secondary x-rays produced in five gases was, within 10 percent, the same as that of the primary. He concluded further that "there is obviously a proportionality between the intensity of secondary radiation and the density of the gas."[95] This supported the impulse concept in Barkla's view. The greater the number of electrons shaken in an atom by the primary pulse, the greater the strength of the expected secondary pulse.

In the course of these studies, Barkla found that a weak tertiary x-radiation could be discerned when the secondary x-rays were directed against some metals. This led to a test for x-ray polarization. A colleague at Liverpool, Lionel Wilberforce, suggested that this tertiary beam should be plane-polarized if the initial impulses were transverse. The electrons in scattering substance *A* can be displaced by the primary beam in all directions except that of the primary beam itself. This follows immediately from the transverse nature of the pulses. Those secondary rays that propagate at right angles to the primary beam should then be able to excite displacement of electrons in scatterer *B* only in a direction perpendicular to both primary beam 1 and secondary beam 2 (see Figure 2.12). The resulting tertiary beam is thus plane-polarized and should

[95] Barkla, *PM, 5* (1903), 696.

show a distinct intensity minimum when viewed in a direction orthogonal to both beams 1 and 2.

Barkla found that the tertiary beam was too weak for the experiments. But he quickly realized that the secondary beam should show the same effect, because the monodirectional cathode ray beam has *already* partially polarized the primary x-rays. Thomson had pointed out in 1898 that the unidirectional cathode beam caused the pulse-producing deceleration to occur in only one direction.[96] He had predicted a factor difference of 2 in x-ray intensity radiated in the line of the cathode beam compared to that seen at right angles to the beam. Experiment disproved the prediction, and Thomson attributed this finding to multiple scattering of cathode ray electrons as they collide with atoms of the anticathode.

Measuring the more completely polarized secondary beam gave Barkla the experimental edge, and he was rewarded with success in 1904.[97] He rotated the x-ray tube itself about the axis defined by the primary x-ray beam 1 and found a 15 percent difference in the ionizing power of secondary x-rays scattered from paper. The result sparked no controversy; as we have seen, there was little opposition to the transverse nature of x-rays by 1904. Although the result did not in fact distinguish between periodic waves and aperiodic transverse impulses, it was widely accepted as further support for the impulse hypothesis.[98]

ERICH MARX MEASURES THE VELOCITY OF X-RAYS

The high point of the seventy-seventh *Naturforscherversammlung* in Meran was the presentation of the first convincing attempt to measure the velocity of x-rays. It had been tried before, notably by René Blondlot, but the results were ambiguous.[99] The method outlined by Erich Marx at Meran seemed to answer all criticism.[100]

---

[96] J. J. Thomson, *PM, 45* (1898), 172–83.

[97] Barkla, *Nature, 69* (1904), 463; *PTRS, 204A* (1905), 467–79. Two years later, Barkla succeeded in showing polarization of tertiary x-rays. *PRS, 77A* (1906), 247–55.

[98] See, for example, Thomson, *Conduction of electricity,* 2nd ed. (1906), 322.

[99] Blondlot, *CR, 135* (1902), 666–70, 721–4, 763–6, 1293–5. An early unsuccessful attempt was made by Brunhes, *CR, 130* (1900), 127–30.

[100] Marx, *VDpG, 7* (1905), 302–21; *PZ, 6* (1905), 768–77; expanded in *AP, 20* (1906), 677–722. The use of electric waves as a comparison had been made for the velocity of cathode rays by Des Coudres, *VDNA, 14* (1895), 86–8; *16* (1897), 157–62. See also Wiechert, *AP, 69* (1899), 739–66.

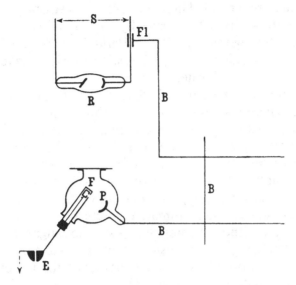

Figure 2.13. Marx' apparatus to measure x-ray velocity. [Franck and Pohl, *VDpG, 10* (1908), 118.]

His apparatus is shown in Figure 2.13. A high negative potential on the curved cathode of x-ray tube *R* causes a burst of cathode rays that impacts at the anode. X-rays are released and travel to electrode *P* of the lower tube. At the same instant, a negative electric impulse is induced in wire *B*, which is attached directly to *P* through the sliding contact on the right.

By 1905 it was well known that x-rays will stimulate the release of electrons from metals. If *P* is negatively charged when the x-rays arrive, electrons will be accelerated to the Faraday cage *F* and thence to the electrometer. This condition is met only if the negative electrical impulse from the wire arrives at *P* simultaneously with the x-rays. Sliding the shunt wire on the right changes the time required for the electric impulse to arrive at *P*. Once he had optimized the shunt for maximum electrometer response, Marx found that the length of wire *B* had to change in step with any alteration of the distance between the discharge tubes. Marx asserted that the x-ray velocity was, within 5 percent, the same as that of the electrical impulse in the wire. It was a common misconception at the time that electrical signals travel through a wire at the velocity of light. For short, uninsulated wires the agreement is very close. Thus, Marx concluded that x-rays also propagate at that velocity. His result was not without opponents, but they were few in number and of little influence.[101] A strong piece of evidence supporting any aether-wave hypothesis of x-rays had been provided. Although this result, like Barkla's, favored neither the pulse nor the periodic-wave hypothesis, it was widely taken as yet more evidence for the former.

By 1905 the impulse hypothesis had achieved virtually unanimous support. There was still a current of thought in Germany that interpreted pulses as if they were constructed of periodic waves. In Chapter 5, we shall see that this was an important influence on early quantum treatments of x-rays in Germany. But in one form or another, the impulse hypothesis of x-rays achieved supremacy in the period 1903–5. The views of H. A. Lorentz, who was widely respected as a consummate theoretical physicist, may be taken to represent those of most scientists. He taught in 1907 that x-rays are "an irregular succession of impacts, each of which persists for a much shorter [time] than does one oscillation of the farthest ultraviolet light yet observed."[102]

---

[101] The most persistent critic was Bernhard Walter, who had a material interpretation of x-rays. *Fortschritte auf dem Gebiete der Röntgenstrahlen, 9* (1905), 223–5. After 1908, Walter's criticism began to find support; see Chapter 6.
[102] Lorentz, *Lehrbuch, 2* (1907), 168.

3

The analogy between γ-rays and x-rays

The γ-rays are very penetrating Röntgen rays [produced
by] the expulsion of the β . . . particle.[1]

The discovery of x-rays reduced much natural reluctance on the
part of physicists to report other discoveries. The most significant
new claim, one recognized within weeks of Röntgen's announcement, was the discovery of radioactivity by Henri Becquerel.[2]
Certain minerals were found to emit rays spontaneously. Not even
the electrical potential difference necessary for the production of
cathode rays and x-rays was required. No human artifice or intervention was needed to produce the Becquerel rays; this was
always their most significant and unique property. For this reason,
it was soon widely believed that the rays were the residue of
spontaneously disintegrating atoms.

However distinct the physical origin of Becquerel rays was from
that of x-rays, telling similarities between the two new emanations
were quickly recognized. Both could penetrate opaque paper to
expose a photographic plate. Like x-rays, Becquerel rays were first
discovered in this way. They seemed to be another invisible radiation, rendered perceivable only through the superhuman sensitivity of photographic plates. It was soon shown that x-rays and
Becquerel rays both induce electrical conductivity in gases through
which they pass. In fact, as we have seen, Becquerel's rays were first

[1] Rutherford, *Nature, 69* (1904), 436–7.
[2] Becquerel, *CR, 122* (1896), 501–3. *Fortschritte der Physik,* the abstracting
journal of the Deutsche physikalische Gesellschaft, first listed research on
Becquerel rays under "Optics" in the section for "phosphorescence and fluorescence." A new subheading in optics for *Becquerelstrahlen* was added in 1900.
Röntgenstrahlen were, from 1896 on, listed as a separate subhead under "electricity and magnetism." Another reorganization in 1903 put Becquerel rays,
together with cathode and x-rays, under "electricity and magnetism."

thought to fill the gap that separated x-rays from ordinary light. Becquerel claimed that they could be refracted and polarized; although soon disproven, these claims served to solidify the view that Becquerel rays resembled light more closely than did x-rays.

For these reasons, Becquerel rays were early on identified as a form of radiation in the aether. Questions were never raised about the generic terms *rays, Strahlen,* and *rayons* used to describe them. Minerals that emitted them were called *radioactive* by the Curies in 1898; *radio,* from the Latin *radix* or *radius,* was selected to emphasize the isotropy of emission and to distinguish it from the ambiguous *rayon,* which in French means both *ray* and *radius.*[3]

Ernest Rutherford began his lifetime research on radioactive phenomena with a study of the conductivity that uranium rays induce in gases. He agreed that the causative agent was likely a true radiation, a view unaltered by his disproof of Becquerel's claims for refraction and polarization.[4] In the course of his first investigations, he found that a sheet of paper or metal foil reduced the ionizing effect of the rays considerably, but additional layers brought no comparable decline. He concluded that two species of ray, one easily absorbed, the other more penetrating, existed side by side; the first he called α to distinguish it from the more penetrating β-component. He then compared the penetrating β-rays to x-rays incident on a metal and described the α-rays in an analogy to the diffuse, and less penetrating, "reflected" x-rays.[5]

The analogy of x-rays and Becquerel rays influenced other discoveries as well. On the heels of Sagnac's demonstration in 1898 that x-rays stimulate the release of electrons from metals, the Curies searched for the same behavior in Becquerel rays. They found that a form of radioactivity could be induced in matter; an ionizing activity was detected that, unlike the secondary emission of electrons, persisted even after the primary irradiation ceased.[6] But this difference appeared to be a consequence of the spontaneous nature of radioactive decay; it did not alter the opinion that the uranium rays were an unusual form of light.

---

[3] P. and [M.] Curie, *CR, 127* (1898), 175–8.
[4] Rutherford, *PM, 47* (1899), 109–63.
[5] The "diffusely reflected" x-rays were largely secondary electrons, but this was not known at the time. See Chapter 4.
[6] P. and [M.] Curie, *CR, 129* (1898), 714–16.

TRANSFORMING WAVES INTO PARTICLES

It came, therefore, as a surprise to many when, in 1899, Becquerel rays were first deflected by a magnet. They bent in precisely the manner expected of negatively charged particles.[7] Soon afterward, it was demonstrated that the ray trajectories also bend in an electrostatic potential.[8] Part of the credit for this recognition must go to the new and more intense sources of radioactivity, polonium and radium, provided by the Curies in 1898. Credit also must go to the widespread interest, particularly in Germany, in techniques for the study of cathode rays by magnetic deflection. In any event, the nature of the Becquerel rays was now in dispute. As one reviewer put it, "not content with attributing to Becquerel rays all the properties of x-rays, physicists and chemists have found that these rays also possess properties hitherto attributed to cathode rays alone."[9]

Pierre Curie showed that Rutherford's division of Becquerel rays into α- and β-components was also expressed in their magnetic deflection. Only the β-component seemed to bend. The α-rays traveled on, apparently unaffected.[10] A photoplate, exposed to uranium rays that had passed through a magnetic field, thus showed two exposed regions. Rutherford's division of Becquerel rays into components offered the first hope of systematic study of uniform properties. No longer were the seemingly variable properties of the rays approachable only through the heroic chemical processes of the Curies, Friedrich Giesel, and others. Around 1899, an effluorescence of research interest occurred in this field, which was increasingly referred to as *radioactivity*.[11] Figure 3.1 shows the number of papers on related topics for the early years cited in an authoritative review by Max Iklé.[12]

[7] Three independent demonstrations were given. The earliest was by Friedrich Giesel at Braunschweig, *AP, 69* (1899), 834–7, following strong hints by Julius Elster and Hans Geitel. See also S. Meyer and von Schweidler, *PZ, 1* (1899), 90–1, 113–14; *CR, 129* (1899), 996–1001.
[8] Dorn, Halle *Abhandlungen, 22* (1901), 39–43, 47–50.
[9] Stewart, *PR, 11* (1900), 155.
[10] P. Curie, *CR, 130* (1900), 73–6.
[11] Becquerel in Guillaume and Poincaré, *Congrès international, 3* (1900), 47–78. Crookes, *PRS, 66* (1900), 409–23.
[12] Iklé, *JRE, 1* (1904), 413–42 is more complete than Wilson, *Nature, 70* (1904), 241–2.

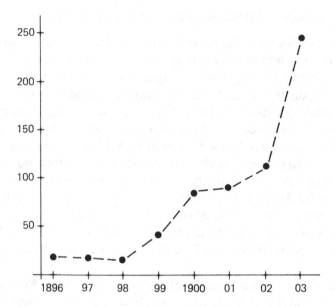

Figure 3.1. Number of studies of radioactivity, 1896–1903.

By 1900 it was clear that the β-rays were high-speed electrons. Measurement of their charge-to-mass ratio (e/m) yielded values on the order of 10^7 emu/g. Their velocities were found to be on the order of 10^{10} cm/sec, higher even than those attained by cathode rays.[13] Reaction to the identification was mixed. Rutherford and Thomson both found it difficult to reconcile the great penetrability of the β-rays with the known properties of the lower-speed corpuscles.[14] R. J. Strutt, the future fourth Lord Rayleigh, undertook a critical reappraisal of x-rays in 1900 to detect a comparable magnetic deflection. He was not able to find one.[15]

Although by 1900 the nature of β-rays had been brought to light, the same was not true for α-rays. Since α-rays seemed to be unaffected by a magnetic field, studies were limited to charting their rapid absorption by matter. Pierre Curie showed in 1900 that any α-ray from radium and polonium lost its ability to ionize after passing through a layer of air whose thickness was characteristic of the radioactive atomic source, that is, independent of the chemical

[13] Becquerel, *CR, 130* (1900), 372–6, 809–15. See also Kaufmann, Göttingen *Nachrichten* (1901), 143–55.
[14] [R. J. Strutt], *Life of Thomson* (1942), 132–33.
[15] R. J. Strutt, *PRS, 66* (1900), 75–9.

compound in which the element was framed.[16] With a qualitative limit thus placed on the penetrability of the undeviated α-rays, the stage was set for the discovery of a third component of Becquerel rays – one that, although not deviated by a magnet, was nonetheless more highly penetrating than the β-rays.

HIGHLY PENETRATING RAYS

Paul Villard, a physicist at the École Normale Superieur in Paris, was a dedicated investigator of cathode rays.[17] His interest had been sparked by study of the chemical effects of both cathode rays and x-rays. Convinced that cathode beams do not actually pass through metal foils, as had been claimed by Hertz and Lenard in Germany, Villard fabricated the implausible hypothesis that cathode rays incident on metal foil stimulate the release of other electrons from the back side of the foil.[18] To test this quaint idea, he developed a research technique of firing cathode beams *along* the plane of a wrapped photographic plate. In this way, one obtained a record of the path followed by the beam both before and after it collided with the metal foil he oriented at right angles to the plate.

This rather recondite research took on new significance in 1900 when Villard added the β-rays from radium chloride to his stock of cathode beam sources. He scattered the beam off a 0.3-mm-thick aluminum sheet and obtained a photograph that he interpreted as showing a "refracted" secondary beam. This, he claimed, had emerged from the back side of the aluminum in a direction normal to the surface of the metal regardless of the angle between the foil and the primary cathode beam. This extravagant but spurious finding was accompanied by more. "I almost always observed," he reported to the Paris Academy in April 1900, "that a straight beam was superimposed on the refracted beam . . . I think it due to the presence of non-deviable rays of lower absorptivity than those described by M. Curie."[19]

Villard corroborated his claim by showing that the new penetrating rays proceed in their straight course even while a magnetic

---

[16] P. Curie, *CR, 130* (1900), 73–6.
[17] Villard, *Rayons cathodiques* (1900); and contribution in H. Abraham and Langevin, *Ions* (1905), 1013–28.
[18] Villard, *CR, 130* (1900), 1010–12.
[19] *Ibid.,* 1012.

field perpendicular to the plate curves the β-beam to the side. Thus, the straight rays could not be the β-component. Nor could they be the α-component; far from being stopped by a few centimeters of air, they passed right through the metal foil, not to mention the 18-cm depth of edge-on photographic emulsion. Soon the find was corroborated by Becquerel in the context of denouncing Villard's wild claim that β-rays are refracted by the foil.[20] Villard himself was somewhat less than pleased about the whole matter because the new rays interfered with his experiments on pure cathode beams. Before he abandoned radium as a cathode beam source, he tried to isolate the properties of what he called the "x-rays from radium," using a strong source borrowed from the Curies for the purpose. He showed that the rays passed through thicknesses of glass and lead that would stop all β-particles (and, of course, all α-rays too). Further, they would expose more photo plates piled in a stack than would either the α- or the β-component.[21]

Villard was not alone in his lack of interest in the new rays; very little effort was expended on their study by anyone in the first few years. One reason was that the discovery came in the midst of attempts to obtain accurate e/m measurements of the β-particle. In addition, the new component was exceedingly weak. Using ionization techniques, Curie had not discerned its existence in his careful studies just prior to Villard's photographic experiment. Rutherford estimated that the new undeviable rays from radium had only 1 percent of the ionizing effect of the accompanying β-rays and only 0.01 percent of that of the α-rays. Finally, there were only two known sources of the rays: thorium and radium. Thorium produced them at a low rate, in keeping with its generally lower activity compared to radium. To study the nondeviable, penetrating rays, one therefore had to have radium, a commodity to which only a handful of physicists had access in the years 1900–5. Although Becquerel, for example, made repeated references to the existence of the nondeviable rays between 1901 and 1903, he studied their properties only once.[22] He showed a curious effect: They darkened a photographic plate more strongly on the

[20] Becquerel, *CR, 130* (1900), 1154–7.
[21] Villard, *CR, 130* (1900), 1178–9.
[22] Becquerel, *CR, 132* (1901), 1286–9; *134* (1902), 208–11; *136* (1903), 431–4.

half covered by a metal sheet.[23] The problem in studying the penetrating rays, he remarked, was "the difficulty one has in stopping them [at all]."[24]

The real topic of research in radioactivity, one that overshadowed any questions about the new rays found by Villard, was the continuing effort to find clues to the nature of α-rays. Many physicists now doubted that these were rays because of their lower penetrability relative to the β-particles. The case for waves had always been weakest for α-rays. "The theory that [both components in] Becquerel rays consist of extremely short ultraviolet light waves is now almost abandoned," wrote Oscar Stewart in 1900.[25] A great effort was made, particularly by Rutherford, to demonstrate magnetic deflection of the α-component. In late 1902, he succeeded.[26] Previously, Rutherford had looked briefly at the as yet unnamed "highly penetrating" rays from radium. He stopped all α- and β-rays with 1 cm of lead, and then tested what remained in the very strong magnetic field he was preparing for his assault on the α-rays. He found no deflection of the third component.[27]

Rutherford had completely changed his view of the nature of radioactive products. By 1902 he was convinced that all three components were streams of particles. The ionization produced in gases by the undeviated rays showed, he claimed, that they "have a closer resemblance to kathode than to Röntgen rays."[28] The rate at which they formed ions in a gas was closer to the rate that had been found for cathode rays and undifferentiated Becquerel rays by J. J. Thomson and by Strutt.[29] Rutherford pointed out the possibility that harder x-rays, as penetrating as the new γ-rays, might match their ionizing properties more closely. But he suggested explicitly that the penetrating γ-rays were electrons; they did not react noticeably to a magnet, he said, because their velocity was essentially equal to that of light itself.

[23] Becquerel, *CR, 132* (1901), 371–3.
[24] Becquerel, *CR, 133* (1901), 710.
[25] Stewart, *PR, 77* (1900), 175.
[26] Rutherford, *PM, 5* (1903), 177–87. For a detailed discussion, see Heilbron, *AHES, 4* (1968), 247–307.
[27] Rutherford, *Nature, 66* (1902), 318–19.
[28] *Ibid.*
[29] J. J. Thomson, *PCPS, 10* (1898), 10–14. R. J. Strutt, *PTRS, 196A* (1901), 507–27.

In 1902 the penetrating rays were given the name γ following the informal but universally accepted taxonomy established by Rutherford.[30] This action signaled the first sustained research interest in γ-rays. Marie Curie, in her thesis on radioactive substances presented in 1903, drew a conceptually crucial analogy between the three components of radioactivity and the known products of electrical discharge: α-rays can be compared to the positively charged molecules called *canal rays,* β-rays to cathode rays, and γ-rays to x-rays. In numerous comparisons between the last pair, she implied strongly that γ-rays arise from the impact of very-high-velocity β-particles upon the atoms within the sample of radium itself.[31] Her diagram, reproduced as Figure 3.2, symbolized the tripartite division of decay products by their relative deviation in a magnetic field. Becquerel was convinced and Rutherford was influenced by Curie's presentation.[32] In a talk to the London Physical Society in June, Rutherford reversed his position yet again, agreeing that γ-rays were "probably" equivalent to x-rays.[33] In his talk to the British Association that September, however, Rutherford demurred; insufficient knowledge, he said, prevented a definite conclusion.[34]

Not everyone agreed with the triumvirate of Rutherford, Curie, and Becquerel. R. J. Strutt recognized only belatedly that his 1901 investigation of the ionizing power of α-rays had been confounded by the presence of γ-rays. Accordingly, he returned to the subject.[35] He thought Curie's interpretation of the γ-rays tempting, but his own data indicated otherwise. He measured the ionizing strength of the rays by the time needed for a given electrical charge to develop or disperse in his electrometer. His table (Figure 3.3) shows clearly that the relative ionization produced in several gases by γ-rays agreed closely with that produced by α- and β-rays, approximately in proportion to the density of the gas. Even the hardest x-rays, those produced by the greatest potential differences, failed to mimic γ-rays. He suspected that γ-rays were parti-

[30] Trenn, *Isis, 67* (1976), 61 – 75.
[31] M. Curie, *Thèses* (1903), 48, 80, 83, 138.
[32] Becquerel, *CR, 136* (1903), 1517 – 22.
[33] Rutherford, *Nature, 68* (1903), 163.
[34] *Nature, 68* (1903), 610.
[35] R. J. Strutt, *PRS, 72* (1903), 208 – 10.

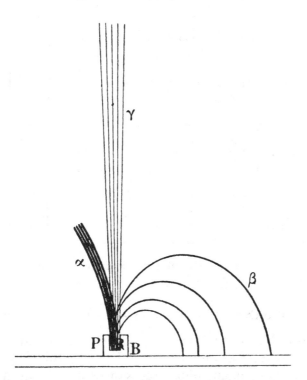

Figure 3.2. Marie Curie's schematic diagram of differential magnetic deflection of radioactive emanations. [*Thèses* (1903), 49.]

cles. Strutt believed that electromagnetic impulses would arise from the β-particle impacts, but he thought them simply too weak to detect. The net β-flux from radium, he pointed out, is some nine orders of magnitude less than the expected cathode ray flux in a typical laboratory electric discharge. Although he offered no empirical evidence for the claim, Strutt explained the lack of magnetic deflection and extraordinary penetrability by suggesting that the γ-particles had lost their charge. By late 1903, discounting the freewheeling speculations of the aged Lord Kelvin,[36] three views were competing for the description of γ-rays: short transverse x-ray impulses, neutral material particles, and extremely high-velocity electrons.

---

[36] Kelvin thought the "simple prima facie view" was that γ-rays are uncharged free atoms of radium. BAAS *Report* (1903), 535–7; *PM, 7* (1904), 220–2.

Relative Ionisations.

Gas.	Relative density.	Relative Ionisation.			
		α rays.	β rays.	γ rays.	Röntgen rays.
Hydrogen	0·0693	0·226	0·157	0·169	0·114
Air	1·00	1·00	1·00	1·00	1·00
Oxygen	1·11	1·16	1·21	1·17	1·39
Carbon dioxide	1·53	1·54	1·57	1·53	1·60
Cyanogen	1·86	1·94	1·86	1·71	1·05
Sulphur dioxide.......	2·19	2·04	2·31	2·13	7·97
Chloroform...........	4·32	4·41	4·89	4·88	31·9
Methyl iodide.........	5·05	3·51	5·18	4·80	72·0
Carbon tetrachloride....	5·31	5·34	5·83	5·67	45·3

Figure 3.3. R. J. Strutt's data on relative ionization of gases by α-, β-, γ-, and x-rays. [*PRS, 72* (1903), 209.]

CORROBORATING THE γ-RAY–X-RAY ANALOGY

J. J. Thomson thought that increasing the density of radioactive matter might increase its rate of decay.[37] He made this suggestion in the course of an enlightening discussion of atomic disintegration as the mechanism of radioactive decay. The longevity of the decay from radium, he proposed, might be explained if only a small fraction of the total number of atoms decay at any instant. The potent ones, he claimed, were those with abnormally high kinetic energy. A decaying atom might raise its closest neighbors to potent energy if the material was packed closely enough. The suggestion appealed to Rutherford, who had moved meanwhile from England to McGill University in Canada. But when Rutherford did some tests, he found that Thomson's predicted increase in activity with sample density did not occur.[38]

A reader of *Nature,* where Rutherford published this result, pointed out that radium γ-rays could not arise from self-bombardment.[39] He alluded to Stokes' opinion that an impulse similar to an x-ray should arise not only where a charged particle is stopped but also where it starts. Might not the γ-rays be due to the acceleration

[37] J. J. Thomson, *Nature, 67* (1903), 601–2.
[38] Rutherford, *Nature, 69* (1904), 222.
[39] Ashworth, *Nature, 69* (1904), 295.

of α- and β-particles *out* of the disintegrating atom? Rutherford agreed, and early in 1904 undertook a critical reevaluation of the mechanism proposed by Marie Curie for β-induced emission of γ-rays. He encouraged his elder colleague, A. S. Eve, to analyze carefully the data Strutt had amassed on the relative ionization of various radioactive species. The following month, Eve reported that the rate of ionization produced by the most highly penetrating x-rays was much closer than previously suspected to that of the γ-rays.[40] (See Figure 3.4 for a comparison of the old and new data.)

Rutherford stepped in immediately to strengthen the new hypothesis by identifying γ-rays as the electromagnetic impulse produced when β-particle electrons exit at very high velocity from a disintegrating atom. "All the experimental evidence so far obtained," he claimed, "is now in agreement with the view that the γ-rays are very penetrating Röntgen rays which have their source in the atom of the radioactive substance at the moment of the expulsion of the β or kathodic particle."[41] "I have found," he went on to explain, "that the γ-rays from radium always accompany the β-rays and are alway proportional in amount to them . . . Active products which give α-rays and no β-rays do not give rise to γ-rays. The β and γ-rays appear always to go together and are present in the same proportion."[42]

If γ-rays are highly penetrating x-ray impulses, they should carry no charge. Supporting evidence for this claim was offered on the heels of the Eve–Rutherford data. James McClelland, a former Cavendish student working in Dublin, passed γ-rays through an insulated cylinder filled with lead shot. The cylinder was long enough so that all the rays were absorbed. With due care to eliminate the effect of ionization in the surrounding air, McClelland showed that no net charge was deposited in the cylinder. "The weight of evidence is certainly in favor of the γ-rays being similar to Röntgen rays," he concluded.[43]

McClelland asked why the γ-rays should be absorbed by gases in approximate proportion to the gas density. For α- and β-particles, indeed for any particle, the law follows directly from the simple assumption that the particle is smaller than the atoms of the target

[40] Eve, *Nature, 69* (1904), 436; *PM, 8* (1904), 610–18.
[41] Rutherford, *Nature, 69* (1904), 436.
[42] Rutherford, *Radioactivity* (1904), 144–5.
[43] McClelland, RDS *Sci trans, 8* (1904), 99–108; *PM, 8* (1904), 76.

Gas.	Relative Density.	Previous values.	Present values.
H	·07	·11*	·42
Air	1·0	1·0	1·0
H_2S	1·2	6	·9
SO_2.......................... ...	2·2	8	2·3
Chloroform	4·3	32	4·6
Carbon Tetrachloride	5·3	45	4·9
Methyl Iodide	5·0	72	13·5

* Strutt ·11; J. J. Thomson ·33; Rutherford ·5; Perrin ·026.

Figure 3.4. Eve's data revising knowledge of ionization by hard x-rays. [*PM, 8* (1904), 613.]

and therefore interacts not with entire atoms but with their constituent electrons. Equal volumes of gases contain equal numbers of molecules, so that the gas density is proportional to the molecular weight. According to Thomson's influential ideas about atomic structure, the number of electrons in an atom is proportional to its atomic weight; thus, it is also proportional to the density of the gas. A particle of subatomic dimension that passes through the gas will lose energy in proportion to the number of electrons it encounters. Therefore, particle absorption should be proportional to the density of the gas.

McClelland was uneasy, however, about whether spreading waves could be absorbed by gases in the same proportion. "It would seem difficult to explain such a law," he said.[44] He did not have in mind the difference one might imagine to exist between the absorption of periodic waves and impulses by bound electrons.[45] Rather, McClelland was echoing reservations that his former teacher, J. J. Thomson, had recently expressed about the isotropic

[44] McClelland, *PM, 8* (1904), 77.
[45] For a periodic wave, not all electrons have natural frequencies sufficiently close to resonance to be appreciably excited. But a sharp impulse will excite all electrons. An organ tone transfers energy only to the closely tuned strings of a harp; a pistol shot excites them all.

character of x-ray pulses; McClelland tried to find justification for Thomson's nascent ideas about the structure of the aether. If a spherical impulse or periodic wave spreads over areas that are large with respect to the size of the atoms, it will no longer, McClelland asserted, interact with constituent electrons. Rather, it must be absorbed by entire atoms, or groupings of atoms. Since the number of molecules per unit volume of a gas is constant for the same pressure and is independent of the gas density, McClelland doubted that spherical impulses could be absorbed in proportion to the gas density. Nor was he alone in his concern. William Henry Bragg in Australia raised the same issue more forcefully at the same time. But Thomson's and Bragg's important views will be considered in the next chapter.

Eve's discovery that hard x-rays follow the same law of absorption in gases as do particles permanently tied the fortunes of the γ-ray to those of the x-ray from 1904 on. The results appeared only in a footnote in Rutherford's book *Radioactivity*, published that year.[46] They brought an end to the cautious balancing of three alternative models for the γ-rays that filled the text itself. But in the second edition of the book, published the following year, Rutherford's discussion of γ-rays is extensively simplified and reorganized around the newly justified impulse hypothesis. The idea of impulse γ-rays was widely circulated in Rutherford's book, which was quickly recognized as a working digest of essential research in radioactivity and was published in several editions over the next decade.

FRIEDRICH PASCHEN AND THE FAST-ELECTRON HYPOTHESIS

At just the time that McClelland showed that γ-rays possess no electrical charge, L. C. H. Friedrich Paschen, noted *Ordinarius* at Tübingen, published what he took to be evidence to the contrary. He had enclosed a sample of radium bromide in a lead box sufficiently thick to retain all α- and β-particles, isolated the box in a vacuum, and found that it developed a spontaneous positive charge.[47] Since the only things able to escape from the box were

[46] Rutherford, *Radioactivity* (1904), 144n.
[47] Paschen, *AP, 14* (1904), 164–71. Paschen was not the first to note the self-charging capability of radium. See M. Curie, *Thèses* (1903), and Dorn, *PZ, 4* (1903), 507–8.

γ-rays, he concluded that they must carry a negative charge. Paschen was convinced that γ-rays were ordinary electrons traveling at speeds greater than even the 2.8×10^{10} cm/sec of some β-particles. The rays had shown no magnetic deflection, he said, only because a strong enough field had not been developed. He cited recent theoretical discussions that proposed a "unit of electricity that moves at the velocity of light," and thought that he had found it in γ-rays.[48]

An elegant but unsound second experiment further confirmed Paschen's view. He used an intense magnetic field to force the β-particles from radium into spiral paths so tight that they could not reach a surrounding lead ring. The ring, shown in cross section in Figure 3.5, is marked *a;* the radium sample is at *b*. Electrons from the sample are forced into paths of circular projection in the plane of the magnet pole faces *N* and *S*. Only the γ-rays could reach the ring, the α-rays having been stopped by the container surrounding the sample. Paschen found that a negative potential developed on the ring, and claimed that it was caused by a negative charge deposited there by the γ-rays.[49] Paschen's clever experimental technique was an extension of prior work by Becquerel and H. Starke.[50] It was later developed into what we now call *β-ray spectroscopy*. But here it led Paschen astray.

Paschen's conclusion was clouded by his total neglect of secondary electrons; he had the misapprehension that only cathode rays could stimulate secondary electrons in matter.[51] He utterly failed to take into account the possibility that γ-rays, like x-rays, might do the same. In his first experiment, secondary electrons might be emitted from the outside of the lead shield, leaving a net positive charge; in the second, electrons released from the surrounding vacuum vessel walls could easily make their way back to the lead ring. The reaction was quick; both McClelland and Eve immediately pointed out Paschen's error.[52] A few of Paschen's results repeated work done in 1899 and 1900 by Becquerel. The latter took an unusual step for a Frenchman and sent a detailed review of his

[48] Paschen, *AP, 14* (1904), 171.
[49] *Ibid.*, 389–405.
[50] Becquerel in Guillaume and Poincaré, *Congrès international, 3* (1900), 63–9. Starke, *VDpG, 5* (1903), 14–22.
[51] Paschen, *PZ, 5* (1904), 502–4.
[52] McClelland, RDS *Sci trans, 8* (1904), 99–108. Eve, *Nature, 70* (1904), 454.

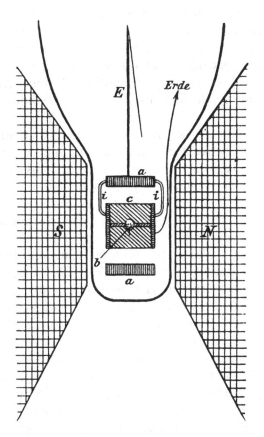

Figure 3.5. Diagram of Paschen's experiment that made him think that γ-rays deposit negative charge. [*AP, 14* (1904), 390.]

work to the editors of the *Physikalische Zeitschrift* for translation into German in the hope that it would "show what the state has been for over a year of the issues that Herr Paschen has taken up again."[53]

Before Paschen's mistake became clear to him, another of his papers appeared; this one contained an estimate of the velocity of the presumed electronic γ-rays. The lack of deflection in a strong magnetic field he took as evidence that their e/m must be less than 10^3 emu/g (in comparison to 10^7 for the β-particle).[54] He cited

[53] Becquerel, *PZ, 5* (1904), 561–3. This was the first of only a few papers Becquerel ever published in German.
[54] Paschen, *PZ, 5* (1904), 563–8.

Abraham's theory and Kaufmann's experiments in order to exploit the idea that the mass of an electron increases as its velocity approaches c. A mass increase by a factor of 10^4 would explain the discrepancy. Paschen concluded that his γ-ray electrons might travel within $10^{-40,000}$ of the velocity of light! The Lorentz transformation predicts only a velocity within 5 parts in a billion (5×10^{-9}) of c.

To support these claims, Paschen turned to measurements of the heat evolved in radioactive decay. If the kinetic energy of each γ-particle is, in fact, as large as that of 40 hydrogen atoms moving very close to the velocity of light, the γ-component should carry away most of the energy. A widely accepted estimate of energy emission held that 1 g of radium evolves 98.5 cal/hr, not counting the γ-rays.[55] So, Paschen attempted the nearly impossible: He tried to determine by ice calorimetry the total heat produced by radium enclosed in so much lead that even the γ-rays were trapped. It was by pure chance that he got any result at all; thermal equilibrium in the massive lead shield could not occur as quickly as calorimetry requires. But miraculously, he found a result, 226 cal/g-hr, that corroborated his conclusion: The γ-component alone radiates more than half of the total energy of radium. Citing data assembled by Wilhelm Wien on β-particles, Paschen concluded that each β-particle carries only 1/74th the energy of a single γ-electron. Thus, he concluded, "the γ-rays cannot be the Röntgen effect of the β-rays" because the latter do not have enough energy.[56]

Rutherford had previously studied the heating effect of the various components of radioactive decay in radium.[57] He had completed a short review of the subject just before Paschen's claims appeared.[58] In light of his experience, Paschen's results seemed simply fantastic. "This result was so unexpected," he wrote, "and was of so much importance in considering the nature of the γ-rays, that we decided to verify the experiments."[59] Before Rutherford could put more than a preliminary result in print, Paschen himself disavowed the issue. He had found that he could not duplicate the calorimetric results during the winter months,

[55] Precht, *VDpG*, 6 (1904), 101–3. Runge and Precht, Berlin *Sb* (1903), 783–6.
[56] Paschen, *PZ*, 5 (1904), 567.
[57] Rutherford and McClung, *PZ*, 2 (1900), 53–5. For a description of Rutherford's research effort, see Heilbron in Bunge and Shea, *Rutherford* (1979), 42–73.
[58] Rutherford, AusAAS *Report* (1904), 86–91.
[59] Rutherford and Barnes, *PM*, 9 (1905), 623.

which was clear evidence that his experiments hid systematic errors.[60] Rutherford's experiments showed that the γ-rays carried only a small percentage of the total heat from radium. Most of the energy resided in the α particles. "No appreciable increase in heating effect was brought about by absorbing the γ-rays," he reported, and the result was corroborated later that year by Knut Ångström.[61]

In the meantime, Paschen's claim that an electrical charge is deposited by γ-rays was destroyed by Thomson, developing McClelland's earlier work. Thomson tested the relative discharge rate of two brass cylinders, both hollow, closed, and electrically isolated in a vacuum.[62] One was empty, the other filled with lead shot. Each was irradiated along its axis with the γ-rays from radium bromide. Secondary electrons stimulated *inside* the metal are quickly reabsorbed. Thus, if the γ-rays can deposit a charge, whether positive or negative, the increase of charge should always be greater for the lead-filled cylinder. If, on the other hand, the γ-rays are uncharged, and affect electrical potential only by the stimulation of secondary electrons *at the surface* of the metal, then the discharge rate of the empty cylinder should be greater. Both ends of the empty cylinder emit secondary electrons, but only the near end of the lead-filled cylinder radiates them because virtually all of the γ-rays are absorbed in passing through the lead. In every case, Thomson found that the discharge rate was greater for the empty cylinder, and it always indicated a loss of negative charge. This showed that Paschen's results were entirely due to secondary electrons, and Paschen acknowledged this fact.

THE STATUS OF THE γ-RAY–X-RAY ANALOGY

In the years after 1905, γ-rays took a permanent place as the radiation component of radioactivity decay. Indeed, the verification of the threefold identification of radioactive decay products with products of electric discharge brought the final severing of ties between the two fields. The editors of the *Annalen der Physik* removed the newer field bodily from the section *Elektricitätslehre*

---

[60] Paschen, *PZ, 6* (1905), 97.
[61] Rutherford and Barnes, *PM, 9* (1905), 626. Ångström, *PZ, 6* (1905), 685–8.
[62] J. J. Thomson, *PCPS, 13* (1905), 121–3.

in their classifactory scheme, elevating *Radioaktivität* to a status equivalent to that of electricity.[63]

R. J. Strutt, who had presented the strongest opposition to the γ-ray–x-ray analogy in Britain, conceded the case in 1904. He described the nature of the γ-rays as "one of the most obscure questions in connection with radioactivity," but concluded his survey with the remark that γ-rays were "probably x-rays."[64] In the second edition of *Radioactivity*, Rutherford rehearsed the arguments against Paschen's interpretation and concluded that "the weight of the evidence, both experimental and theoretical, at present supports the view that the γ-rays are of the same nature as the x-rays but of a more penetrating type."[65] In Britain, where the impulse hypothesis for x-rays had been accepted most firmly, the extension of the hypothesis to include the γ-rays was complete in 1904.

In Germany there was greater diversity of opinion. Paschen's debacle is indicative of this fact. Aside from his eminent position at Tübingen, Paschen had held an international reputation since 1895 for his work in optical and infrared spectroscopy. That he was simultaneously ignorant of possible secondary electrons, and willing to exhibit that ignorance in defense of a corpuscular hypothesis of γ-rays, is symptomatic of the greater resistance in Germany to the facile acceptance of the γ-ray–x-ray analogy. It also indicates the difficulty of isolating the purely wavelike properties of the penetrating γ-rays. For many in Germany, the requisite hard evidence had simply been lacking before the reaction to Paschen. Even after the barrage of evidence from Britain, occasional voices of opposition were heard. Otto Wigger, for example, refused to adopt the analogy in his 1905 review of γ-ray absorption.[66] But by then his was a minority view, even in Germany.

There was no serious resistance to the identification of γ-rays with x-rays following Paschen's experience. The reaction to Paschen's advocacy showed beyond doubt that negatively charged particles are released from metals by the action of γ-rays. Becquerel had noted in 1901 that a wrapped photoplate exposed to

[63] See the tables of contents of the *Beiblätter zu den Annalen der Physik* for the relevant years.
[64] R. J. Strutt, *Becquerel rays* (1904), 86.
[65] Rutherford, *Radioactivity*, 2nd ed. (1905), 186.
[66] Wigger, *JRE*, 2 (1905), 391–433.

γ-rays was more heavily exposed if the plate was covered by a metal sheet.[67] He suggested that this result was probably caused by the stimulated emission from the metal of some photoreactive ray that was more easily absorbed by the emulsion than were the γ-rays; it was clear now that these secondary emissions were electrons. Eve thought in 1904 that, like x-rays, γ-rays stimulate both negatively charged particles and secondary γ-rays.[68] McClelland jumped to the conclusion that the particles are β-electrons, "the mass and the charge being no doubt the same," although he had no evidence to corroborate the claim.[69] Indeed, no serious study to determine the e/m ratio for the secondary particles was undertaken until 1906.[70] The result corroborated McClelland's claim, but by then it was widely expected. X-rays stimulate secondary electrons; so should γ-rays.

The γ-ray – x-ray analogy was settled by 1905. Shortly thereafter, continental leaders such as Lorentz in theoretical physics and Walther Nernst in physical chemistry accepted the analogy as an accurate pedagogical classification, if not an explanation, of γ-rays.[71] There was a marked decline in research on γ-rays between 1905 and 1908, and this too may be explained by the orthodox analogy. As soon as one could unambiguously associate γ-rays with the impulse model of x-rays, they lost allure as a research topic. Experimental study of either ray offered data on both, and x-rays were considerably easier to study.

The strength of the γ-ray – x-ray analogy held quite generally until the early 1920s. Albeit based more on similitude than on hard data, it remained a rock in the sea of uncertainty that would surround attempts to specify the precise nature of both forms of radiation in the intervening years. What came under discussion first in 1907 and increasingly thereafter was no longer the similarity of x-rays to γ-rays, but rather the impossibility of describing either in any electromechanically consistent manner.

[67] Becquerel, *CR, 132* (1901), 371–3.
[68] Eve, *PM, 8* (1904), 669–85.
[69] McClelland, RDS *Sci trans, 8* (1904), 169–82; *PM, 9* (1905), 230–43.
[70] Allen, *PR, 23* (1906), 65–94.
[71] Lorentz, *Lehrbuch, 2* (1907), 587. Nernst, *Theoretische Chemie*, 5th ed. (1907), 410. Rutherford inserted a new chapter on γ-rays in the second edition of *Radioactivity* (1905).

PART II
Ionization and the recognition of paradox
1906–1910

4

Secondary rays: British attempts to retain mechanism

> There is . . . a reasonable argument that the γ and
> x rays are also material.[1]

When x-rays or γ-rays strike atoms, electrons and secondary x-rays are emitted. Experiments to sort out the differing properties of the two secondary components produced new and perplexing observations in the first decade of this century. The most perplexing problems concerned the effect of the rays in producing secondary electrons. First, x-rays and γ-rays ionize only a small fraction of the total number of gas molecules through which they pass. Spherically expanding pulses should affect all molecules equally; manifestly, they do not. Second, the velocity that x-rays impart to electrons is many orders of magnitude higher than one would expect to come from a spreading wave. The energy in the new radiations seemed to be bound up in spatially localized packages, available to an electron *in toto.*

British physicists responded to these paradoxes with attempts to revise, rather than replace, classical electromechanics. J. J. Thomson suggested that old ideas about the microscopic structure of the aether might have to be reformulated. William Henry Bragg concluded that x-rays and γ-rays are not impulses at all but rather neutral material particles. Bragg and Thomson tried at first to find explanations using models based on human experience with machines. Each sought a resolution within the context of classical mechanics; each ultimately failed. But the discussion of basic concepts that resulted enriched physics in two ways. Materially, it stimulated Charles Barkla to identify a new form of secondary x-ray, one that is homogeneous in penetrating power and charac-

[1] W. H. Bragg, *TPRRSSA, 31* (1907), 97.

teristic of the atom in which it is stimulated. Conceptually, the discussion directed concern to the difficulty of formulating *any* theory of x-rays and γ-rays that answered the dual paradoxes encountered in the study of their ionizing power, as we shall see in this chapter.

THE ENERGY OF SECONDARY ELECTRONS

Early attempts to determine the energetic properties of x-rays focused on the heat they produce in irradiated bodies. Röntgen had predicted but not investigated the effect.[2] An unfairly forgotten physicist, Ernst Dorn, professor at Halle, irradiated metal foils enclosed within one of two connected glass vessels in 1897. The enclosed air was heated, the pressure rose, and the expansion could be read by the movement of a fluid plug in the tube that connected the vessels. By comparing the effect caused by a known electric current, Dorn derived approximate values for heating by x-rays.[3]

Ernest Rutherford recognized the importance of the heating effect to his earliest studies on radioactivity. In his first year in Canada, he found a capable assistant in Robert McClung. Together they measured the change in electrical resistivity of a platinum strip bolometer heated by x-rays.[4] They then measured the fraction of ionizing power lost by x-rays in traversing a unit thickness of air. The two results taken together gave an estimate of the total energy extracted from the beam by a known volume of air. Measurement of the total charge released then allowed Rutherford to calculate the approximate energy needed to ionize a single nitrogen atom: 3.8×10^{-10} erg. Their goal was to estimate from the measured effect of radium on the same gas the total energy available in the radioactive rays. It was this prior work that gave Rutherford the instant conviction that Paschen's calorimetric results must have been wrong.

All of these early experiments measured macroscopic effects, the net result of a large number of interacting x-rays and secondary products. None was capable of giving information on the detailed energy balance of single x-ray impulses. Moreover, early investiga-

---

[2] Röntgen I, p. 5.
[3] Dorn, *AP, 63* (1897), 160–76.
[4] Rutherford and McClung, *PTRS, 196A* (1901), 25–59.

tions were confounded by a mixture of two secondary effects; in addition to electrons, bodies irradiated by x-rays also emit secondary x-rays. At first, both were referred to as *diffuse reflection* of x-rays. The dual nature of these secondary effects was unrecognized, but it complicated experimental study. One had to learn first to distinguish the secondary x-rays from the particle flux, at the beginning called *secondary rays,* that only later came universally to be identified as electrons.

During his initial researches, Röntgen discovered that x-rays discharge electrified bodies, but he did not mention this in his first paper. Immediately, a number of physicists lodged claims for priority in showing that the rays neutralize electrically charged objects.[5] Röntgen's second paper was devoted to this effect.[6] Most people agreed that the action bore a remarkable similarity to the photoelectric effect, which was at the time incompletely understood. But the difference was even more telling: Whereas ultraviolet light discharged only a negative electrical potential, x-rays always acted, apparently oblivious to the sign of the initial charge.[7]

Only after Thomson's isolation of the electron in cathode rays in 1897 did the study of microscopic energy relations become feasible. The mixture of electrically active particles indiscriminately called *ions* produced in gases by the passage of x-rays suggested that ions, like the cathode rays, were mostly electrons.[8] By 1900, investigations on both sides of the English Channel had corroborated the fact that x-rays and ultraviolet light stimulate the release of electrons from solids and gases.[9] The difference lies in the velocity with which each is emitted. Lenard had shown that a typical velocity for photoelectrons was 10^8 cm/sec.[10] In 1900 Dorn magnetically deflected the electrons released by x-rays and inserted the value of e/m known for cathode rays.[11] The derived velocity

---

[5] Among the claimants were J. J. Thomson, *PRS, 59* (1896), 274–6; Righi, Bologna *Memorie, 6* (1896), 231– 301; and Benoist and Hurmuzescu, *CR, 122* (1896), 235–6.

[6] Röntgen II.

[7] For contemporary understanding, see Wheaton, *Photoelectric effect* (1971).

[8] J. J. Thomson, *PM, 46* (1898), 528–45.

[9] The major study besides Thomson's was by P. Curie and Sagnac, *CR, 130* (1900), 1013–15.

[10] Lenard, *AP, 2* (1900), 368.

[11] Dorn, *Archives Néerlandaises* (2), *5* (1900), 595–608. The original measurements appeared in Dorn, Halle *Abhandlungen, 22* (1900), 37–43.

exceeded 10^9 cm/sec, which was equivalent to the recently determined velocity of cathode rays.[12]

Dorn's result did not arouse the interest that in retrospect might have been expected. He had shown that the velocity of secondary electrons produced by x-rays is of the same order of magnitude as that of the electrons whose impacts produce the x-rays in the first place. This is surprising because the energy density in the x-ray would be expected to decrease rapidly as the x-ray spreads spherically and traverses macroscopic distances. Nonetheless, the *full* energy of the ray seemed to be concentrated on an individual electron in the irradiated body. Dorn did not call attention to this issue, nor is there reason to suspect that he recognized its peculiarity. He admitted that irregularities in the output of his x-ray tube might have clouded the result. The work was not published in a major journal. Dorn himself was not considered to be a leading experimentalist – competent, but not distinguished.

THE TRIGGERING HYPOTHESIS

By the time that Dorn's empirical result was widely recognized to be accurate, another way of explaining it was available; thus, the potential problem went unrecognized for another five years. Philipp Lenard completed important experiments on the photoelectric effect in 1902.[13] He uncovered the surprising fact that the maximum velocity with which electrons are ejected by ultraviolet light is entirely independent of the intensity of the light. This result convinced him that there could be no transformation of light energy into electron kinetic energy. Instead, he proposed that electrons in an atom *already* possess their photoelectric velocity, or the potential energy equivalent, by virtue of their membership in the atomic system. The light only triggers the release of selected electrons; it does not add energy to them. Until 1911 this *triggering*

[12] Wiechert, *AP, 69* (1899), 739–66.
[13] Lenard, *AP, 8* (1902), 149–98. Despite claims to the contrary, even by historians, Lenard did not here demonstrate that photoresponse varies with light frequency. He only claimed that it varies with the type of light used – arc light, spark light, or the type of electrode metal. The word *frequency* appears nowhere in his study; *wavelength* appears only once, and in a different context. Compare Jammer, *Conceptual development* (1966), 35; Kleinert and Schönbeck, *Gesnerus, 35* (1978), 318; Hendry, *Annals of science, 37* (1980), 64.

hypothesis formed the basis of almost all physicists' understanding of the photoelectric effect.[14]

The triggering hypothesis owed some of its appeal to its compatibility with contemporary hypotheses of atomic disintegration. It seemed to promise a means of investigating the mysterious decay of radioactive elements. Thomson adopted triggering as an explanation of the photoelectric effect in 1905.[15] Between 1905 and 1907, his student, J. A. McClelland, devoted a series of studies to the corroboration of the triggering hypothesis for the emission of β-particles by γ-rays. McClelland's chief weapon in this study was the observed monotonic increase he found in the total ionizing power of the secondary rays, including what he thought were secondary γ-rays, with the atomic weight of the scattering material. He took this as evidence that the primary beam, a mixture of β- and γ-rays from radium, could induce atomic disintegration in matter.[16] If the electrons were free before interacting with the rays, McClelland explained, "it is difficult to see how such remarkable relations should exist between the atomic weight and the intensity of the secondary radiation."[17]

It was no great step to expand the capabilities of the triggering hypothesis to include the excitation of electrons by x-rays. Wilhelm Wien in Würzburg concluded in 1905 that the triggering mode offered the only realistic explanation.[18] Thomson publicly announced his adoption of the triggering hypothesis for x-ray ionization in 1906.[19] His student, P. D. Innes, found direct supporting evidence the following year; like the velocity of photoelectrons, the velocity of the secondary electrons from x-rays appears to be fully independent of the intensity of the incident radiation.[20]

For a brief period in 1906, the influence of the triggering hypothesis was at its height. It was widely thought to explain the observed velocities of electrons ejected from matter by light, by x-rays, and

[14] Wheaton, *HSPS, 9* (1978), 299–323.
[15] J. J. Thomson, *PM, 10* (1905), 584–90.
[16] McClelland, RDS *Sci trans, 9* (1905), 1–8, 9–26, 37–50. Righi also found evidence of a correlation with atomic weight. Lincei *Atti, 14* (1905), 556–9.
[17] McClelland, RDS *Sci trans, 9* (1905), 36.
[18] Wien, *AP, 18* (1905), 991–1107. See Chapter 5.
[19] J. J. Thomson, *PCPS, 13* (1906), 322–4; *Corpuscular theory* (1907), 320–1. Some of the evidence that convinced Thomson came from Bumstead, *PM, 11* (1906), 292–317.
[20] Innes, *PRS, 79A* (1907), 442–62.

by γ-rays. Indeed, for a while it seemed as if the study of these velocities would reveal the distribution of electron energy within the atom and shed light on the mysterious process of radioactive disintegration. These were two of the most pressing fields of physical research in the first decade of the century.

THE PARADOX OF QUANTITY

J. J. Thomson had one serious reservation about the impulse explanation of x-ray ionization. His early experimental studies had convinced him that only a vanishingly small fraction of gas atoms is ionized by x-rays.[21] It is difficult to see how any spreading wave could fail to affect all atoms in its path equally. Each x-ray impulse spreads outward from its source. In any particular direction along an x-ray "beam," the diverging pulse should disturb all atoms to the same extent. Indeed, the lateral area of the x-ray beam is of macroscopic dimension, and all atoms within the beam should be affected the same way. Röntgen had shown in his third study that the intensity of the x-ray beam did not decrease appreciably until one reached very large angles from the normal to the anticathode.[22]

Thomson had quickly exploited the discovery that x-rays can induce conductivity in a sample of gas through which they pass. The earliest means of measuring the ionizing strength of x-rays was to determine the rate at which a given initial charge decayed in an electrometer. In their early study of the process of ionization by x-rays, Thomson and Rutherford had charted the course of the induced conductivity by measuring charge per unit time as a function of the increasing potential difference applied between two electrodes in a gas (see Figure 4.1). X-rays produce ions in the gas; when a small potential difference is applied, current immediately begins to pass. As the applied potential increases, the current increases too, but only up to a point. At values of a few hundred volts per centimeter, the current settles down to a *saturation* value.

The discharge is a dynamic equilibrium. X-rays tear electrons from atoms, but the electrons spontaneously recombine with stripped atoms if given the chance. A high potential forces more

[21] J. J. Thomson and Rutherford, *PM, 42* (1896), 392–407. See also Dorn, *AP, 63* (1897), 160–76.
[22] Röntgen III, pp. 21–3.

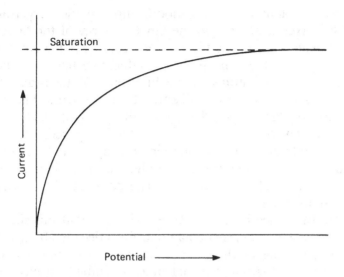

Figure 4.1. Saturation of current produced by ionizing a gas.

free charges to reach the electrodes before recombination occurs. The existence of a saturation current shows clearly that the total number of ionization events is limited. If the space between the electrodes is increased, the saturation current also increases because more gas atoms are now able to contribute electrons. It was clear from electroscopic measurement that not all atoms are treated equally by the x-ray beam; only a small fraction of gas atoms is ionized. In 1896 Thomson had estimated it to be a maximum of 1 in 10^{12} atoms.[23]

The puzzling fact that an x-ray beam ionizes only some of the atoms over which it passes I shall call the *paradox of quantity*. Note that Thomson's discovery was partly due to his recent determination of the charge on the electron. Without knowledge of this unit of electricity, it would not have been possible to count the number of ionized atoms and learn that there were so few. The paradox of quantity was created in the wake of the unitary electron.

In 1903 Thomson offered several potential explanations of why x-rays ionize so few atoms. He raised the unsettling possibility that different atoms of the same substance are, in some undetermined way, different from each other. Perhaps, he suggested, the x-rays only affect electrons already free of their parent atom. His

[23] J. J. Thomson and Rutherford, *PM, 42* (1896), 392–407.

most realistic explanation was a modification of the triggering hypothesis; it fixed attention on the kinetic energy of the target atoms. Thomson suggested that ionization occurred only in atoms with velocity greater than some threshold value.[24] Perhaps only the very small fraction of atoms at the high end of the Maxwell velocity distribution was able to discard electrons. Were this the case, heating the target should increase the yield of electrons. Thomson enlisted the services of Robert McClung, recently arrived from Montreal, to test the matter. In May 1903, McClung reported that there was no measurable difference in the ionization of air over a range of 14° to 200°C.[25] The paradox of quantity remained unanswered.

Thomson was concerned, but not despairing, about this situation. He briefly revived ideas he had had for a decade about the microscopic structure of the electromagnetic field. As early as 1893, Thomson had speculated that lines or "tubes" of electric force might be more than just mathematical abstractions. If they were thought to be finite in number, an electromagnetic theory of light would result with some "characteristics of the Newtonian emission theory."[26] An oncoming pulse, traveling on the tubes, would not be continuously distributed in space. In 1903 Thomson described the pulse front in his Silliman Lectures at Yale as a pattern of "bright specks" on a dark background.[27] This view directly explained the small number of ionized atoms; there was only a limited number of tubes. Thomson submitted this opinion largely in the spirit of medieval *speculationes;* for four years, he hesitated to raise the radical hypothesis again. When he did so, it was in response to more than simply the paradox of quantity. In the intervening years a more formidable difficulty had arisen from closer study of the ionizing energy, or "quality" of both x-rays and *γ*-rays.

THE QUALITY OF X-RAYS

In the first years of x-ray research, the very newness of the rays made it difficult to know how best to characterize their strength.

[24] J. J. Thomson, *Conduction of electricity* (1903), 255–8.
[25] McClung, *PCPS, 12* (1903), 191–8.
[26] J. J. Thomson, *Notes* (1893), 3; *Conduction of electricity* (1903), 258. See McCormmach, *BJHS, 3* (1967), 362–87.
[27] J. J. Thomson, *Electricity and matter* (1904), 62ff.

There were differences of opinion about which attribute of the rays should be quantified as a measure of x-ray quality. Röntgen had tried to quantify the fluorescent activity of a detecting screen using photometric techniques. He found "from three closely agreeing measurements" that the inverse-square law seemed to hold for the decrease in x-ray intensity.[28] Rutherford too attempted to quantify the brightness of x-ray fluorescence.[29] But formidable problems confronted the accurate measurement of fluorescent or photographic effectiveness of x-rays. One of the reasons for the dearth of useful empirical work on x-rays on the Continent in the early years was the widespread use there of photography to detect the rays. Accurate photometry was still in its infancy.

In Britain, where until 1907 the most useful research on x-rays was done, the approach was quite different. The method devised by Thomson to measure the saturation current passing through an irradiated gas was widely applied. For a given gas at a fixed pressure and for a standard electrode separation, a fairly reproducible measure of x-ray ionizing strength could be obtained. In turn, this method made possible the calibration of a new, and ultimately more direct means to measure x-ray quality. By using the electrometer to determine saturation current, one could measure the relative intensity of the rays that passed through an absorbing sheet and express this as a fraction of the incident intensity. One could then fit the result to an anticipated exponential decay in the effective intensity of the rays,

$$I = I_0 \exp(-\mu d). \tag{4.1}$$

Here I_0 is the original measure of what was called the *intensity* of the rays; I will return in a moment to this interesting term. The quantity d is the thickness of the absorbing sheet, and μ is the *absorption coefficient* of the rays relative to the absorbing substance. The coefficient μ is a measure of the penetrating power of the rays; large μ implies low penetrating power. The value of μ measured in aluminum came soon to be accepted as a standard.

Each of these methods measured something distinct in the rays, and it was unclear which, if any, ought to be adopted as the standard means to specify x-ray intensity. Indeed, there was a certain novelty in defining the intensity of a train of electromag-

[28] Röntgen I, p. 7.
[29] Rutherford and McClung, *PTRS, 196A* (1901), 25–59.

netic impulses. The accepted definition was clearly insufficient. Classically, intensity expresses the energy flux in periodic waves. Poynting had shown that this directed quantity has a magnitude equal to the vector product of the amplitudes A_E and A_H. Considering that both A_E and A_H decrease as $1/r$ (cf. equations 2.2 and 2.3), their product will decrease as $1/r^2$, as expected.

But impulses, unlike periodic waves, require an additional statement about quantity. To fix the net energy radiated in a given direction, one needs to know the number of impulses as well as the electromagnetic amplitude of each. As the number of periodic radiating sources increases, their additional output can simply be added to the varying net electromagnetic wave amplitude. But the temporal discontinuity of impulses requires that their quantity be explicitly taken into account in measuring intensity. Thus, intensity as determined by Thomson's ionization methods included a major new emphasis on the *number* of pulses radiated past a point, a quantity that does not vary with the azimuthal angle or the distance from the source. The result was to modulate the definition of *radiant intensity* in a subtle and historically significant manner.[30]

This change in emphasis, unrecognized then, was to have far-reaching ramifications. One of its first influences was to give the pulse width a role in determining the energy of the impulse. As Thomson had shown (equation 2.5), the total energy carried by a single pulse is inversely proportional to the pulse width. Thus, an impulse x-ray of given width should transport a fixed quantity of energy. Although it was widely anticipated from Thomson's analysis that different pulse widths would be manifested as different penetrating powers in the rays, there was no real evidence of either a correlative or a causative relation. Sommerfeld's analysis of diffraction was, at the time, the only direct experimental means to determine x-ray pulse width.

The research of Barkla, above all, served to fix x-ray penetrability in aluminum as the standard means to characterize x-ray quality. From its introduction in 1903 until 1912, no other way existed of measuring the temporal duration of x-ray impulses. To the extent that a given beam could be assigned a single coefficient of absorption, one could speak of the beam as *homogeneous* in

[30] Wheaton, *HSPS, 11* (1981), 367–90.

ionizing power. If a second thickness of absorber failed to remove the same fraction of ionizing capability from the beam as the first, the beam was declared inhomogeneous, and effort was devoted to separating out the homogeneous components. This means of characterizing x-rays gained a great deal from Rutherford's work to separate the α- and β-rays.

W. H. BRAGG AND THE PROCESS OF IONIZATION

William Henry Bragg was Elder Professor of Mathematics and Physics at the University of Adelaide in Australia. He was born in England and educated at Cambridge, but he left for the colony in 1885, a year after Thomson had assumed the Cavendish Professorship.[31] Bragg did very little research in physics until 1904 and then undertook a critical analysis of the explanation of ionization produced in gases by various forms of radiation. He took special objection to what he thought was the unwarranted assumption that the exponential law of absorption of radiation could be applied to ionization by particles. In the course of this study, he convinced himself that γ-rays could not possibly be the spreading impulses that, after 1904, most European physicists took them to be.

Bragg delivered his results early in 1904 as his presidential address to the section on physics, astronomy, and mathematics of the Australasian Association for the Advancement of Science.[32] The talk is a testament to Bragg's remarkable physical intuition and is rich in the use of mechanical analogies. His conclusions are based on simple calculations using data culled from a wide, but not exhaustive, knowledge of relevant experiments. Already discernible is his tendency to overstate his position; his eloquence and persuasiveness occasionally made the evidence in support of his claims seem to be greater than it was. But he succeeded in giving his audience a clear understanding of each step in the mechanics of ion formation in gases.

Bragg's comments were directed largely to the ionization induced by particles, but early in his talk he established an important distinction between the properties of x-rays and particles. When

[31] Forman, *DSB, 2,* (1973), 397–400.
[32] W. H. Bragg, AusAAS *Report, 10* (1904), 47–77. Bragg was not inactive before 1904; see AusAAS *Report, 3* (1891), 57–71; *4* (1892), 31–47; *6* (1895), 223–31.

electrons or α-particles traverse matter, they interact with the individual electrons that comprise the atoms of the absorbing body. Therefore, they transfer energy through collisions in proportion to the number of electrons they encounter, and are thus absorbed roughly in proportion to the density of matter. As pointed out by Röntgen and demonstrated by Strutt, x-rays are absorbed according to a different law, and they therefore constitute, as Bragg said, an "exception to the rule that penetration . . . depend[s] only on the density of the substance penetrated."[33] Impulses or periodic waves in the aether, Bragg said, would not interact with single electrons, but rather with the "molecule as a whole." Consequently, x-rays transfer energy less readily to matter than do electrons of comparable energy.

One of the significant contributions of Bragg's early study was his recognition that, strictly speaking, the exponential law of absorption does not hold for streams of particles. The exponential law is a direct consequence of absorption in proportion to absorber density; Bragg claimed that for particles this was valid only as a first approximation. As α- and β-particles slow down, their energy is more easily lost because collisions with atoms become more frequent. Therefore, a particle conforms to the exponential law only in the high-velocity first part of its trajectory; toward the end of its path, an accelerating increase in its absorption coefficient is to be expected. Other observers had deluded themselves that the exponential law holds because they experimented on beams containing a fairly broad distribution of particle velocities.

Bragg therefore had particular reason to differentiate waves from particles. Whatever the difference in the relationship of absorption coefficient to absorber density, for homogeneous waves the absorption coefficient should remain constant. For particles, which are able to slow down, the coefficient increases with time. When one is confronted with a new sort of ray, the characteristic way that it is absorbed by matter offers a means to identify it as particulate or aetherial. Bragg evaluated the γ-rays according to this condition, and although he could not demonstrate that their absorption rate was constant, he could nonetheless see clearly that even in the first approximation their absorption was substantially different from that of periodic waves or impulses.

[33] W. H. Bragg, AusAAS *Report, 10* (1904), 72–3.

REVERSING THE γ-RAY – X-RAY ANALOGY

Bragg discussed Strutt's experiments on the ionization produced by γ-rays, and he emphasized that γ-rays appear to follow the laws governing the absorption of α- and β-particles. They do not match the x-ray ionization rate. Bragg accordingly came to the conclusion that γ-rays act on the individual electrons in atoms. Therefore, in some sense, they must be localized in space. He was at first reluctant to attribute to them a purely corpuscular nature because of their great penetrability and because they seemed to suffer no electric or magnetic deflection. But, he said, if they are waves, "they are waves so small as to be unable to act [on] a whole molecule or atom at once."[34]

Bragg's statement is remarkably obscure. A wave that, after traversing macroscopic distances, can affect only a region smaller than an atom is not a wave in the classical sense. Bragg was at a loss to characterize it accurately. In his ambiguous statement, one perceives an unconscious attempt to combine wave and particulate characteristics. But soon he resolved his indecision by attributing to γ-rays all the essential properties of particles. Bragg therefore reacted quite differently than did most European physicists to Eve's discovery that hard x-rays also follow the absorption law characteristic of particles. What Rutherford took as proof in 1904 that γ-rays must be impulses like x-rays was interpreted by Bragg as evidence that x-rays must have the same particulate qualities as γ-rays.

Bragg's recognition that the absorption coefficient of a particle will increase along its trajectory opened the way to new studies of α-particles. In the years 1904–7 he was able to study the range of α-particles thanks to the gift of some radium bromide. The work was done with the assistance of various students at Adelaide. Richard D. Kleeman was the first and most important of the collaborators.[35] With his help, Bragg soon found that the total distance in a substance or gas over which the ionizing effect of the α-particles persists is directly related to the velocity with which the particle left the disintegrating atom.

---

[34] *Ibid.,* 77. The passage actually reads "unable to act as a whole molecule," but from the next sentence it is clear that this is a misprint.

[35] These ten papers and one letter may be found under "Bragg, W. H." in the indexes of *PM, 8, 10, 11, 13,* and *14* (1904–7).

In most of these experiments, Bragg adopted Thomson's means to measure the ionizing strength of the α-beam: A standard potential difference is applied between electrodes enclosed in the irradiated gas. The potential is then increased until the current stimulated by the beam achieves its fully saturated value. In 1906 Bragg discerned that some source other than the incident α-particle beam was ionizing the gas. He quickly dismissed the possibility that the electrode potential alone could be the cause. "If it were," he explained, "there would be no saturation current"; potential gradients substantially greater than the few hundred volts per centimeter at which saturation occurs give ions sufficient energy to ionize atoms themselves.[36] An avalanche effect occurs, and the current increases very rapidly. Rather, Bragg suspected that the new ionizing emanation might be the x-rays expected to arise when the charged α-particles strike gas molecules.

But herein lay a significant problem. The ionization caused by α- or β-particles *should* saturate. Each particle can interact with only a limited number of molecules. But the x-rays produced by collisions should radiate in all directions (in the orthodox view) and the impulses should affect all molecules in their vicinity equally. Since large numbers of atoms would therefore be affected, the ionization current should not saturate. Bragg, like Thomson, knew that the current *does* saturate at microscopically small levels. It appeared to Bragg that far too few electrons were being released. Thus, Bragg confronted the paradox of quantity, only hinted at in his 1904 claim for "waves so small." It was reaffirmed in a particularly clear operational form for Bragg, and he could not ignore it without upsetting his ongoing α-particle studies.

Bragg suggested a possible explanation for the ionization due to x-rays produced by α-rays. Perhaps, he thought, the secondary effect of the x-ray is fully expended within a microscopically small distance from its source. Within that short distance, the impulse might retain sufficient energy to ionize an atom. But this explanation clearly could not be complete. The current induced by ordinary x-rays from a discharge tube also saturates. This occurs even after the rays have traveled distances equal to billions of atomic diameters. Even if one allowed the impulse to be severely restricted in its spatial width, after traversing these long distances it would

[36] W. H. Bragg, *PM, 11* (1906), 623.

have diffracted to encompass regions much larger than atoms. The ionization data indicated that soft x-rays might do this, but hard x-rays and γ-rays acted like particles: They did not appear to diffuse their ionizing power in space at all. By 1906 it was clear to Bragg that something was seriously amiss in the widely accepted hypothesis of x-ray impulses, and he reaffirmed the extension to x-rays of the corpuscular status he had assigned to γ-rays in 1904.

THE PARADOX OF QUALITY

Bragg's research interests shifted in 1906 from α-particles to γ-rays. He soon came to recognize a new problem that the pulse hypothesis of x-rays seemed incapable of explaining. It was the same issue that the early researches of Dorn had raised but that had been glossed over by the adoption of the triggering hypothesis in Europe. Bragg, however, was not in Europe. His relative isolation in Australia gave him the courage to question the impulse hypothesis of x-rays.

Bragg apparently had little firsthand knowledge of the photoelectric effect. In 1904 the relevant evidence he cited came from Townsend's study of ionization of gases by ultraviolet light, not Lenard's work.[37] The triggering hypothesis was not mentioned. Although Bragg was certainly familiar with the triggering hypothesis, his understanding derived from his work on radioactive disintegration, and he was not convinced that triggering needed to be extended to include ionization by x-rays. Nor did he take seriously McClelland's claim that the triggering hypothesis also explains the high velocity of secondary electrons stimulated by γ-rays. Bragg was convinced that γ-rays are particles, and he suspected as much for the hard x-rays. He had also become aware of the high secondary electron velocities that Dorn measured after irradiating matter with x-rays.[38] The high velocity of secondary electrons was easily understood if one assumed that x-rays and γ-rays are particles. The evidence that others used to show that the triggering mechanism operated convinced Bragg that the rays could not be aether pulses at all. Viewed this way, the evidence of secondary electron velocity constituted a new paradox for the pulse theory.

[37] W. H. Bragg, AusAAS *Report, 10* (1904), 47–77.
[38] W. H. Bragg, *TPRRSSA, 31* (1907), 95.

Since the difficulty deals with the quality, or energy content, of x-rays and electrons, I shall call it the *paradox of quality*. It is a test of the microscopic validity of energy conservation. Each impacting electron brings kinetic energy in an amount determined by its velocity. That energy reappears in the form of a transverse impulse that propagates, it would seem, isotropically outward. The impulse eventually passes over an atom or molecule. One would expect that the amount of pulse energy available to the cross-sectional area of an atom is exceedingly small. The atom subtends an insignificant angle when viewed across the macroscopic distances over which the potency of the x-rays is known to persist. The pulse apparently dissipates itself in all directions. How can the *full* energy in the impulse be concentrated on the atom? Yet it seems to be. The atom releases an electron with an energy virtually equal to that of the electron whose impact alone created the impulse!

Bragg had already clearly perceived the distinction of quantity and quality in 1904 when he discussed the energetics of particle streams. He had objected that the simple phrase *amount of [parti-cle] radiation* made no sense.[39] He insisted that one had to consider separately the energy transferred by the particles and the particle flux itself. In 1904 he transferred this way of thinking to γ-rays and by 1906 was convinced that it also applied to x-rays.

One of the influences on Bragg in this critical period was the work of his former colleague, Richard Kleeman. Kleeman left Australia in 1906 for Emmanuel College, Cambridge. There he began research work at the Cavendish Laboratory under J. J. Thomson. His first experiments, perhaps still under the influence of Bragg's concerns, treated an issue closely related to the conservation of quality in γ-ray interactions. He showed that the velocity of secondary electrons produced by the combined action of β- and γ-rays from a radioactive source far exceeded that due only to the α-component.[40] Interpreted by Bragg, this evidence was significant.[41] One knew that the largest part of the energy in radioactive decay is carried by the α-particles. Nevertheless, even the small fraction of energy carried by the γ-component seemed to be concentrated on individual electrons.

In March 1907 Kleeman provided more evidence that the action

[39] W. H. Bragg, AusAAS *Report, 10* (1904), 69.
[40] Kleeman, *PM, 12* (1906), 273–97.
[41] W. H. Bragg, *TPRRSSA, 31* (1907), 96.

of the γ-rays is concentrated on individual atoms. He showed that the ionization produced by all three components of radioactive decay appeared to be strictly an atomic, not a molecular, property.[42] One first measures the rate of ionization caused by the rays in a gas consisting only of atoms A, and denotes the result by a; that found in gas B is called b. Then the ionization rate produced by the same radiation in a gas consisting of molecules $A_n B_m$ is approximately $na + mb$. To Bragg, Kleeman's result signified again that all three component radioactive emanations, γ included, interact with the electrons that comprise the atoms, not with entire molecules.[43]

In May 1907, Bragg submitted a paper outlining his new thoughts to the Royal Society of South Australia. It was titled "A comparison of some forms of electric radiation." In it Bragg explained: "when x-rays were first investigated, and again when γ-rays were discussed, it was often suggested, in each case, that the radiation might consist of material particles."[44] The "often" expressed Bragg's boundless optimism rather than historical fact; the great penetrating power of the rays had always stood in the way of a corpuscular interpretation, and few had suggested it. "The difficulty of accounting for the great penetration of these radiations on the basis of a material interpretation," Bragg explained, "was quite exaggerated, and even imaginary."[45] And Bragg proposed an entirely new view of x-rays and γ-rays.

THE NEUTRAL-PAIR HYPOTHESIS

To replace the "overrated" impulse, Bragg proposed the *neutral pair*. It was a dipole consisting of an electron coupled with a bit of positive charge to make the whole electrically neutral.[46] In 1907

---

[42] Kleeman, *PRS, 79A* (1907), 220–33.

[43] W. H. Bragg, *TPRRSSA, 31* (1907), 96.

[44] *Ibid.,* 90.

[45] *Ibid.,* 90. Bragg implied that Röntgen had proposed a particulate hypothesis of x-rays, but this was a convenient falsehood. Röntgen had pointed out in Röntgen III that x-rays are similar to cathode rays. Bragg's own corpuscular model of cathode rays led him to project a similar interpretation of x-rays on Röntgen, a view that the latter never held.

[46] Andrade has suggested, in RSL *Obituary notices, 4* (1943), 280, that Bragg's idea of neutral pairs may have been derived from Lenard's dipole "dynamid," proposed in 1903. Bragg was aware of Lenard's hypothesis but commented negatively on it in 1904.

Bragg selected the α-particle for the neutralizing role; later investigations showed that its charge was too large, and other candidate particles took its place. The easy penetration by the pair was attributed to its electrical neutrality and consequent lack of interaction with matter. Some ionization events not only disturb the gas atom but also disrupt the pair. When this happens, the heavy α-particle moves off slowly, but the less massive electron continues with essentially the same high velocity it possessed within the x-ray pair.

At one blow, Bragg solved the paradoxes of both quantity and quality. In his own words:[47]

If the x-ray is an aether pulse, it is difficult to understand . . . why the spreading pulse should only affect a few of the atoms passed over [in ionization], why the secondary cathode rays are ejected with a velocity which is independent of the intensity of the pulse which weakens as it spreads, and why [the pulse] should be able to exercise ionising power when its energy is distributed over so wide a surface as that of a sphere of, say, ten or twenty feet radius.

Like the triggering hypothesis, Bragg's new hypothesis required no transformation of radiant to mechanical energy. The energy of the neutral pair is mechanical to begin with. The high velocity of the electron exists before ionization of an atom. But instead of arising from the internal dynamics of the atom, the electron is released to move independently, where before it had constituted half of a pair.

Bragg had good reason to suggest that the high-velocity electrons were not secondary products of ionization. Not just α-, β-, and γ-rays from radioactive decay, but also their electrical analog canal, cathode, and x-rays induce conductivity in gases. The vast majority of charge-carrying particles produced in gases by these radiations are electrons moving at about 10^8 cm/sec. So ubiquitous are these slow electrons that Bragg adopted the special name δ-*rays* applied to them by others.[48] "In all cases," Bragg said, "the bulk of the ionization which the rays effect is of the same character. [It] consists in the displacement of slow moving electrons, or δ rays, from the atoms of the gas."[49] In this sense, all six forms of radiation

[47] W. H. Bragg, *TPRRSSA, 31* (1907), 91–2.
[48] J. J. Thomson, *PCPS, 13* (1904), 49–54. This use of δ has, of course, no connection with my use of δ to denote pulse width.
[49] W. H. Bragg, *TPRRSSA, 31* (1907), 79.

were "electric," and this is how Bragg intended the term in his title. That the velocity is common to electrons released by all forms of radiation convinced Bragg that it was an atomic property. The slow electrons are the true product of atomic ionization, he claimed; the high-speed electrons are more likely remnants of the incident neutral-pair x-rays.

Bragg's study appeared in two parts in Australia. They were published back-to-back but were separated by a month's careful consideration. In the first part, where he discussed the evidence regarding δ-ray electrons, Bragg made it clear how seriously he took Marx's demonstration in 1905 that x-rays travel with the velocity of light. "If we attempt to explain the properties of the X-rays on the supposition that it is a neutral pair," he complained, "we meet with a difficulty which does not occur in the case of the γ-ray . . . Unless some way out is found [the hypothesis] must remain simply an interesting comparison."[50] During the month between the two papers, Bragg realized that only slow electrons would be susceptible to the detailed control exercised by the electric oscillations in Marx's experiment. The very fast secondary electrons would easily travel from the cathode P to the electro-meter F in the apparatus shown in Figure 2.13, even if the ambient field momentarily acted to decelerate them.

In the second part of his study, called "The nature of Röntgen rays," Bragg proposed that two entities inhere in any x-ray beam. First, there are electromagnetic impulses that travel at the velocity of light. Combined with these are the neutral pairs that travel at somewhat lower velocity. Upon collision, the latter can break into their constituent parts and release high-velocity electrons. The pulses, Bragg suggested, might be those produced when the neutral pairs themselves break away from the anticathode of the x-ray tube.[51] The actual velocity of the pairs could be anywhere within the limits set by the minimum velocity required to penetrate matter and the minimum velocity of the released free α-particles. Based on Thomson's results, he estimated the range as 10^8 to 10^9 cm/sec.

Bragg's hypothesis applied only to x-rays and γ-rays. He never

---

[50] W. H. Bragg, *TPRRSSA, 31* (1907), 91–3.
[51] Note the influence of Rutherford's hypothesis that γ-rays are the impulses produced when β-electrons are ejected from a disintegrating atom.

intended it to apply to light. The ionizing action of x-rays, he pointed out, "must be entirely different from that of ultra-violet light."[52] Although Einstein had proposed his lightquantum hypothesis two years before, Bragg was unaware of it and would not likely have seen the connection to his own ideas in any event.[53] Indeed, in disclaiming any connection between x-rays and visible light, Bragg betrayed the essentially conservative nature of his appeal; it was intended as even more of a throwback to purely mechanical theory than the impulse hypothesis. To Bragg, all six types of radiation – α, β, γ, canal, cathode, and x – were particles, although some were accompanied by unavoidable aetherial effects such as impulses.

After satisfying himself that he could answer possible criticism based on Marx's experiment, Bragg allowed the entire study to appear in the *Philosophical Magazine*.[54] It was a broadside attack on the British pulse theory, which at the time was most strongly supported by Charles Barkla. While Bragg had been developing his novel hypothesis, Barkla's experiments had turned up new and remarkably homogeneous secondary x-rays. In its fully interpreted form, this evidence would prove to be virtually unanswerable by Bragg. And it was Barkla who took on the responsibility of defending the impulse hypothesis from Bragg's assault.

HOMOGENEOUS X-RAYS

Barkla had shown in 1903 that x-rays scattered off light elements reradiate with a quality roughly equivalent to that of the primary rays. In this way, he sought corroboration of Thomson's simple picture of the interaction of pulses and matter. In the course of his polarization studies in 1904 and 1905, he extended his survey to some of the heavier elements – calcium (Ca), iron (Fe), copper (Cu), zinc (Zn), platinum (Pt), and lead (Pb) – and found that the secondary x-rays are consistently less penetrating than the primary ones.[55] He invoked Thomson's "thunder and lightning" clause in

[52] W. H. Bragg, *TPRRSSA, 31* (1907), 96.

[53] "When I first put forward the neutral pair theory I was ignorant of the work of Einstein . . . I did not think of carrying over the idea to the theory of light; on the contrary, I had hopes of proving that no connection existed between the two kinds of radiation." W. H. Bragg, *Studies in radioactivity* (1912), 192–3.

[54] W. H. Bragg, *PM, 14* (1907), 429–49.

[55] Barkla, *PTRS, 204A* (1905), 467–79.

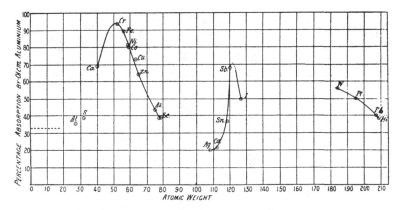

Figure 4.2. Barkla's graph of periods in the penetrating power of x-rays scattered off elements. [*PM, 11* (1906), 820.]

explanation: Interionic forces in the solid are presumably responsible for lowering the penetrating strength of the secondary x-rays by broadening the secondary impulses.

J. A. McClelland began a fruitful line of investigation in 1905 by surveying a wide variety of elements for their response to the β- and γ-rays from radium.[56] He did this as part of his investigations, mentioned earlier, to demonstrate that atomic disintegration is the mechanism of secondary electron emission. He measured the intensity of secondary electrons by the ionization rate they induced in a gas sample; he was therefore measuring the number of electrons released from different elements by the same combination of β- and γ-rays from radium. He discovered that the number of secondary electrons increases monotonically with the increasing atomic weight of the scattering substance. Moreover, the rate at which secondary electrons increase with atomic weight appeared to change from period to period among the elements. This was the behavior that convinced McClelland that the secondary emission must be an "atomic" property caused by a mechanism analogous to the triggering hypothesis.

Barkla was encouraged by McClelland's work, and he tried to discover a similar reaction between the penetrability of the secondary x-rays (as distinct from secondary electrons) produced in matter by a primary x-ray beam. A survey in 1906 turned up the rough relationship illustrated in Figure 4.2: Increases in the pene-

[56] McClelland, RDS *Sci trans, 9* (1905), 1–8.

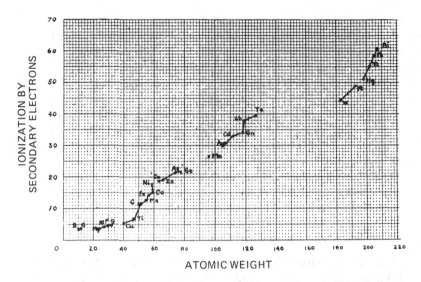

Figure 4.3. J. J. Thomson and Kaye's graph of ionizing strength of secondary electrons released from elements under x-ray bombardment. [*PCPS, 14* (1906), 113.]

trating power correspond to the periodic properties of the elements.[57] Barkla admitted that the instability of his primary x-ray beam rendered the results somewhat unreliable, but he was firmly convinced that he had found a means to determine the order of the elements in a way independent of the atomic weight.

Thomson quickly saw the power of the new survey, and with G. W. C. Kaye he studied the secondary electrons produced by x-rays.[58] They discovered that the ionizing strength of the secondary electrons increases monotonically with the atomic weight of the scattering substance. Figure 4.3 illustrates their result. McClelland had found a similar behavior of electrons produced by β- and γ-rays. Thomson found that, according to x-ray analysis, nickel was improperly placed when ranked according to atomic weight in the periodic table.

The next year, 1907, Kleeman carried the investigation into the realm of γ-rays and was careful to separate the primary gammas from their accompanying betas. He then repeated McClelland's experiment of two years before by measuring the relative ioniza-

[57] Barkla, *PM, 11* (1906), 812–28.
[58] J. J. Thomson, *PCPS, 14* (1906), 109–14.

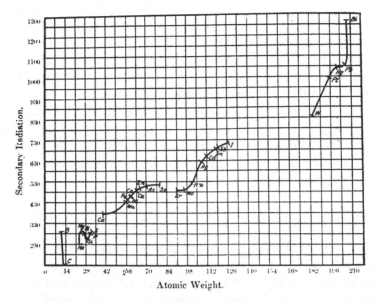

Figure 4.4. Kleeman's graph of ionizing strength of all secondary radiation from elements bombarded by γ-rays. [*PM, 14* (1907), 624.]

tion caused by the secondary electrons.[59] His graph, Figure 4.4, is remarkably similar to Thomson's, and Kleeman took this as yet another proof that x-rays and γ-rays are equivalent. Fully under the Thomson influence by this time, he concluded in August 1907 that both radiations were "probably" electromagnetic impulses.[60]

In the meantime, Barkla enlisted the assistance of his Liverpool physics demonstrator, Charles Sadler, in order to investigate further the anomalous case of nickel found by Thomson. They discovered that most of the secondary x-rays spent their ionizing power after traversing a few layers of aluminum. But among those that persisted was a component of relatively high intensity that always lost the same fraction of its ionizing power in passing through successive test thicknesses of aluminum.[61] This component followed the absorption law, equation 4.1, with a constant value of μ, clear evidence of homogeneous penetrating power. Not only were some secondary rays homogeneous, but they showed a

---

[59] Kleeman, *PM, 14* (1907), 618–44.
[60] *Ibid.,* 643.
[61] Barkla and Sadler, *PM, 14* (1907), 408–22. Barkla, *Nature, 75* (1907), 368.

particular transparency for the material of their origin. Secondary rays from copper, for example, passed with lower than normal loss through copper. Barkla called this a "property which appears to be unknown among x-ray beams hitherto experimented upon."[62]

The announcement of this new sort of x-ray came in September 1907. But the full significance of Barkla's discovery only emerged slowly over the next year as his investigation proceeded. Barkla was encouraged to develop and extend his interpretation of the result by the controversy that quickly developed over Bragg's paper on neutral pairs. It appeared in England just a month after Barkla's paper on homogeneous x-rays.

THE BRAGG–BARKLA CONTROVERSY

The conflict between Bragg and Barkla arose largely from the fact that the protagonists approached the common domain of x-rays from opposite directions. Bragg's view had developed out of detailed studies of the properties of α-particles, the results of which were first applied to γ-rays and by extension to x-rays. Barkla's research was concerned exclusively with x-rays, and not particularly hard x-rays at that. The two had very little common ground from the very start of their discussion. Barkla was dismayed that Bragg could suggest what appeared to him to be so implausible an idea as neutral pairs. Bragg frequently exploited the impassioned tone in Barkla's defense of the impulse hypothesis. The debate was carried on in the pages of *Nature* in 1908; the details have been presented elsewhere, and I will discuss only those of direct concern to the broader issues of our study.[63]

Bragg had attempted to reinterpret Barkla's experimental results according to the neutral-pair hypothesis. He had suggested that the penetrating power of a pair was, at least in part, determined by the separation of its constituent charged particles, which in turn fixes the dipole moment of the pair.[64] Soft x-rays, produced at relatively

---

[62] Barkla and Sadler, *PM, 16* (1908), 560.

[63] The interchange fills many letters to *Nature, 77* and *78*. Barkla's are virtually all titled "The nature of x rays" and Bragg's "The nature of γ and x rays." See Stuewer, *BJHS, 5* (1971), 258–81. Barkla had all the evidence except unpolarizability for characteristic x-rays before the debate with Bragg started. The controversy led him to reinterpret data that already existed.

[64] Bragg left ambiguous whether the speed of the pair or its moment determines its penetrating power. He wished to attribute it solely to velocity, but had to invoke the change in moment to explain Barkla's scattering results.

low discharge potentials, are poor penetrators because, said Bragg, they have large moments. Hard x-rays do not ionize gas atoms readily because they possess small moments. On striking a relatively light atom able to absorb some of the collision energy by rebounding, a neutral pair might simply change its course with no change in dipole moment. But on striking a heavy atom, it could suffer either an increase in moment or be "shattered altogether." If shattered, it releases its captive high-speed electron. In this way Bragg explained Barkla's discovery that x-rays scattered from light atoms exhibit roughly the same penetrability as the primary rays, whereas secondary x-rays from heavy atoms are less penetrating than their primaries.

Bragg's dipoles also rotated, and the plane of rotation of the pair was conserved. Bragg exploited this capability to explain Barkla's polarization data, and it was here that the opening shots in the debate were fired.[65] Bragg assumed that electric oscillations within an atom occur only in a fixed plane. An atom can absorb a neutral pair, but only if the planes of rotation of the atom and of the pair are parallel. The polarization of secondary x-ray beams followed directly from this constraint between the two interacting planes of rotation. An atom, Bragg said, can eject a pair only into the plane of the atom's oscillation.

Barkla carried the consequences of Bragg's position further to argue that an x-ray beam from an anticathode could consist only of pairs rotating in planes parallel to the beam direction. The secondary radiation stimulated by this beam, when viewed at right angles to the primary beam, could then consist only of pairs rotating in the plane defined by both beams. If the atoms in the scattering material are randomly oriented, the net flux of secondary neutral-pair x-rays should be "proportional to the density of the lines of longitude" (like those running from pole to pole in Figure 4.5), that is, as the secant of the latitude. So, Barkla reasoned, the secondary rays seen at *A,* only slightly off the beam axis, should be many times more intense than those scattered at 90° to the beam, seen at *B.* He estimated the difference to be about 8:1, then allowed for possible mismatch between the planes of atom and pair, and revised the ratio to 4:1.

On the other hand, Thomson's proposed mechanism for the stimulation of secondary x-rays predicted that an unpolarized

[65] Barkla, *Nature, 76* (1907), 661–2.

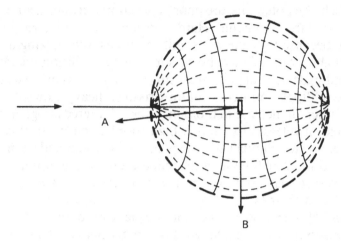

Figure 4.5. Barkla's prediction of x-ray scattering according to Bragg's neutral-pair hypothesis.

primary x-ray beam would produce electron displacements only in the planes perpendicular to that beam. Secondary x-ray impulses radiated by these electrons in the direction of the primary beam would have the benefit of all electron accelerations as sources. But impulses viewed from a direction normal to the primary beam are caused only by the component of electron displacement that is confined to the single plane normal to *both* directions. A moment's reflection will show that the intensity of secondary x-rays directed at right angles to the primary beam is just half that directed back along the primary beam. Barkla concluded that the intensity distribution of secondary impulses should approximate that shown in Figure 4.6, and that the ratio for observations at A and B could not exceed 2:1.

Experiment showed that soft x-rays yield ratios approaching 1.85:1. For hard x-rays, the ratio was less. Barkla described this result as "most conclusive evidence against the 'neutral pair' hypothesis."[66] But it scarcely supported the impulse hypothesis any better for hard x-rays. The test also raised questions about the recently discovered homogeneous secondary x-rays from heavy atoms. When, in defense of the pulse hypothesis, Barkla extended his analysis to heavy atom scatterers, he found that the ratio

[66] Barkla, *PM, 15* (1908), 288–96.

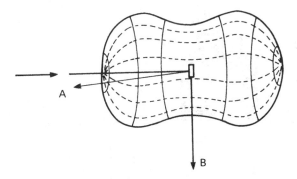

Figure 4.6. Barkla's expectation of scattered x-ray intensity according to Thomson's hypothesis.

approached 1:1. The homogeneous secondary x-rays are isotropically scattered, in accord with neither prediction!

Bragg was not at all deterred by Barkla's argument. He pointed out that one was not entitled to assume isotropic scattering of pairs into the plane defined by atomic vibration. To do so would require detailed knowledge of atomic structure, "as to which it is scarcely possible to do more than speculate."[67] He responded with a "decisive argument" based on his recent experiments with J. P. V. Madsen on the intensity distribution of secondary rays following irradiation by γ-rays.[68] If γ-rays are transverse impulses, Bragg argued, the secondary rays, whether material electrons or aetherial impulses, should reradiate with equal probability in the forward and backward directions. (Thomson had made a similar assumption the previous year.[69]) Neutral pairs, on the other hand, would be expected to conserve their incident momentum; secondary rays should preferentially scatter in the forward direction.

Bragg and Madsen's experiments showed unmistakable evidence for an asymmetric emission of secondary electrons. Bragg used an ionization chamber (Figure 4.7) closed at the top and bottom by double plates of lead (Pb) and aluminum (Al). Bragg and Madsen showed that with the aluminum sheet against the

[67] W. H. Bragg, *Nature, 77* (1908), 560.
[68] W. H. Bragg, *Nature, 77* (1908), 270–1. Bragg and Madsen, *TPRRSSA, 32* (1908), 1–10; *PM, 15* (1908), 663–75.
[69] J. J. Thomson, *Conduction of electricity,* 2nd ed. (1906), 405ff. Asymmetry had been observed for scattered β-electrons by MacKenzie, *PM, 14* (1907), 176–87.

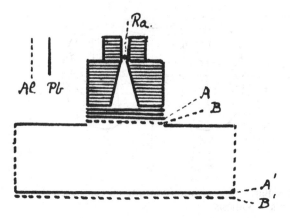

Figure 4.7. Apparatus used by W. H. Bragg and Madsen to demonstrate asymmetric emission of secondary electrons after γ-irradiation. [*PM, 15* (1908), 665.]

chamber at the top, the ionization was slightly greater than when the two top plates were interchanged. This was so regardless of the order of plates on the bottom. Kleeman's finding that a metal emits secondary rays in rough proportion to its atomic weight thus implied that the net ionizing radiation scattered in the "through" direction was greater for aluminum than for lead.

When the order of plates was changed on the bottom with no change on the top, 44 percent larger currents occurred with lead closest to the chamber. Thus, the number of rays scattered back, against the direction of the primary beam, was greater for lead than for aluminum. Since all four layers were too thin to absorb appreciable γ-rays, either aluminum or lead or both emitted more secondaries in the forward compared to the backward direction. This, they claimed, was "fatal" to any transverse impulse or wave theory of γ-rays.

Madsen soon claimed that what he took to be secondary γ-rays also are asymmetrically emitted from matter.[70] Bragg completed the quatrain with an investigation of the weaker effect in secondary x-rays. There, deep within Barkla's own x-ray research domain,

---

[70] Madsen, *TPRRSSA, 32* (1908), 163–92; *PM, 17* (1909), 423–48. For the conditions under which Madsen carried on his researches, see his correspondence with Bragg in Home, *Historical records of Australian science, 5:2* (Nov. 1981), 1–29.

Bragg found the asymmetry he sought.[71] But Bragg, always confident and somewhat parochial, had quite neglected the momentum carried by an electromagnetic wave. The error was first mentioned by Charlton Cooksey at Yale when he presented evidence of an equivalent asymmetry in the distribution of secondary electrons stimulated by x-rays.[72] For the moment, nothing was made of the oversight.

Bragg's commitment to the neutral-pair hypothesis led to a second important discovery in 1908. In May he listed six properties of the γ-rays that, he claimed, were best interpreted in terms of the neutral-pair view. One, on the transformation of γ-rays, had previously been buried in raw experimental data: "the penetration and therefore the speed of the β radiation . . . produced [by γ-rays] increases with the penetration of the γ radiation to which it is due."[73] All neutral pairs of equivalent penetrating power should release electrons of equal speed. The velocity of the secondary betas should be independent of the number of neutral pairs and should depend only on their penetrability.

This result was difficult for any impulse hypothesis to explain. If one adopted the triggering hypothesis, the velocity of secondary electrons should depend on the type of scattering material. Bragg's conviction led directly to a test, and he was vindicated.[74] As with the photoelectric effect, measured velocities of the secondary β-particles were independent not only of the intensity of the incident γ-radiation but also of the nature of the scattering material.

Bragg was convinced that the neutral-pair theory explained "all the known properties of the γ-rays much more simply and completely" than did the impulse hypothesis.[75] He conveniently summarized his position in a series of largely rhetorical questions related to the possible origin of secondary betas: (1) both the particle and the energy arise from the atom (in essence, the trigger-

---

[71] W. H. Bragg and Glassen, *TPRRSSA, 32* (1908), 301–10; *PM, 17* (1909), 855–64.

[72] Cooksey, *Nature, 77* (1908), 509–10. Thomson had pointed out in 1907 that x-ray impulses would carry momentum. In the "bright speck" hypothesis of light, to which we shall return in Chapter 6, he claimed that x-rays would "have all the properties of material particles." *PCPS, 14* (1907), 424.

[73] W. H. Bragg, *Nature, 78* (1908), 271.

[74] W. H. Bragg and Madsen, *TPRRSSA, 32* (1908), 35–54; *PM, 16* (1908), 918–39.

[75] W. H. Bragg, *Nature, 78* (1908), 271.

ing hypothesis); (2) the particle comes from the atom but its energy is transformed from the incident radiation (the impulse or wave hypothesis) and (3) both the electron and the energy come from the primary γ (the neutral-pair hypothesis). Of (1) he inquired, how could the penetrability of the γ-ray affect the energy of the β-particle at all? Also, why does the energy not vary with the type of atom? Finally, how could one explain the asymmetry in secondary emission, as if the γ-rays found the "guns pointed in the direction in which they are travelling themselves?"[76] Explanation (2) hardly fared better. To use one of Bragg's favorite expressions of later years, it was as if a plank dropped from a height of 100 meters into the sea sent out a circular impulse that, after spreading over thousands of kilometers, concentrated its effect on another plank, giving it sufficient energy to impel it 100 meters into the air! The γ-ray or x-ray somehow delivered its energy bundle to a single electron. But replace the bundle by a neutral pair, he exclaimed, "and the whole affair seems simple enough."[77] A cathode-ray electron, incident on the anticathode, picks up a neutralizing positive charge and becomes an x-ray or γ-ray. On collision it breaks into its component parts, and the electron continues on with close to its original velocity.

CHARACTERISTIC SECONDARY X-RAYS

Throughout the debate, Barkla stood firm by the x-ray evidence and refused to be drawn into a discussion of more penetrating radiations. He believed his position to be secure for soft x-rays. Bragg's data were largely drawn from the behavior of the highly penetrating γ-rays. Each felt confident in his own domain. Barkla never answered the paradoxes of ionization, and Bragg always admitted the difficulty presented by the partial polarization of x-rays.

In response to Bragg's second attack, Barkla retreated to his strongest evidence: homogeneous x-rays scattered from matter. In 1908 he made it clear that there are two distinct types of x-rays. "One," Barkla explained, is "a scattered radiation produced by the motion of electrons controlled by the electric force in the primary

---

[76] W. H. Bragg and Madsen, *PM, 16* (1908), 934.
[77] *Ibid.,* 936.

Röntgen pulses," and is therefore inhomogeneous. These scattered x-rays are polarizable; this is a consequence of the unidirectional cathode beam that produced the primary x-rays. The other type of x-ray is, according to Barkla, "a homogeneous radiation characteristic of the element emitting it, and produced by the motion of electrons uncontrolled by the electric force in the primary pulses."[78]

The emphasis on the characteristic nature of the secondary x-rays was new and was offered largely as a challenge to Bragg. Barkla had found that these homogeneous rays characteristic of heavy elements are emitted isotropically, and they show no inclination to polarize. Bragg was hard-pressed to find any explanation using neutral pairs for the characteristic nature of the secondary rays. Neutral pairs are, by definition, independent of the atoms that scatter them. The hypothesis denies detailed interaction between a pair and an atom; it consequently cannot explain how the pairs scatter so differently depending on the weight of the atom.

The discovery of homogeneous x-rays forced careful study of their properties; it was quickly recognized that the results of all prior experiments might have been affected by their selective but strong influence.[79] They also provided a potential source of x-rays of definite penetrating power needed for detailed analysis of x-ray absorption. Anticipating that fine distinctions in penetrating power might be discerned, Barkla discovered that the secondary rays from some elements – silver (Ag), tin (Sn), antimony (Sb), and iodine (I) – each possess two distinct components. He called the less penetrating component A, the more penetrating one B, and realized that the periodicity he had charted the year before (Figure 4.2) was due to the overlap of the two species of secondary x-ray (cf. Figure 4.8). Within two years he had renamed the species L and K; he thought it "highly probable that series of radiations both more absorbable and more penetrating exist," and he wanted to leave room on both ends for additions.[80]

Unlike Bragg, Barkla could at least hint at a plausible mechanism for the production of the characteristic secondary x-rays. When a pulse disturbs an atom, electrons within it should oscillate

[78] Barkla and Sadler, *PM, 16* (1908), 576.
[79] Barkla and Sadler, *PM, 17* (1909), 739–60. Barkla and Nicol, *Nature, 84* (1910), 139. Barkla, *PCPS, 15* (1909), 257–69.
[80] Barkla, *PM, 22* (1911), 406n.

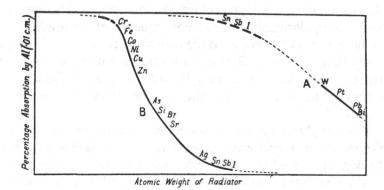

Figure 4.8. Barkla's graph showing two components of different penetrating power in secondary x-rays. [*PCPS, 15* (1909), 259.]

at frequencies determined by the atomic structure. The radiated
impulse that results can be thought of as compounded of a large
number of secondary partial oscillations. The net effect could be a
secondary impulse with a width determined entirely by the atom.
This particular pulse width is relatively intense because each atom
in the sample gives rise to it.

The most remarkable property of the new homogeneous secondary x-rays, only hinted at in earlier studies, became clear after
Barkla moved from Liverpool to London in 1909. His former
colleague, Charles Sadler, continued the investigation of precise
conditions for the stimulation of the characteristic x-rays.[81] Using
the fine distinction in the penetrating power of the secondary
x-rays to excite selected tertiary characteristic x-rays, he corroborated in detail his suspicion that characteristic rays can be stimulated only by an incident radiation of lower absorption coefficient
than the anticipated secondary. If the primary beam does not meet
this requirement, only the inhomogeneous scattered radiation is
detected.

Barkla's hypothesis for the origin of the characteristic secondary
x-ray impulses partially supported the uncorroborated association
of pulse width with penetrating power. The rays should have equal
pulse widths because all rays are produced from the same mixture
of inhomogeneous pulses by identical atoms. To Barkla and Sadler
this meant that each of the new homogeneous pulses carries the

[81] Sadler, *PM, 18* (1909), 106–32.

same energy; x-ray intensity for homogeneous rays is then proportional to the simple number of impulses. Here was a reaffirmation of the reinterpretation of radiant intensity provided by the impulse hypothesis. Bragg had defined the quality of neutral pairs in terms of the dipole moment and the rotational velocity of a pair; the quantity was simply the flux of pairs. Many pulse theorists came to define the intensity of an x-ray beam (whether composed of characteristic rays or not) in terms of only the number of pulses passing a point in space, the quality of each impulse as a function solely of pulse width.

Seen in this light, the characteristic x-rays took on new significance. Their stimulation required a primary beam of greater penetrating power than they themselves had. The necessary threshold could now be expressed in terms of pulse width. This, in turn, was analogous not to the *amplitude* of a periodic wave but to the *wavelength*. The condition found by Sadler for the stimulation of characteristic x-rays appeared to many, especially those outside Britain, to be analogous to Stokes' law of fluorescence. Just as light of a given wavelength can be stimulated only by light of a shorter wavelength, so characteristic x-rays can be stimulated only by other x-rays of shorter pulse width. Some physicists began to speak of the characteristic rays as *fluorescent* x-rays for this reason, but Barkla resisted the identification before 1911. He admitted that this "extension of Stokes' law of fluorescence" was "the most difficult point to explain" about secondary x-ray pulses.[82]

Barkla's influence was at its peak in the years 1908 – 11. He was confident that the sum of the evidence on homogeneous secondary x-rays "strikingly verified Thomson's scattering theory" of pulses.[83] Thomson himself thought that the continued study of characteristic x-rays was "of the utmost importance."[84] Three of Barkla's most comprehensive studies were translated into German and published in the *Jahrbuch der Radioaktivität und Elektronik*.[85] The interpretation placed on his results by some Germans, most notably the editor of the *Jahrbuch,* Johannes Stark, was quite different from Barkla's, as we shall see next.

[82] Barkla, *PCPS, 15* (1909), 269.
[83] *Ibid.,* 257.
[84] J. J. Thomson, *Encyclopaedia Britannica,* 11th ed., *23* (1911), 695.
[85] Barkla, *JRE, 5* (1908), 246–324; *7* (1910), 1–15; *8* (1911), 471–88.

5

The appeal in Germany to the quantum theory

The unit of light energy does not travel in all directions away from an oscillator, but only in a single [direction.][1]

German-speaking physicists arrived at an explanation for the problems facing radiation theory different from that of their British colleagues. Moreover, their attempt to discover a consistent theory formed a complete contrast to the British attempt. The problems themselves were perceived differently. It would be an exaggeration to say that the problems ever reached the same status as paradoxes in Germany as they did in Britain. Physics as practiced in Germany was not as dependent on conceptual pictures, and physicists there were much more willing to adopt formal principles to solve the difficulties without demanding a consistent physical interpretation.

By 1905 the very word mechanics meant to many Germans something essentially different from its meaning in Britain. Influential voices had proclaimed that matter itself is only a construct of the mind, fashioned out of the more ontologically significant electromagnetic forces that give rise to human perceptions of mass and extension. The more influential Germans were not closely tied to the logical requirements of mechanistic thought. When the electromagnetic impulse hypothesis of x-rays was accepted in Germany, it was interpreted by many as a further example of the versatility of an electromechanical ontology. Electromagnetic impulses were considered by some, most notably those who first tried to make sense of x-ray behavior, as a form of wave with extended properties in space. Consequently, as we shall see in this chapter, the sharp distinction between x-rays and periodic light was never made as strongly in Germany as it was in Britain.

[1] Stark, *Elementare Strahlung* (1911), 262.

The most far-reaching proposal to reformulate classical ideas about radiation was not concerned with x-rays at all. Albert Einstein, an unknown patent clerk lacking advanced degrees and totally outside the profession of academic physics, proposed in 1905 that ordinary light behaves as though it consists of a stream of independent localized units of energy that he called *lightquanta*. He was led to this revolutionary view by a statistical analysis of the properties of an idealized sample of radiation in thermodynamic equilibrium. The statistical mode of reasoning was not common in physics at the time, and the resulting "heuristic" description of lightquanta seemed to virtually all physicists to be simply untenable.

Before reading about Einstein's lightquantum, Wilhelm Wien introduced a significant change in the way energy is reckoned in electromagnetic impulses. Building on an analysis by Max Abraham of the energy radiated from uniformly accelerated electrons, Wien recognized that impulse width as well as pulse amplitude varies with the angle from the direction of the source of acceleration. Wien began to use the impulse width, rather than the classically orthodox field amplitude, as the parameter by which to specify the energy content of the impulse.

Shortly thereafter, Wien and Johannes Stark each applied the new quantum principle to x-rays. To do so, they had to describe the impulse energy in terms of a frequency; this was possible because of the previous revision Wien had suggested in the significance of the pulse width. Arnold Sommerfeld came to the defense of classical electrodynamics, but he also justified the nonclassical use of pulse width to fix the penetrating power of x-rays radiated in a given direction. The impulse hypothesis of x-rays, as interpreted in Germany, was a critical conceptual stepping stone between classical and quantum definitions of the intensity and frequency of radiation.[2]

PHYSIK OHNE DEN ÄTHER

There were two opposing theoretical programs in early-twentieth-century Germany designed to resolve the inconsistencies confronted in the interaction of aether and matter. The most widely

---

[2] Wheaton, *HSPS, 11* (1981), 367–90.

accepted and developed was the electromagnetic view of nature in its various forms. As mentioned in Chapter 1, the most radical framers of this program wished to dispense entirely with the concept of matter and to reformulate mechanical properties solely in terms of electrodynamics. The primacy of force was asserted; many German physicists believed that progress was being made when the gross concept of *matter* could be traced back along the Kantian chain of perception to the forces that underlay it. Diametrically opposed to this aether ontology was the theory of relativity proposed by Albert Einstein in 1905.[3] Einstein reformulated Newtonian dynamics and resolved the inconsistencies that complicated understanding of the aether–matter interaction by eliminating the aether altogether. Einstein's solution was ultimately the more successful.

An integral part of Einstein's rejection of the medium for light waves was his suggestion of the lightquantum hypothesis.[4] He began by questioning the baseless assumption that there must be a strong conceptual distinction between light and matter – light being thought to be waves spreading through a medium, matter to consist of localized particles.[5] His suggestion of the lightquantum hypothesis arose from the close analogy he perceived between the behavior of radiation and the behavior of a gas.[6] His point is easily understood today, although the statistical form of his argument was unusual at the time.

Einstein showed that the entropy of an idealized sample of radiation enclosed in a cavity of known volume could be obtained directly from the energy density. In the mid-1890s, Wien had given a possible form for energy density as a function of frequency and temperature.[7] Einstein recognized that Wien's law was valid only for high-frequency light. But he used it to calculate the change in entropy S to be expected in the radiation within the frequency range v to $v + dv$ if the occupied volume in the cavity was decreased from V_0 to V:

$$S - S_0 = \frac{E}{bv} \ln (V/V_0). \tag{5.1}$$

[3] McCormmach, *HSPS, 2* (1970), ix–xx, 41–87. Miller, *Relativity* (1981).
[4] Einstein, *AP, 17* (1905), 132–48.
[5] See Klein, *Natural philosopher, 2* (1963), 59–86; *3* (1964), 3–49.
[6] Klein, *Science, 157* (1967), 509–16.
[7] Wien, *AP, 52* (1894), 132–65; *AP, 58* (1896), 662–9.

Here E is the total energy contained in the sample and b is a constant.

According to Ludwig Boltzmann, the entropy change at constant temperature in a gas of particles can be expressed as a function of the gas constant R and Avogadro's number N. It is

$$S - S_0 = \frac{R}{N} \ln W \tag{5.2}$$

where W is the probability that a state of the particles of entropy S will occur relative to the likelihood of a state of entropy S_0. Clearly, the probability that, at a random instant, one molecule will be in a subvolume V smaller than the cavity volume V_0 is the ratio V/V_0. The relative probability that n molecules will be found in volume V is thus $W = (V/V_0)^n$. Einstein therefore rewrote equation 5.2 as

$$S - S_0 = \frac{Rn}{N} \ln (V/V_0) \tag{5.3}$$

Noting the similarity in form between equations 5.1 and 5.3, Einstein remarked that "the entropy of a monochromatic radiation of sufficiently low density follows the same laws of variation with volume as does the entropy of an ideal gas or a dilute solution."[8] The application of Boltzmann's statistical formulation of the second law of thermodynamics led directly to the surprising result that the energy of the radiation might be quantized. Noting that E/bv for the radiation corresponded to Rn/N for the gas, Einstein concluded that monochromatic radiation "behaves in a thermodynamic sense as if it consisted of mutually independent energy quanta of magnitude $[Rbv/N]$."[9]

Einstein went on in this epochal paper to sketch the observable consequences of his hypothesis. He pointed out that Stokes' law of fluorescence, a relationship that had never been fully understood, followed immediately from the fact that the energy of each light-quantum is proportional to its frequency. Light of a given frequency can never stimulate light of a greater frequency because of simple conservation of energy. He also drew some verifiable conclusions about photochemical reactions. But for our purposes, the prediction of the functional form of the photoelectric relation is

[8] Einstein, *AP, 17* (1905), 139.
[9] *Ibid.*, 143.

the most significant consequence of the lightquantum. If light consists of localized quanta of energy Einstein argued, an electron in an atom will receive energy from only one lightquantum at a time. Monochromatic light of frequency v can therefore grant electrons only energy Rbv/N, or, as we express it today, hv, where h is Planck's constant. If one supposes that some small part p of that energy must be used to release the electron from the metal itself, all electrons of charge e so released will be stopped by a decelerating potential P, following the relation

$$Pe = hv - p \qquad\qquad (5.4)$$

At the time Einstein made this prediction, very little was known about the precise form of the photoelectric relation. The triggering hypothesis was gaining wide acceptance, and it effectively rendered close experimental study of the relation unnecessary.[10] Note that Einstein's formulation also explains the observation that had led Philipp Lenard to the triggering hypothesis in the first place. The maximum velocity of photoelectrons must be independent of the light intensity according to the lightquantum hypothesis; only the frequency of the light determines how much energy any electron may receive. The intensity of light becomes, in Einstein's view, equivalent to the total lightquantum flux: the number of lightquanta that pass through a unit area in unit time. The size of the photocurrent should vary with light intensity, but the stopping potential should not. A full decade of uncertainty about the experimental form of the photorelation was to ensue before Einstein's predicted equation 5.4 was finally and completely verified.

There are two other aspects of Einstein's study that we should note. First, Einstein used Wien's distribution function to make the calculation; he did not use Planck's. The following year, Einstein showed that Planck had erred in his derivation of the blackbody law, and could have arrived at his new, successful relation only by implicitly adopting the lightquantum hypothesis.[11] Recent scholarship has suggested that Einstein's lightquantum theory owed very little to Planck's work. It has been shown that Planck placed no proscription on the energies of material oscillators in 1900; they could possess and transfer energy in a continuum of values. Ein-

---

[10] Wheaton, *Photoelectric effect* (1971); *HSPS, 9* (1978), 299–323.
[11] Einstein, *AP, 20* (1906), 199–206.

stein's paper of 1905 therefore marks the origin of the quantum theory as we understand it today.[12]

Second, Einstein's paper contains no reference to x-rays or to γ-rays. Einstein did note that the properties of light he wished to emphasize were most pronounced at high frequency, where Wien's law was valid. But he did not suggest, nor is he likely to have believed, that similar considerations should be applied to x-rays. X-rays and periodic light were then thought to be essentially different forms of radiation. Viewed retrospectively, Einstein's lightquantum hypothesis solved the dual paradox just then being recognized for any classical theory of x-rays. But it also introduced serious difficulties for radiation theory that most physicists justifiably thought insurmountable.

Einstein's hypothesis of lightquanta was not taken seriously by mathematically adept physicists for just over fifteen years. The reasons are clear. It seemed to be an unnecessary rejection of the highly verified classical theory of radiation. Most physicists rejected the idea as merely the resuscitation of the discredited emission theory of light. In the domain of visible or ultraviolet light, not even a hint of the x-ray paradoxes then existed. How lightquanta could possibly explain interference phenomena was always the central objection. Einstein offered very little explanation; reconciliation of the lightquantum with interference was a problem with which he wrestled for the rest of his life. In 1921 he characterized the conflict to his friend Paul Ehrenfest as something fully capable of driving him to the madhouse.[13]

Although Einstein's theory of relativity was soon accepted by leading physicists, his lightquantum hypothesis was not. Einstein's own study of the problems of radiation had almost no influence on the work done by others before the early 1920s. For this reason, we shall not discuss it in further detail here. Some very useful historical discussions have been provided, but it is a story yet deserving scrutiny.[14]

---

[12] Kuhn, *Black-body theory* (1978), argues that Einstein and Ehrenfest first quantized the energy of oscillators in 1905–6. For another view, see Klein, *AHES, 1* (1962), 459–79.

[13] Einstein to Ehrenfest, 15 March [1921]. AHQP 1, 77.

[14] Klein, *Natural philosopher, 2* (1963), 59–86. McCormmach, *HSPS, 2* (1970), 41–87.

WILHELM WIEN AND THE ENERGY OF X-RAYS

The first person to perceive the connection of Einstein's unorthodox lightquantum hypothesis with the nature of x-rays was Wilhelm Wien, since 1900 *ordentlicher* professor of physics at the University of Würzburg. Wien was a proponent of the electromagnetic view of nature. He was used to thinking that the properties of matter are artifacts of the continuous electromagnetic field, not consequences of actual material particles. He believed that there was great value in the idea that the electron mediates between the events in the electromagnetic field that we call radiation and those that we ascribe to atoms. Therefore, his concepts were closely allied to the electron theory of H. A. Lorentz, in which interaction between matter and radiation occurs only by means of the electron.[15]

But Wien deviated from Lorentz in two important respects. First, Wien thought that the electron itself has no material existence; its apparent mechanical properties are due exclusively to electrodynamic effects. Second, he believed that the electron is compressible, and it was this idea that first directed his attention to x-rays.[16] He gave the electron a dipole moment that could change with time. In 1904 he treated x-ray impulses as if they arose from impacts of this deformable electron.[17] He thought the impulse hypothesis was the "most probable" explanation of x-rays but later that year admitted that "theoretical conceptions of x-rays can in no way be considered certain."[18] He hypothesized that some 3×10^{-13} ergs of energy should be radiated away in each electron impact, and to test this prediction he turned to experiments on the heat produced in material bodies by x irradiation.

The results appeared in a *Festschrift* dedicated to Adolph Wüllner in 1905.[19] Wien found by experiment that no more than one part in a thousand of the heating capacity of the cathode rays actually appears as x-ray energy. Joseph Larmor had shown in 1897 that the rate at which energy radiates from an accelerated particle with charge e is:[20]

[15] Hirosige, *HSPS, 1* (1969), 151–209.
[16] Wien, *Archives Néerlandaises* (2), *5* (1900), 96–107.
[17] Wien, *PZ, 5* (1904), 128–30.
[18] Wien, *JRE, 1* (1904), 215–20.
[19] Wien, *Festschrift Adolph Wüllner* (1905), 1–14.
[20] Larmor, *PM, 44* (1897), 503–12.

$$\frac{2}{3}\frac{e^2 a^2}{c} \tag{5.5}$$

where a is the mean acceleration. Wien used a more recent form for the total energy derived by Max Abraham to take account of what we would now classify as relativistic effects from a rigid spherical electron accelerating over a time interval τ:[21]

$$\frac{2}{3}\frac{e^2}{c^3}\int_0^\tau a^2 \frac{dt}{k^6} \tag{5.6}$$

where $k = \sqrt{1 - v^2/c^2}$. Wien integrated this expression over the time needed to decelerate the electron uniformly in a distance l. The total emitted energy thus obtained had to be $\frac{1}{1000}$th of the kinetic energy of the electron, which in turn may be determined directly from the discharge tube potential.

To turn this result into a determination of the pulse width, Wien had to settle one more matter, and in so doing he made a significant new observation. "The impulse width depends on the direction of the emitted wave," he pointed out.[22] This is certainly true. Along the line of electron motion, "the radiated electromagnetic energy is zero." Classically, the energy radiated along the line of acceleration is zero because the directed field amplitudes A_E and A_H both drop to zero. But Wien implicitly ascribed the penetrating power, intimately related to the energy content in the pulse, to the pulse width, not the pulse amplitude. "Even for quite uniform deceleration and completely equal electron velocities," Wien pointed out, "the radiation will be inhomogeneous."[23] The physical analog of the pulse width is the wavelength; although the amplitude of a periodic wave varies with the azimuth, its wavelength remains constant with the changing direction from which it is viewed. The pulse width, on the other hand, varies according to the azimuthal angle. Wien showed that at an angle θ with respect to the axis of electron motion, the pulse width is

$$\delta = l(\cos\theta + 2c/v) \tag{5.7}$$

[21] M. Abraham, *AP, 10* (1903), 156.

[22] Wien, *Festschrift Adolph Wüllner* (1905), 8.

[23] *Ibid.*, 8–9. Wien's unorthodox views reflect the relative lack of interest in x-ray research in Germany at the time. Röntgen never seriously returned to the study of x-ray properties, and a perusal of the papers given at the *Naturforscherversammlungen* between 1900 and 1908 turns up none concerned with the nature of x-rays apart from those by Marx.

From this relation, Wien found that in the direction perpendicular to the electron acceleration, the pulse width δ had the value 10^{-10} cm, about 3 percent of that calculated by Sommerfeld from his analysis of diffraction.

After making this comparison, Wien turned from the production of x-rays to an analysis of their absorption. He had expected that all of the x-rays would "convert themselves to secondary [cathode] rays."[24] At first, he thought that he had shown this to be true. But before his expanded study appeared in the *Annalen der Physik,* he realized that something was amiss. "If one looks more closely at the mechanism of this transformation," he pointed out, "difficulties arise in seeing just where the high velocities of the secondary electrons originate."[25] Using the classical definition of energy content in the pulse, that is, the product of electric and magnetic field amplitudes, Wien calculated the maximum velocity that each x-ray impulse could produce in an electron. For a typical x-ray tube potential, it was on the order of 10^{-4} cm/sec; Dorn had shown that typical x-rays release secondary electrons that move as fast as 10^9 cm/sec. The electrons moved some thirteen orders of magnitude faster than Wien expected! All the x-ray impulses over a very long time interval, on the order of 0.01 sec, would have to conspire to contribute their energy to the *same* electron. This Wien could not believe, and he concluded, "it is therefore impossible that secondary [electrons] obtain their velocity through direct acceleration by the x-ray waves."[26] Wien had run full tilt into the paradox of quality.

Wien found only temporary solace in the triggering hypothesis, suggesting that the incident x-rays simply release electrons that *already* possess high velocity because of their membership in the atomic system. But Wien's allegiance to the triggering hypothesis was not strong. The photoelectric effect required a set of electrons in the atom with velocities on the order of 10^8 cm/sec; x-rays now required another set of even greater velocity. Furthermore, Wien believed that serious problems would arise for this explanation once radiative losses were taken into account. To explain the high secondary-electron velocities, an atom must contain electrons moving at about 10^9 cm/sec. Wien imagined that an electron

---

[24] Wien, *Festschrift Adolph Wüllner* (1905), 7.

[25] Wien, *AP, 18* (1905), 1003-4.

[26] *Ibid.,* 1005.

revolves in the inverse-square potential surrounding a positive charge; the orbit is of atomic radius 10^{-7} cm. An electron moving 5×10^7 cm/sec, slower even than electrons released by light in the photoelectric effect, should, according to equation 5.6, radiate its 5×10^{-13} erg at the rate of 10^{-6} erg/sec. The total energy of the electron is gone in 10^{-7} sec![27] As early as 1905, Wien had concluded that grave difficulties confronted the triggering hypothesis, not simply for x-rays but for visible light as well. But he had no better alternative to offer at the time.

Actually, the problem is not as severe as Wien implied. In the article in which Larmor derived expression 5.5, he went on to suggest that radiative losses might drop almost to zero for atomic systems with more than one electron.[28] This would be true as long as the vector sum of all electron accelerations remains zero, a situation most easily achieved when two electrons describe the same circular orbit at opposite ends of a diameter. J. J. Thomson developed this idea in 1903 for radiation from multielectron orbits, finding that the energy radiated per electron drops by a factor of roughly 1,000 for each additional electron in the ring when the particles move at a velocity of $0.01c$.[29] Each of six electrons in a ring radiates only 10^{-17} of the energy predicted by expressions 5.5 or 5.6. But of this work Wien was apparently unaware.

Wien was convinced very early that there is very little difference between x-rays and periodic light. His use of the term *Röntgen-welle* to describe x-ray pulses, as quoted above, indicates the close connection in his mind of the two forms of radiation. At the same time as he assumed the similarity of x-rays and light, most British physicists were emphasizing their differences. Very likely, Wien's conflation of periodic wave and discontinuous pulse was a product of his experience in conjuring the localized properties of atoms out of the continuous electromagnetic field. It is, of course, possible to describe an impulse as the sum of many periodic waves, each of which extends to infinity in space. In any event, the connection Wien saw between x-rays and light, combined with his concern to find an explanation for x-ray ionization, increased his interest in any attempt to explain the photoelectric effect on some basis other than the triggering hypothesis.

[27] The radiated power is given on p. 1007 as 10^6 ergs/sec, but the sign of the exponent is clearly a misprint.
[28] Larmor, *PM, 44* (1897), 503–12.
[29] J. J. Thomson, *PM, 6* (1903), 681.

QUANTUM IMPULSES

In the spring of 1907, Wien found what he had been looking for: Einstein's unorthodox treatment of the photoelectric effect using the lightquantum. In all likelihood, Wien discovered Einstein's paper by himself. He was fully conversant with the literature on cavity radiation, particularly Planck's work, and Einstein's paper was closely related. Furthermore, Wien was aware of Einstein's new relativity theory and recognized early the profound nature of Einstein's thought. A direct influence may also have contributed to Wien's awareness of Einstein's quantum studies. Wien's student, Jakob Laub, completed a dissertation on secondary electrons early in 1907.[30] At about the same time, he prepared a seminar talk at Würzburg on the theory of relativity.[31] Wien was sufficiently impressed to suggest that the younger man travel to Bern to meet and learn from Einstein.[32] This Laub did. After a short visit, he returned in the spring of 1907 to Würzburg. The following year, he went back to Bern to collaborate with Einstein.[33] It is not at all unlikely that Laub's close early contact with Einstein widened Wien's appreciation for Einstein's work, even if Laub's enthusiasm for Einstein's lightquantum hypothesis provided nothing new, and was not entirely acceptable, to Wien.

Wien's first use of the quantum hypothesis, the *Planck theory*, as he called it, arose out of his interest in the problems of cavity radiation. It was an attempt to extend Planck's quantum treatment of heat radiation to the discrete optical emission spectrum.[34] Wien sought the conditions under which emission lines might, like cavity radiation, be expressed solely as a function of intensity and frequency for a given temperature. But Wien soon recognized another application of the quantum transformation rule, one that he saw clearly only after reading about Einstein's treatment of the photoelectric effect. It might hold the key to the difficulties yet unresolved for x-ray ionization. "A simple generalization of the Planck radiation theory to x-rays," he suggested in November 1907, "would not only explain, but demand, the production of the

[30] Laub, *AP, 23* (1907), 285–300.
[31] Laub, *AP, 23* (1907), 738–44; *AP, 25* (1908), 175–84.
[32] Seelig, *Albert Einstein* (1954), 85–6; *Albert Einstein, Leben und Werk* (1960), 120ff. For more on the Laub–Einstein connection, but only with regard to relativity theory, see Pyenson, *HSPS, 7* (1976), 83–123.
[33] Einstein and Laub, *AP, 26* (1908), 532–40; *AP, 26* (1908), 541–50.
[34] Wien, *AP, 23* (1907), 415–38.

high velocities of secondary electrons. [It would] as well provide yet another way to calculate the wavelength of the active x-rays."[35]

Note Wien's use of the word "wavelength" to characterize x-ray impulses. The lack of distinction in his mind between impulses and waves here yielded a benefit. The quantum transformation relation requires that radiation be expressed in terms of a frequency. An impulse has no frequency. But Wien could now interpret the inverse pulse duration τ as analogous to a frequency. Thus, the circle was complete; by 1907 Wien characterized the energy in x-rays not in terms of pulse amplitude but in terms of pulse width alone. He set the kinetic energy of the electron $mv^2/2$ equal to h/τ and then derived δ from the relation $\delta = c\tau = 2hc/mv^2$. The result was $\delta = 6.75 \times 10^{-9}$ cm, considerably closer than his 1905 estimate to Sommerfeld's derived value for the x-ray pulse width. In support of his use of the quantum relation, Wien cited the experiments by P. D. Innes showing that the maximum electron velocity stimulated by x-rays, like that stimulated by periodic ultraviolet light, does not depend on the intensity of the x-rays but only on their quality.[36] But neither, in the first approximation, did it seem to depend on the material from which electrons are ejected, so Wien explained just where the analogy to light breaks down.[37]

Wien employed the quantum relation to achieve a specific result; he was not particularly concerned to find a *modellmässig* interpretation for why the relation should be employed. The quantum transformation relation offered quantitative success without recourse to the suspect triggering hypothesis. Wien's extension to x-rays of Einstein's relation for visible light was made possible by his close identification of periodic light with impulse x-rays, an identification that followed in turn from Wien's acceptance of the electromagnetic view of nature. In December 1907, a month after the new calculation of δ, Wien declared in an important review of radiation theory, "there seems to me to be no basis for any essential distinction between x-rays and light."[38]

The quantum transformation relation was not synonymous

---

[35] Wien, Göttingen *Nachrichten* (1907), 599. Laub soon repeated Wien's experiment. *AP, 26* (1908), 712–26.

[36] Innes, *PRS, 79A* (1907), 442–62.

[37] Wien neglected the energy required to release an electron from the metal after it has left the atom. He felt that this was negligible relative to x-ray energies but important for visible light.

[38] Wien, Würzburg *Sb* (1907), 107.

with Einstein's lightquantum hypothesis. Wien accepted the former but not the latter. One could not assume that the energy of x-rays exists in "indivisible quanta [in the way that] matter consists of atoms."[39] His analysis had been directed solely to the production of x-rays by electron impacts; he did not treat the conceptually more demanding case of x-ray absorption. Only in the latter case need one explain how an isotropic pulse can concentrate its total energy on an individual electron.

JOHANNES STARK AND THE QUANTUM HYPOTHESIS

Wien's identification of x-rays with light and his recourse to the quantum relation were both developed and extended by Johannes Stark. Like Wien, Stark did not distinguish x-ray impulses from periodic waves. In a review article written in 1905 on electric discharges, he repeatedly interchanged the terms *wavelength* and *pulse width,* and used *wavelength* where he explicitly considered the properties of discontinuous impulses.[40] The mere equivalence of these terms carried with it an implicit connection of x-ray energy with frequency. It is a short step from this to the attribution of a frequency to the x-ray, a frequency that measures x-ray energy. Stark explained, freely mixing incompatible concepts, that among the various pulse widths in an x-ray beam, "one definite wavelength will dominate corresponding to the velocity of the stimulating cathode rays."[41]

Thus, Stark was already conceptually primed for the quantum interpretation of x-rays when Wien dropped it in his lap. Wien's first use of the quantum relation had been for spectral radiation from positively charged ions, the so-called canal rays. Stark had discovered in 1905 that the light radiated by these ions shows a Doppler shift in frequency. It was Wien who had suggested that canal rays were molecular ions, and his work on the subject was of continuing interest to Stark.[42] Moreover, Stark read widely in the relevant literature because of his position as editor of the *Jahrbuch der Radioaktivität und Elektronik.* When Wien applied the quan-

[39] *Ibid.,* 106.
[40] Stark in Winkelmann, *Handbuch, 4* (1905), 642–8.
[41] *Ibid.,* 643.
[42] Wien, *AP, 5* (1901), 421–35, *8* (1902), 244–66; *9* (1902), 660–4.

tum relation to canal ray spectra in May 1907, Stark could hardly miss its importance. Wien suggested that secondary electrons produced by cathode rays and canal rays might themselves stimulate fluorescence "in much the same manner as in that connection presumed by Einstein [to hold] between the elementary quanta of energy and the velocities of photoelectrons."[43] The quantum relation, and explicit reference to Einstein's lightquantum, passed from Wien to Stark.

In the earlier study, in which Wien calculated the pulse width of x-rays, he pointed to two cases in which the quantum relation might be profitably applied: the photoeffect and canal rays. Independently of Wien, Stark recognized a third case: the ionization produced by x-rays. In October, Stark published his first quantum paper, a dilettante's attempt to connect a hypothetical unit of positive charge and the electron with a synthesis of quantum and relativistic concepts.[44] Among the chaff, he inserted an irrelevant but inspired footnote outlining a way to determine the minimum "wavelength" of x-rays produced by electrons decelerating from a known velocity. Electrons accelerated by potential P have energy $Pe;$ if they stop in a distance taken to be one-half the "wavelength" of the spherical pulse that results, then $\lambda_{min} = 2hc/Pe$. For a typical potential of 60,000 V, λ_{min} is 6×10^{-9} cm. As we have seen, Wien reported the same result the following month and compared it favorably to the value 5×10^{-9} cm obtained in Haga and Wind's second detailed study of the x-ray diffraction photograph.[45]

But unlike Wien, Stark went on to treat the inverse case of ionization by x-rays. In extending the quantum treatment to the absorption of radiation, he opened the door that led to the lightquantum itself. The transfer of an electron's kinetic energy in quanta to a spherical wave or impulse may have no classical rationale, but it is conceptually understandable. How the full quantum of energy contained in that spherical wave can be passed back to an individual electron defies understanding; it makes credible the spatial localization of the radiant energy proposed by Einstein. The maximum velocity of the electrons stimulated by light or x-rays would, in the quantum hypothesis, be independent

[43] Wien, *AP, 23* (1907), 433.
[44] Stark, *PZ, 8* (1907), 881–4. Stark persistently used λ and referred to the "wavelength" of x-ray impulses.
[45] Haga and Wind, *AP, 10* (1903), 305–12.

of the radiation intensity and, Stark claimed, independent of the chemical nature of the metal as well. Of course, Einstein had already pointed this out for ultraviolet light; in a subsequent paper, Stark gave Einstein credit for a "similar treatment" one that he said he had read only after the fact.[46] This may be true; all that Stark really needed had been contained in Wien's paper. Although Wien had cited Einstein's lightquantum paper, Stark's first plunge into the quantum theory was quite possibly without direct knowledge of that seldom cited source. Stark did not take into account the possibility that energy might be lost in the removal of an electron from the structure of the metal. This was likely a holdover from his initiation to the subject by Wien, who made a similar assumption. The work function is negligible in the case of x-ray ionization, but not in the photoelectric effect.

Stark was certainly acquainted with Einstein. The two were corresponding over a review article on relativity for Stark's journal. Einstein, perhaps protective of his ideas because of his lack of academic position, or perhaps simply encouraged to hear a respected physicist employ the quantum condition, shipped off reprints of all his papers to Stark and wrote to express his happiness that Stark had "become aware of the lightquantum issue."[47]

Stark eagerly exploited the quantum relation. His first topic after x-rays was his own discovery, the Doppler shift in canal-ray spectra.[48] The problem was to explain why a discrete line appeared at a slightly higher frequency for each original, rather than a continuous broadening of each line toward higher frequency. Presumably a continuous range of ion velocities exists in the discharge tube from zero up to some maximum. The lack of a corresponding continuum of shifted frequencies cast doubt on the Doppler nature of the effect. The quantum principle offered a possible explanation for the exclusion of lines below a certain threshold. Although later shown to be incorrect, Stark's quantum treatment sparked wider interest in a quantum discussion of topics other than cavity radiation and specific heats.[49] Next, Stark used the

[46] Stark, *PZ, 8* (1907), 914n; *9* (1908), 767–73.
[47] Einstein to Stark, 7 December 1908. Quoted in A. Hermann, *Sudhoff's Archiv, 50* (1966), 272.
[48] Stark, *PZ, 9* (1908), 767–73.
[49] Einstein to Stark, 2 December 1908. A. Hermann, *Sudhoff's Archiv, 50* (1966), 276; *Frühgeschichte* (1969), 88.

quantum hypothesis to calculate the high-frequency absorption limit for the stimulation of molecular spectra.[50] He soon went on with a student to analyze the relation between photoelectric sensitivity and fluorescence in some fifty-seven organic molecules.[51]

In the course of these studies, Stark returned repeatedly to the close connection he perceived between ionization, the photoelectric effect, and fluorescence. Even before recourse to the quantum in 1907, he had expressed his confidence in the interconnection.[52] With the success of the quantum principle demonstrated, his confidence increased. "Absorption of light," he said, "has as its consequence the ionization of the absorbing atom or molecule, that is, the excitation of photoelectric cathode rays, accompanied by a fluorescence conforming to Stokes' law."[53] Stokes' law, which stated that the frequency of incident light must equal or exceed that of emitted fluorescent light, lacked a classical explanation. In the spring of 1908, Stark expressed his appreciation of Einstein's success in "illuminating another dark point of fluorescence phenomena" by giving the straightforward quantum theoretical basis for Stokes' law. This prompted his request of Einstein in 1908 for an elaboration of the subject, suitable for chemists' understanding, to appear in the *Jahrbuch*.[54]

A review article by Charles Barkla, which spelled out the detailed properties of characteristic x-rays for the first time in the German language, appeared early in 1908 in Stark's *Jahrbuch*.[55] Stark, as editor, had known of its content for some time. In that paper, Barkla revealed that the characteristic secondary x-rays possess precisely the trait that Stark had emphasized in his long study of optical absorption: "properties of the chemical elements that are periodic functions of atomic weight."[56] Barkla emphasized that "without exception" the stimulation of a homogeneous radiation requires a primary beam of greater penetrability than that produced. Further, the intensity of the primary beam could be

[50] Stark, *PZ, 9* (1908), 85–94, 356–8.
[51] Stark and Steubing, *PZ, 9* (1908), 481–95; 661–9.
[52] Stark, *PZ, 8* (1907), 81n.
[53] Stark, *PZ, 9* (1908), 92.
[54] Stark to Einstein, 11 February 1908. Hermann, *Sudhoff's Archiv, 50* (1966), 272–3.
[55] Barkla, *JRE, 5* (1908), 264–324.
[56] Stark, *JRE, 5* (1908), 124–53.

altered almost at will without affecting the hardness of the secondary.[57]

Stark had already converted in his own mind x-ray "hardness" or "penetrability" to frequency. He recognized immediately that Barkla had found the x-ray analog to optical fluorescence. The hardness threshold for the stimulation of characteristic x-rays was a simple case of Stokes' law of fluorescence. Stark was very early in this realization; it was to be three years before Barkla would speak of fluorescent x-rays as "an extension of Stokes' Law." Stark was quick to ensure that his quantum view of characteristic x-rays appeared in English in *Nature*.[58]

Stark was unusual among German physicists in his early recognition of the problems surrounding the paradox of quantity. Before adopting the quantum relation, he puzzled over the "small fraction of the total number of molecules" that are placed in a "chemically different state" after absorbing light energy.[59] Following contemporary work by Lenard on fluorescence, Stark came to the conclusion in 1908 that the effect of light is "localized in definite atomic groups."[60] Then Barkla raised the analogous issue for x-rays. The primary impulse "passes over every single electron" in all atoms of the scattering substance, but "only a small fraction of the total number of electron [radiation] sources produce the homogeneous [secondary] radiation."[61] Not only did Barkla provide the evidence for Stark's extension of fluorescence to x-rays, he pointed out the single most difficult observation to be explained in a theory of spreading pulses, whether treated with the quantum or not.

SPATIALLY LOCALIZED X-RAYS

Stark was the first influential physicist after Einstein to take the lightquantum seriously. He was led to this step from the analogy of x-rays to light. Because he therefore was able to assign a wavelength to the x-rays, he could apply the quantum relation. This

[57] Barkla, *JRE*, 5 (1908), 229, 306, 308, 309, 317, 318, 319.
[58] Stark, *Nature*, 77 (1908), 320.
[59] Stark, *PZ*, 8 (1907), 82n.
[60] Stark, *JRE*, 5 (1908), 149.
[61] Barkla, *JRE*, 5 (1908), 320, 323.

worked so well to explain observed frequency limits and secondary electron velocities that little doubt remained in his mind about the validity of the relation. Although Stark adopted the quantum relation, he tended more than most German physicists to interpret the elementary processes in terms of conceptual mechanisms. Unlike most of his German colleagues, Stark took very seriously the observation that x-ray ionization affects only a very small fraction of atoms. Sometime before 1909, this forced him to adopt the lightquantum as a physical reality.

Stark may have adopted the lightquantum by the time of his talk to the *Naturforscherversammlung* held in Cologne in 1908. There was a marked shift in his terminology. What had formerly been the *Planck'sche Elementargesetz,* or Planck's law, became the *Lichtquantenhypothese.* His talk at the meeting was entitled "New observations on canal rays with respect to the lightquantum hypothesis," and he considered the significant new case of spectral lines radiated after ion collisions.[62] He had undertaken the experiments, he claimed, "as a proof of the lightquantum hypothesis," and expressed the hope that so suspect a motive might not discredit his results. "This hypothesis is so unusual and so strongly opposes the customary concept of emission and absorption of light that the general reservations against it are perfectly understandable," he said.[63] All he actually used in the study was the quantum transformation relation, which by that time had been adopted by Planck, Wien, Einstein, and others. That Stark so clearly recognized the controversy into which he headed suggests strongly that he had already accepted the spatially localized lightquantum.

If Stark had previously been ignorant of Einstein's work, he was no longer. He cited all three relevant papers, doubtless assisted by the reprints Einstein had sent him. He spoke freely about the transfer of radiant energy in units of lightquanta, although such a view clearly did not require accepting the lightquantum per se.[64] In an *Anhang* added to the printed version of his talk, he reviewed the successes of the hypothesis, significantly crediting it solely to Einstein. His playing down of Planck's contribution may have

[62] Stark, *PZ, 9* (1908), 767–73.
[63] *Ibid.,* 768.
[64] As Einstein is reputed to have put it: "Although beer is sold only in pint bottles, it does not follow that it exists only in indivisible pint portions." Frank, *Einstein* (1953), 71.

been due to his acceptance of Einstein's 1906 claim that Planck's work was a consequence of the more basic lightquantum assumption. There was at that time general misunderstanding of just what Planck had done in 1900.[65] Einstein, who had hoped to attend the meeting in Cologne at which Stark spoke, decided instead to spend his short vacation elsewhere. But he was naturally exceedingly interested in Stark's remarks because he saw how close Stark had come to accepting the "revolutionary" lightquantum.[66]

I suggest that Stark had already done so. The terminology of his Cologne talk strongly suggests it; but radiation from canal rays – light produced by mechanical impact – did not *require* spatially localized rays in the same way as did absorption of light. Stark had only to express his results in a manner that would allow him to refer back to this work as consistent with lightquanta. Stark had other reasons to pause before taking so bold a step. He knew how controversial the topic was. And he was at that moment embarking on a bit of political maneuvering to obtain a regular appointment at Aachen. For this he was counting on, and getting, strong support from the former resident physicist, Arnold Sommerfeld.[67]

Sommerfeld was the foremost German proponent of the impulse hypothesis for x-rays, precisely the object of Stark's quantum assault. Immediately after he received the post, with Sommerfeld's endorsement, Stark offered Einstein a position as his assistant. Einstein wisely declined.[68] The next month, Einstein published a masterful review of the state of radiation theory and began to formulate a statistical treatment of radiation based on large numbers of lightquanta.[69] Stark wrote that the new view interested him greatly.[70] Four months later, Stark proclaimed publicly that he had adopted the spatially localized lightquantum for x-rays.[71]

His paper was entitled "On x-rays and the atomic constitution of radiation." The critical transition to the lightquantum was de-

[65] Klein, *AHES, 1* (1962), 459–79. Kuhn, *Black-body theory* (1978).
[66] Einstein to Stark, 2 December 1908. A. Hermann, *Sudhoff's Archiv, 50* (1966), 276.
[67] Stark to Sommerfeld, 12 October 1908. A. Hermann, *Centaurus, 12* (1968), 41–2.
[68] Einstein to Stark, 6 April 1909. A. Hermann, *Sudhoff's Archiv, 50* (1966), 278.
[69] Einstein, *PZ, 10* (1909), 185–93. See also the discussion between Einstein and Ritz, *ibid.*, 224–5, 323–4.
[70] Stark to Einstein, 8 April 1909. A. Hermann, *Sudhoff's Archiv, 50* (1966), 278.
[71] Stark, *PZ, 10* (1909), 579–86.

signed to win converts directly from the impulse theory. First, he calculated the mean time between electron impacts in the x-ray tube; typically operating at 30,000 V and passing 3×10^{-4} amps, the tube produced 10^{14} collisions every second. On the other hand, the accepted pulse width, 10^{-8} cm, implied a mean pulse duration of some 10^{-19} sec. It was extremely unlikely, Stark claimed, that two or more pulses would combine their effect on the same atom. The impulses were separated by an interval 10^5 times longer than the duration of any pulse. Note that Stark implicitly assumed that ionization begins instantaneously when the radiation arrives; he had explicitly claimed this before, but did not do so now.[72]

Stark sketched a vague mechanism that he thought might explain interference for visible lightquanta. Optical lightquanta are radiated from "centers" in atoms that are generally far closer together than the wavelength. Thus, a "space–time superposition" of lightquanta arises that has a "complex of specifically ordered properties." It was not clear what Stark meant by this, and it was likely that Stark himself did not know. In some unexplained manner, the superposition of lightquanta would apparently direct quanta toward regions of constructive interference and away from regions of destructive interference. For wavelengths that are short compared to the separation of the centers, on the order of 10^{-8} cm or less, the complex does not arise and the radiation acts like more or less directed units.[73] Stark concluded that the "oscillating energy" of each impulse must therefore be contained within a volume of space equal to $(c/v)^3$. He no longer bothered to qualify his use of the then standard frequency symbol to characterize x-rays, while at the same time his argument depended on the rays' spatial and temporal limitations.[74] This article formally opened Stark's attempt "as quickly as possible to bring the conflict between the lightquantum hypothesis and its older opposing concepts from the sphere of speculation and theoretical discussion to the firm ground of experiment."[75]

---

[72] Stark had recognized the significance of an immediate response in optical fluorescence. He thought that his chemical interpretation explained "the fact that the onset and cessation of fluorescence occurs almost at the instant the illumination start and stops." Stark and Steubing, *PZ, 9* (1908), 491.

[73] Stark, *PZ, 9* (1909), 582.

[74] *Ibid.*, 583. The standard symbol for frequency was *n*, not *v*.

[75] *Ibid.*

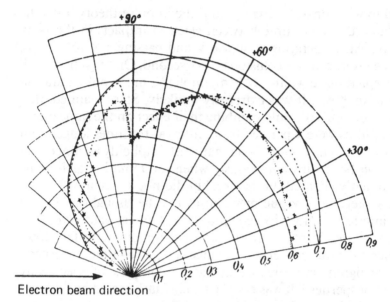

Electron beam direction

Figure 5.1. Stark's graph of forward scattering of x-rays. [*PZ, 10* (1909), 911.]

The weapon he chose was the asymmetry of emitted x-rays. He thought he had found an *experimentum crucis*. In November he applied his substantial gifts as an experimentalist to a careful demonstration that x-rays are not emitted isotropically from the anticathode. He found that they show a distinct maximum intensity in the forward direction. The unbroken line in Figure 5.1 records his corrected measures of photographic blackening produced by x-rays scattered from an aluminum plate at different angles from the horizontal direction of the cathode beam toward the right.[76] For a typical cathode beam velocity of 10^9 cm/sec, the maximum effect lies within a cone some 60° from the forward direction. For once, Stark's knowledge of the foreign-language literature failed him. Bragg had emphasized Cooksey's demonstration of the asymmetry in 1908 and earlier in 1909 had himself repeated the result for x-rays.[77]

Stark claimed that the asymmetry of emission could not be

[76] Stark, *PZ, 10* (1909), 902–13.
[77] Cooksey, *Nature, 77* (1908), 509–10. W. H. Bragg and Glasson, *TPRRSSA, 32* (1908), 301–10.

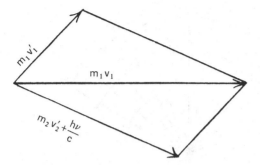

Figure 5.2. Stark's vector diagram of x-ray emission. [*PZ, 10* (1909), 905.]

explained by what he called the *aether wave hypothesis* of x-rays. By this he seems to have meant the view that each electron radiates isotropically and that the energy density in the impulse decreases as the square of the distance traveled. On the other hand, he asserted, lightquanta carry momentum in quantity $h\nu/c$. Since the electron beam is unidirectional, each lightquantum released on impact should have a net forward component of momentum, as indicated in the vector diagram, Figure 5.2. Here m_1 is the mass of the electron, v_1 its velocity before impact, and v_1' its velocity after impact; m_2 and v_2' are the mass and the final velocity of the atom that is struck; and $h\nu/c$ is the "electromagnetic momentum" of the lightquantum.

In addition to the variation in x-ray intensity, Stark claimed that there should be a decrease in the penetrating power or, as he put it, the frequency ν, of the emitted x-rays as the angle from the forward direction increases. Thus, he expected that the absorption coefficient of the rays should increase the farther one looked from the forward direction. This experimental result he also demonstrated, as shown in Figure 5.3 where, again, the beam direction is horizontal toward the right. It was soon made clear to Stark that the classical impulse theory was fully capable of explaining the azimuthal intensity variation of x-rays and the shift with the azimuth in the hardness of x-rays. Arnold Sommerfeld pointed it out. Sommerfeld felt that when a man as influential as Stark made a mistake of such magnitude, the correction of the error should be made public.[78]

[78] Sommerfeld to Stark, 4 December 1909. Hermann, *Centaurus, 12* (1968), 45.

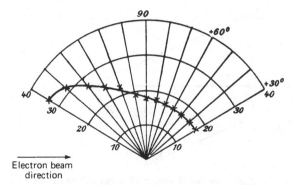

Figure 5.3. Stark's graph of variation of absorption index of x-rays emitted at various angles to the direction of the incident cathode-ray beam. [*PZ, 10* (1909), 912.]

THE SOMMERFELD–STARK EMBROGLIO

Arnold Sommerfeld, since 1907 *ordentlicher* professor of theoretical physics at Munich, had a personal stake in x-rays. His treatment of diffraction by impulses in 1900 had convinced many in Germany that x-rays were discontinuous impulses in an electromagnetic continuum. Although he had not written on x-rays since then, at Munich he was in close contact with Röntgen, whose students studied their properties. In fact, even before Stark's claims for the lightquantum, Röntgen had asked Sommerfeld to consider the asymmetry that his student, Bassler, had observed in x-ray intensity.[79] Moreover, Barkla's characteristic homogeneous x-ray radiation appeared to corroborate Sommerfeld's early suggestion that a continuum of x-rays exists from the periodic to the aperiodic. Stark provided the opportunity for Sommerfeld to express considerations on which he had been reflecting for a "long time."[80]

Before addressing Stark, Sommerfeld commented on the new homogeneous x-rays to distinguish them explicitly from the old.[81] The isotropic, unpolarized, homogeneous secondary x-rays, he said, were likely due to emission of energy by electrons bound

[79] Bassler, *AP, 28* (1909), 808–84. Sommerfeld, *Scientia, 51* (1932), 48. Other important evidence was provided by Kaye, *PCPS, 15* (1909), 269–72.
[80] Sommerfeld, *PZ, 10* (1909), 969.
[81] *Ibid.*

within the atom. He thereafter called these *fluorescent rays.* The partially polarized, aperiodic transverse impulses created by electron impacts he called the *Bremsanteil* or *braking part* of the rays. This division was not original; the British and Stark had made it before.[82] But although Sommerfeld allowed the homogeneous fluorescent rays a wavelength, he categorically denied it to the *Bremsanteil*. "For this, as before," he said, "one would more correctly speak of pulse width rather than wavelength."[83] Although Sommerfeld here first admitted that the Planck relation might be needed to treat the fluorescent x-rays, he was certain that it had "nothing to do with the *Bremsanteil*." The division between the two types of x-rays seemed so strict to Wien that he quickly wrote to be assured that Sommerfeld did not conceive of either type as "nonelectromagnetic" in origin.[84]

Then Sommerfeld turned to the defense of the electromagnetic theory of pulses. "Recent speculations on the lightquantum," he said, "have tarnished our trust in the validity of electromagnetic theory."[85] To set the picture straight, he showed that the directed field amplitudes A_E and A_H at a great distance from an electron, stopped instantaneously from a velocity v, are both proportional to

$$\frac{(v/c) \sin \theta}{1 - (v/c) \cos \theta} \tag{5.8}$$

Here θ is the angle between the direction of the acceleration and the direction of interest. For the more realistic case in which the deceleration occurs over a finite time interval, he argued that the field amplitudes become proportional to

$$\frac{\sin \theta}{[1 - (v/c) \cos \theta]^3} \tag{5.9}$$

Energy radiates instantaneously at a rate given by the product of these vectors; the total energy radiated comes from integration over the duration of the impulse.[86] Sommerfeld found that the total energy flux varies with θ as

[82] Stark had suggested that there was a distinction between impulse and fluorescent x-rays, *PZ, 10* (1909), 579–86. He was one of the first physicists to point up the close parallels between x-rays and light, their identity being a viewpoint that, he said, was "gaining ground." See also Stark, *PZ, 10* (1909), 614–23.

[83] Sommerfeld, *PZ, 10* (1909), 970.

[84] Wien to Sommerfeld, 27 December 1909. Sommerfeld papers; AHQP 34, 13.

[85] Sommerfeld, *PZ, 10* (1909), 976.

[86] Poynting, *PTRS, 175* (1884), 343–61.

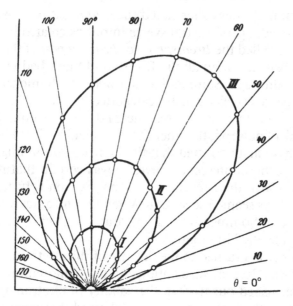

Figure 5.4. Sommerfeld's graph of forward scattering expected of classical x-ray impulses. [*PZ, 10* (1909), 972.]

$$S(\theta) = \frac{\sin^2 \theta}{\cos \theta} \left(\frac{1}{[1 - (v/c)\cos\theta]^4} - 1 \right) \qquad (5.10)$$

The maximum of this expression occurs at an angle with cosine equal to v/c. The closer the initial velocity is to c, the greater is the fraction of energy radiated toward the front. The resulting intensity distribution is shown in Figure 5.4. Roman numerals denote the initial electron velocities $c/10$, $c/5$, and $c/3$ in increasing order. The maximum in the intensity at about $\theta = 60°$ for case III agreed well with Stark's measurement for an initial electron velocity near $c/3$.

Sommerfeld commented that the asymmetry he had derived classically could be visualized in a simple diagram. In Figure 5.5, point i marks the beginning of the electron's deceleration. Point x marks the position that the electron would have occupied had it not slowed at all. The expanding impulse front is centered on i. The trailing edge of the impulse emanates from point f, where the deceleration ends. The resulting pulse width varies with the azimuthal angle θ, as shown exaggeratedly in the figure. The field vectors are maximum where the density of the radial force lines is greatest. This will occur in the plane perpendicular to the instanta-

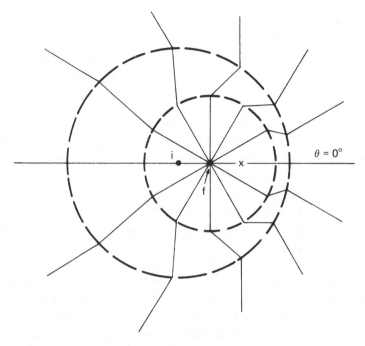

Figure 5.5. The electric field in the vicinity of a rapidly decelerated electron, showing the azimuthal variation in impulse width.

neous velocity vector of the electron. But it is not until after the pulse has passed a point in space that the field, so to speak, knows that any acceleration has occurred. The maximum energy flux will always be directed through the region of the impulse that intersects an imaginary plane perpendicular to the acceleration vector and centered on the moving point x. Consequently, the maximum x-ray intensity will always be in the forward direction, and will occur at smaller angles θ the greater the initial electron velocity.

But the net energy flux is compounded of both the quantity of pulses and the quality of each. Sommerfeld turned to Wien's analysis of 1905 to point out something "that appears to be important in connection with recent observations."[87] Figure 5.5 directly answers Stark's claim that the penetrating power of the x-rays decreases for increasing azimuth. The impulse width increases with angle θ up to 180°. Thomson had shown that the energy carried by a pulse varies as $1/\delta$; but he had not noticed that δ will

[87] Sommerfeld, *PZ, 10* (1909), 975.

vary with the azimuthal angle around the pulse. In pointing this out explicitly, Wien and Sommerfeld were providing a significant reformulation of the definition of radiant intensity.[88]

If an electron oscillates periodically about a fixed point, it radiates a periodic wave that varies in amplitude with the azimuth, but does not vary in wavelength or in frequency. An impulse acts differently. Both its amplitude and its pulse width vary with the azimuth. The electromagnetic impulse hypothesis offered a conceptual route, unavailable with periodic waves, whereby the properties associated classically with wave amplitude could be transferred to pulse duration or frequency. When it became clear that a quantum theory of x-rays was needed, Sommerfeld's and Wien's analyses justified assigning a significant role in defining the energy content of impulse radiation to the impulse's duration rather than to its amplitude.

Stark had placed himself in so indefensible a position that an essential point he wished to convey was obscured. In describing the radiation from a single electron, Stark had said that its energy density far from the electron "has equal magnitude in all directions."[89] Sommerfeld showed without question that Stark's claim was simply wrong. Stark had confused the equality of energy in all directions with the constancy of energy in a given direction. When Sommerfeld pointed out his error, Stark retreated to a position somewhat closer to his intent. What he now sought to explain with the lightquantum was the fact that "the electromagnetic radiation from a single electron does not spread to a volume that grows as the square of the separation from the emission center."[90] He now said that what he had called the *Ätherwellenhypothese* was not compatible with Maxwell's electromagnetic equations.[91] He seems to have meant, in accordance with the successes of the quantum hypothesis, that a true aether wave must be able to transfer energy in this non-Maxwellian way.

Stark's position was no clearer to others. Einstein commented privately that Stark had "once again produced pure dung."[92]

[88] Wheaton, *HSPS, 11* (1981), 367–90.
[89] Stark, *PZ, 10* (1909), 903.
[90] *Ibid.*
[91] Stark to Sommerfeld, 6 December 1909. Hermann, *Centaurus, 12* (1968), 46.
[92] "Stark hat wieder einmal gediegenen Mist produziert, Sommerfeld wohl die Beweiskraft jenes Phänomens überschätzt." Einstein to Laub, 16 March 1910. From a transcription by Laub in the Einstein papers.

Sommerfeld was unable to fathom what Stark meant and concluded that his caricature of classical electromagnetic theory had only been a straw man set up especially for the kill. "Your aetherwave hypothesis," he wrote, "is electromagnetically impossible and has neither followers nor interest; it makes no sense [for you] to oppose it by experiment."[93] In the ensuing discussion, Stark deleted all reference to his unfortunate phrase "equal magnitude in all directions." Instead, he emphasized only that energy will remain constant in a given direction.[94] Sommerfeld knew perfectly well that Stark had changed his stance, and called the later exposition an "artfully dissected transcription" of the original version.[95] Stark's final paper on the x-ray asymmetry reported only empirical findings that were no longer in doubt. No interpretation of the results, which would have brought him back into the dispute, was provided.[96] In any given direction from the source of an x-ray impulse, the electromagnetic amplitude decreases as the pulse spreads over larger volumes of space; the pulse width likewise changes. But the frequency of a periodic disturbance remains constant in a given direction and is equal in all directions. Already used to interpreting pulse duration as an intrinsic frequency for x-rays, Stark saw its directed constancy as the key to the preservation of the energy units required by the quantum relation. Sommerfeld's impulse did not fulfill that condition. When Stark tried to point this out after the fact, it appeared only that he was switching to a new argument to save face. That may have been the case, but Stark had also raised an important question. Not only did the electromagnetic theory have to show that energy is localized, it also had to demonstrate how that energy *stays* localized as the "wave" propagates.

In the first volume of a long exposition on the *Principles of atomic dynamics,* published in the midst of the controversy with Sommerfeld, Stark stated his general concern. He opposed what he called the uncritical application of traditional continuum mechanics to the new field of elementary processes. "All objections raised to date against the lightquantum hypothesis are based on

---

[93] Sommerfeld to Stark, 10 December 1910. A. Hermann, *Centaurus, 12* (1968), 48–9.
[94] Stark, *PZ, 11* (1910), 30.
[95] Sommerfeld, *PZ, 11* (1910), 101.
[96] Stark, *PZ, 11* (1910), 107–12.

incorrect or unprovable assumptions," he said.[97] He began to do experiments on visible and ultraviolet light.[98] Even stronger claims for the lightquantum appeared in the second volume of Stark's *Atomic dynamics,* entitled *Elementary radiation,* which appeared in 1911. There Stark reformulated the lightquantum in terms more suited to the temperament of German physicists. It should be thought of as a "deformation in the energetic medium" that remains localized.[99] Stark could afford to come somewhat closer to Sommerfeld's position because by 1911 it was clear that the energy of x-rays is distributed far less isotropically than had been thought. From this point on, the issue was no longer whether x-rays diverge spherically from a source, but whether they diverge at all. And it was far from clear that any electrodynamic theory, with or without the new definition of frequency and intensity, could achieve the degree of localization that ionization experiments implied the energy of x-rays possessed.

[97] Stark, *Electrischen Quanten* (1910), vii.
[98] Stark, *PZ, 11* (1910), 179–87.
[99] Stark, *Elementare Strahlung* (1911), 283–4.

PART III
Seeking an electrodynamic solution
1907–1912

6

Localized energy in spreading impulses

On the basis of . . . pure undulatory theory, one is
led . . . to consequences that approximate the
Newtonian emission theory.[1]

The discussions between Bragg and Barkla in Britain, and between Stark and Sommerfeld in Germany, encouraged attempts to determine how much spatial concentration of radiant energy was demanded by ionization, and how much was compatible with classical electromagnetic theory. Mirroring the different perceptions of and approach to the problems of x-ray absorption in Britain and in Germany, there were two parallel responses. J. J. Thomson recognized that the paradoxes applied equally to all forms of radiation. He influenced important experimental tests of energy localization, and concluded that the classical ideal of symmetry of the electric field around an electron would have to be abandoned. Arnold Sommerfeld took no refuge in the alteration of the physical interpretation of Maxwell's theory. He pressed formal electrodynamics to its extreme to show that the energy of γ-rays might be localized to an extent compatible with recent experiments.

But neither Thomson's nor Sommerfeld's treatment addressed the most compelling problem raised by Bragg and Stark. The energy transferred from a wave to an electron does not seem to change when the distance from the source changes. No matter how severely limited in angular extent the energy of x-rays and γ-rays might be according to Thomson or Sommerfeld, that energy is still diluted in strength as the rays traverse space. Neither of them therefore confronted the issues in quite the same way as did Bragg, who had never intended his neutral-pair hypothesis to apply to ordinary light. As evidence increased in 1910 and 1911 that x-rays

[1] Sommerfeld, München *Sb, 41* (1911), 4.

share fundamental properties of light, Bragg began to call for a truly dualistic interpretation of radiation, one in which wave and particulate properties were to be combined in a radically nonclassical fashion. To some, predominantly the British physicists, the paradoxes of radiation had reached crisis proportion by 1911.

THE VICISSITUDES OF THE TRIGGERING HYPOTHESIS

William Bragg had not been the only British physicist to express reservations about the impulse hypothesis of x-rays in 1907. J. J. Thomson, long uneasy about the paradox of quantity, rejected the triggering hypothesis as an explanation of x-ray ionization in 1907, as we shall see. For Thomson, the paradox of quality was held at bay no longer. He instituted a systematic search for evidence on the seemingly contradictory properties of all forms of radiation. Unlike Bragg, who consistently excluded light from his concerns, Thomson realized as early as 1907 that the problems involved *any* spreading wave, periodic or not. It had become clear to him that even for the photoelectric effect with visible light, the triggering hypothesis was open to serious question.

In September 1907, Erich Ladenburg reported in Berlin that ultraviolet light releases electrons with a continuum of velocities by photoelectric action.[2] He found that the maximum electron velocity increased monotonically with the frequency of the light; there seemed to be no resonance peaks. Although Ladenburg interpreted his results as providing support for Lenard's triggering hypothesis, Thomson saw it differently. If every incident frequency of light can induce an electron to leave an atom with a corresponding velocity, the triggering hypothesis demands that an unreasonably large number of different electron motions must be contained in an atom. Recall that in this view each electron possesses a velocity in the atom that is uniquely related to its rotational or vibrational frequency. If all frequencies of incident light release electrons, it would seem that the photocathode atoms contain a continuum of electron oscillation frequencies.

Thomson was wary. Was it possible that different atoms of the same element have electrons of quite different frequency; that only

---

[2] E. Ladenburg, *VDpG*, 9 (1907), 504–14.

the large number of atoms provided the apparent continuum effect? This proposal was disagreeable to anyone familiar with chemistry and conflicted with spectroscopic evidence for atomic homogeneity. Could it possibly be that within a single atom so large a number of vibrational or rotational frequencies are represented? To account for this, extremely large numbers of electrons would have to reside in each atom.

Just the year before, Thomson had challenged the prevailing notion that atoms contain large numbers of electrons.[3] Based on results of scattering experiments, Rutherford and Bragg had supposed that most of the atomic weight of any atom consists of electrons; a single atom of hydrogen would possess a swarm of 1,000 electrons. But Thomson, combining evidence from β-particle absorption, x-ray scattering, and optical dispersion, revised the estimate downward by a factor of 10^3. This left about as many electrons per atom as the atomic weight in units of hydrogen atoms.[4] It was difficult to see how so few electrons could execute mechanical oscillations of enough different frequencies to explain Ladenburg's findings. The paradox of quality, eclipsed by the triggering hypothesis since 1902, began to exert its influence on Thomson.

The year before, Thomson had suggested to P. D. Innes that some further tests be made on x-rays by analogy to the photoeffect. Innes had found that the velocities of secondary electrons due to x-rays are independent of the intensity of the x-rays and depend only on their quality.[5] The similarity to photoeffect behavior had convinced Innes that the triggering mechanism should be extended to x-rays. And Thomson, like Wien before him, temporarily adopted the triggering hypothesis for x-ray ionization. Now Thomson realized that, whatever the theoretical difficulties for a triggering explanation of the photoeffect, the same considerations would apply to ionization by x-rays. Within a month of Ladenburg's announcement, and just as Bragg's paper on neutral-pair

[3] J. J. Thomson, *PM, 11* (1906), 769–81.
[4] Heilbron, *AHES, 4* (1968), 269–74.
[5] Innes used a variable spark gap to measure the x-ray tube potential and, consequently, the x-ray quality. He found that changing the gap from 5.5 to 16 cm produced a 7 percent increase in the maximum velocity of released electrons, *PRS, 79A* (1907), 442–62.

x-rays appeared in England, Thomson returned to the problem of radiation.[6]

The problem was especially acute for x-rays. Whereas the maximum velocity of electrons released by ultraviolet light is on the order of 10^8 cm/sec, Innes had corroborated Dorn's figure of several times 10^9 cm/sec for the velocity of x-ray electrons. The energy differs by a factor of about 10^3; Thomson believed it to be 10^4. Velocities of electrons released by γ-rays are even greater. Even if the atom encompassed the profusion of electron motions required by the triggering explanation of the photoelectric effect, an entirely new set of motions, with energy 10^3 times greater, would be required by the x-ray data. The γ-rays needed an even more energetic set. It was clear to Thomson that, at least for x-rays and γ-rays, the energy could not come from the atom but must be transformed from the radiation itself. The double paradox of quantity and quality convinced him in 1907 that the energy of the new radiations comes in *units,* the "bright specks on a dark background" that he had first discussed in 1903.[7]

Thomson had already calculated the size of the x-ray unit. In 1906 he had used the absurdity of the result to support the triggering hypothesis.[8] The transverse electric field amplitude A_E acts on a charge e to induce a velocity measured to be at least 10^9 cm/sec. The net change in electron velocity must be $(A_E e/m)dt$ where dt is the time the pulse takes to pass the electron. The known value of e/m implies that $A_E dt$ is approximately 60 g cm/emu/sec. But the energy density in the expanding impulse at the position of the ionized electron is

$$A_E^2\delta/4\pi c^2 = A_E^2\, dt/4\pi c \tag{6.1}$$

Substituting 60 for $A_E dt$, and putting an upper limit of 10^{-5} cm for δ, Thomson found that almost 1 cal/cm² has to be provided by the x-ray impulse.[9] The results of his earliest analysis showed that even in the direction of maximum radiant intensity, $\theta = 90°$, a large

[6] J. J. Thomson, *PCPS, 14* (1907), 417–24. This answers McCormmach's query on why Thomson should have revived his 1903 speculations on light at this time. McCormmach, *BJHS, 3* (1967), 372, 374.

[7] J. J. Thomson, *Electricity and matter* (1904).

[8] J. J. Thomson, *PCPS, 13* (1906), 321. It was, he said, "an altogether inadmissible amount" of energy. The following year, he interpreted the result to indicate the extreme coarseness of the distribution of energy "units" in x-rays.

[9] J. J. Thomson, *Conduction of electricity,* 2nd ed. (1906), 320–1.

enough energy density occurs only within 10^{-20} cm of the origin of the x-ray pulse.[10] This is inside the decelerating electron itself!

In 1907 Thomson turned the argument around. If the energy was bound up in units, he said, the spatial density of those units needed only to be correspondingly low. For the faint but visible light intensity of 10^{-4} erg/cm²/sec, there would be only 1 unit/liter of space – "exceedingly coarse," Thomson said.[11] "As these units possess momentum as well as energy they will have all the properties of material particles," he continued. "Thus we can readily understand why many of the properties of the γ-rays resemble those of uncharged particles moving with high velocities."[12]

The implicit reference to Bragg's neutral-pair hypothesis was clear enough. Thomson tested Bragg's hypotheses: Were the slow electrons emitted by different causes of ionization actually of uniform velocity? It appeared that they were. Thomson found that the electrons released by cathode rays had about the same velocity as did those elicited by canal rays. The velocity was independent of the type of irradiated gas and of its pressure.[13] Early in 1908, Thomson concluded that this common effect could be explained if "a doublet consisting of a corpuscle and a positively charged particle" was removed from the atom.[14] He soon agreed, at the height of the Bragg–Barkla controversy, that Bragg's hypothesis had much to offer, not just for x-rays but for γ-rays as well.[15]

But Thomson did not believe that x-rays are particles. He sought a solution within quasi-classical electromechanics. The coarse-grained aether, he thought, would rescue the advantages of a wave model of radiation and still explain the dual paradoxes. The energy transmitted in x-ray and γ-ray impulses is, in his interpretation, concentrated in specific directions, so that few atoms are hit, and those receive close to the full complement of radiant energy. He expected to be able to explain the similar behavior of periodic light waves in much the same way.

[10] J. J. Thomson, *PM, 46* (1898), 528–45.
[11] J. J. Thomson, *PCPS, 14* (1907), 423.
[12] *Ibid.,* 424.
[13] Thomson, *PCPS, 14* (1908), 541–5. Füchtbauer, *PZ, 7* (1906), 748–50.
[14] *Ibid.,* 545.
[15] J. J. Thomson, *PCPS, 14* (1908), 540. Just what the effect of this was on Barkla, who saw himself as the defender of Thomson's views, is not clear. But the message was limited to the small readership of the *PCPS*. Not until 1910 did Thomson raise these issues in the widely distributed *PM*.

THOMSON'S RESTRUCTURED AETHER

In 1910, after two years of silence on the nature of light, Thomson announced that prior concepts about the microscopic structure of the aether must be in error.[16] According to his "coarse-grained aether" hypothesis, radiant energy does not propagate spherically from its source because the structure of electric field lines surrounding each electron is microscopically discontinuous. Now he went even further, suggesting that each electron is the seat of only a single "tube" of electric force; its energy is radiated only into one narrow cone. Macroscopic charged bodies appear to exert electric force isotropically only because of the large number of electrons they contain. Each of these electrons broadcasts its influence in only one specified direction in space.

The impulse produced by a rapid displacement of the electron is represented as a transverse kink in the electric field, as in Figure 6.1. The kink travels outward along the cone of solid angle Ω defined by the tube. The transverse electric force is approximately

$$\frac{2\pi e v}{r\,\Omega c^2\, dt} \tag{6.2}$$

where v is the total velocity lost by the radiating electron and dt is the time interval over which the deceleration occurs. The total energy carried by a kink is therefore

$$\frac{\pi e^2\, v^2}{\Omega c^3\, dt} \tag{6.3}$$

As the kink moves outward, the transverse displacement as well as the angular width of the cone increase in proportion to the distance r. Thus, the total energy of the kink remains constant. Thomson concluded that "the resulting Röntgen radiation is concentrated into small patches which possess momentum and energy . . . we have a condition of things much more closely represented in many respects by the old emission theory of light than by the wave theory."[17]

One impediment to the extension of this idea to visible light was the potential difficulty of explaining interference phenomena. In

---

[16] J. J. Thomson, *PM, 19* (1910), 301–13. Hints of Thomson's radical proposal had been given in the BAAS *Report* (1909), 3–29; *Nature, 81* (1909), 248–57.
[17] J. J. Thomson, *PM, 19* (1910), 308.

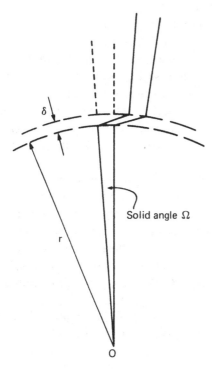

Figure 6.1. Thomson's concept of x-ray impulses as kinks in an electromagnetic tube of force.

1909 Thomson had encouraged his student, Geoffrey Taylor, to seek interference effects in light of extremely low intensity. Perhaps one might find evidence of the "grainy" structure of light if the number of radiating sources was sufficiently low. In January 1909, Taylor reported the results of, among others, a test that spanned 2,000 hr of illumination by a subvisible source.[18] He found clear evidence of ordinary interference fringes.

Thomson was reluctant in 1910 to extend his revision of the electromagnetic field beyond x-rays and γ-rays. He still held out hope that the photoeffect would prove tractable for the triggering hypothesis. This required an atom that would support a continuum of electron frequencies. In 1910 Thomson offered such an atomic model to the readers of the *Philosophical Magazine*.[19]

[18] Taylor, *PCPS, 15* (1909), 114–15.
[19] J. J. Thomson, *PM, 20* (1910), 238–47.

Electrons follow orbits in the effective inverse-cube force surrounding a dipole such as one of Bragg's neutral pairs. Thomson showed that electrons describing orbits on a cone extending from the negative end of the dipole possess kinetic energy in proportion to their orbital frequency. Because an orbit may occur at any distance from the dipole, a full continuum of frequencies awaits a passing light wave. Electrons released by resonance fly off with an energy proportional to the triggering frequency. Thomson even showed that the constant of proportionality was tolerably close in value to Planck's h.

Thomson was not receptive to the quantum hypothesis and set no great store in a quantum of light. He compared his photoeffect constant to Planck's not because it was required by the empirical data then available but in an attempt to show how one might explain so-called quantum effects on the basis of classical concepts. His continued attempts to develop the idea after 1911 must be read as opposition to the growing acceptance of the quantum theory, which, he thought, achieved mere agreement with experimental results without providing a conceptual explanation of the paradoxes he perceived in the absorption of radiation.

STATISTICAL FLUCTUATIONS: CAMPBELL'S TEST OF LIGHT

Thomson's concerns about light influenced a Cambridge student to undertake a remarkable and significant test of its nature. Unlike Thomson and virtually all of his compatriots, Norman Campbell was convinced of the value of Einstein's lightquantum. His early adoption of this unorthodox view was the result of two influences: Thomson and Einstein. In 1907, as Thomson made the first tentative proposal that the unit hypothesis might apply even to ordinary light, Campbell showed his detailed understanding of the paradoxes that had led to that state of affairs. His book, *Modern electrical theory*, published that year, reviewed the peculiar ionizing behavior of radiation with depth and originality.[20] By 1909, under the sway of Einstein's relativistic abandonment of the aether, Campbell suggested that the "confusions" about light arose from continued reliance on the outmoded concept of an aether. The aether, he said, should be relegated "to the dust heap where

[20] Campbell, *Theory* (1907), 221–9.

now 'phlogiston' and 'caloric' are mouldering."[21] The new "atomic" theories of radiation, free of aetherial taint, should be adopted. "The trend of modern theory," he explained, "is everywhere to replace by discontinuity the continuity which was the basis of the science of the last century."[22]

Casting about for a suitable formalism to treat discontinuous events, Campbell selected the statistical methods used to describe radioactive decay of individual atoms. Egon von Schweidler, a Viennese expert in the field, had realized in 1905 that a statistical interpretation of atomic decay offered a means to determine the absolute number of decaying atoms. One needed only to measure the size of the fluctuations from the ideal exponential law of decay. At the First International Conference on Radiology and Ionization held in Liège, Schweidler considered how the resultant ionization current changes as Z, the number of undecayed atoms, decreases.[23] The fluctuation ϵ in the current increases as Z gets smaller. But what, precisely, is the form of this deviation from an exponentially declining current?

The number, Z, of Z_0 initial atoms still undecayed after an elapsed time interval t is given in the ideal case by

$$Z = Z_0 \exp(-\zeta t) \tag{6.4}$$

where ζ is a constant characteristic of the element used. This relation is shown by the central dashed line in Figure 6.2. The measured ionization current (mean value proportional to Z) will fluctuate about the ideal case, as shown schematically by the irregular line. But the limits of the fluctuation, shown by dotted lines, should increase as the number of undecayed atoms falls. Simply applying Bernoulli's calculation of expected error, Schweidler found the mean value of the limit ϵ expressed as a fraction of Z_0:

$$\bar{\epsilon} = \left\{ \frac{1 - [1 - \exp(-\zeta t)]}{[1 - \exp(-\zeta t)]Z_0} \right\}^{1/2} \tag{6.5}$$

If t is taken to be small enough so that $\exp(-\zeta t) \simeq 1$, an adequate approximation is:

$$\bar{\epsilon} = 1/\sqrt{Z} \tag{6.6}$$

---

[21] Campbell, *PM, 19* (1910), 181–91.

[22] *Campbell, PCPS, 15* (1909), 117.

[23] *von Schweidler, Premier congrès, Liège* (1906), paginated section 3, pp. 1–3. The time rate of change of Z is equal to $-\zeta Z$ because $d/dx \exp(ax) = a \exp(ax)$.

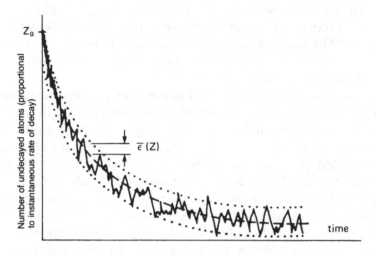

Figure 6.2. Schematic representation of the limit of fluctuations of actual radioactive decay about the ideal law of decay.

Schweidler thought that a measure of $\bar{\epsilon}$, from the ionization current produced by the decaying element, would directly yield Z. But the application was not as easy as it appeared. In 1905, errors in methods used to obtain relative values of Z were larger than the anticipated fluctuations. The method was applied in 1906 by K. W. F. Kohlrausch to determine the number of α-particles ejected from polonium;[24] an experiment repeated the following year by Edgar Meyer and Erich Regener in Berlin.[25] Kohlrausch had found the fluctuations to be larger than anticipated, and Meyer located the difficulty in the slow response of the electrometer used to measure the ionization current. Hans Geiger at Rutherford's laboratory in Manchester criticized Kohlrausch for failing to ensure that his currents were at saturation, and went on to develop a new means of counting individual α-particles. This bypassed the problem by allowing individual α-particles to induce electrical discharges large enough to be detected.[26]

Thomas H. Laby at Cambridge developed a sensitive quartz fiber electrometer that he hoped would allow measurement of

[24] Kohlrausch, Wien *Sb, 115:IIa* (1906), 673–82.
[25] E. Meyer and Regener, *VDpG, 10* (1908), 1–13; *AP, 25* (1908), 757–74.
[26] Geiger, *PM, 15* (1908), 539–47. Rutherford and Geiger, *PRS, 81A* (1908), 141–61.

rapid transient effects. He apparently tried to test the device in 1908 on the ionization produced in a gas by the γ-rays from radium.[27] Conversations with Norman Campbell may have provided him with the technique he was seeking. Shortly thereafter, Campbell began to apply Schweidler's method to the paradoxical problems of light then under scrutiny by J. J. Thomson. First, Campbell reformulated Schweidler's derivation using finite differences and combinatorial analysis, as befit its application to discontinuous processes.[28] But the real intent of his work became clear in a two-part study of the photoelectric effect in 1909.[29]

Campbell saw in the fluctuation method a means to test the dependence or independence of the light radiated in different directions from a single point source. He split the light from a Nernst lamp into two beams, each incident upon one of a matched pair of sensitive photocells. He then balanced the two output signals so that only the difference between them was observed. If the light spreads spherically, one would expect that the "elementary" ionization events in the two cells would coincide and produce few differences. However, if the radiation in each beam is directed, and there is no correlation of microscopic effect, then a measurable fluctuation in the null current would arise.

The experiment failed because Campbell was unable to find a light source simultaneously intense and steady enough. He carefully calibrated the quartz fiber electrometer and removed the oxidized impurities from the photocells. He also developed a new form of high resistance across which to read the electrometer current, one that was itself free of fluctuations.[30] But he required high-intensity light. The only suitable source was a 50-watt Nernst lamp, and, because it was run off the Cambridge ac mains, its output varied by 5 percent, considerably greater than the fluctua-

---

[27] Laby, *PCPS, 15* (1909), 106–13. See also Laby and Burbidge, *Nature, 87* (1911), 144; Campbell, *PCPS, 15* (1909), 136.

[28] Campbell, *PCPS, 15* (1909), 117–36. By "atomic" theories of radiation, Campbell had in mind "at least two:" Thomson's unit hypothesis and "Planck's theory (as interpreted by Stark)," *PM, 19* (1910), 190.

[29] Campbell, *PCPS, 15* (1909), 310–28, 513–25.

[30] Others, including Meyer and Regener, typically used a "Bronson" resistance – a gas cell in which low conductivity is induced by a source of ionizing radioactive material. Campbell understandably criticized Meyer for introducing an additional source of fluctuation and replaced the Bronson resistance by a liquid measure of xylol and ethanol, calibrated at 9×10^{10} Ω.

tions of interest.[31] A few months later he reported failure, for the same reason, in a test on x-rays.[32]

Although Campbell reached no experimental conclusion, he considerably advanced the analysis of the problem. The mean square fluctuation in the number of electrons registered from a single photocell is

$$\overline{x^2} = N(\overline{\omega^2} + \overline{\eta^2})dt, \tag{6.7}$$

where N is the mean number of "light disturbances" (spherical impulses or localized units) that appear in a unit of time.[33] The period of illumination is dt, and ω is the number of electrons released by each disturbance; today we would call ω the *quantum efficiency*. But ω can vary from one ionization event to another, and η measures the instantaneous deviation of ω from its mean value characteristic of the cathode atoms, the light frequency, and light intensity. Campbell fixed attention on the change in $\overline{\eta^2}$ when the illuminated area of the photocathode becomes larger and the photocurrent consequently increases. The larger area serves to integrate irregularities in electron receptiveness to the light over a larger sample of atoms. According to the classical wave theory of light, waves spread to encompass *all* electrons in the illuminated surface, so that fluctuations in ω should approach zero when larger areas are illuminated. For the rival theory, this is manifestly not the case. A directed light unit affects electrons only along a specific direction from the source; $\overline{\eta^2}$ should remain roughly constant as the cathode area increases.

If one increases the intensity of the light while holding the illuminated area of the photocathode fixed, the current also changes. Under all circumstances the current is proportional to $N\omega$, the total number of electrons released per unit time. For

[31] The Nernst lamp was designed to compete with carbon-filament light bulbs. It needed no surrounding vacuum and produced a higher ratio of illumination to electrical power. Nernst sold the patent rights to his lamp to the Allgemeine Elektrizitäts Gesellschaft in the mid-1890s for a million marks, a good-sized fortune at the time. But the zirconium rare-earth compound used as the filament requires preheating to bring its internal resistance down to allow steady operation, and the lamp was never produced commercially. See Mendelssohn, *Walther Nernst* (1973), 44–6.

[32] Campbell, *PZ, 11* (1910), 826–33.

[33] Campbell, *PCPS, 15* (1909), 314, shows η^2 rather than $\overline{\eta^2}$, but this is clearly a misprint.

spreading waves, the number of impulses N is independent of the distance from light to cell. Thus, the decrease in current with light intensity must be due to a corresponding decrease in the potency of each event, ω. Neglecting the variation in ω, the mean square fluctuation in ionization current is then proportional to ω^2. Since the light intensity must be proportional to ω, $\overline{x^2}$ should be proportional to the square of the light intensity. But it was Campbell's stated "hope" to find evidence against the "spherical wave theory" of light.[34] For the unit hypothesis, ω remains constant; each ionizing event, if it occurs at all, delivers the full unit of energy. Thus, it must be N that changes with the intensity of the light. A smaller solid angle of light containing fewer energy bundles will hit the photocathode as the source is moved away. In this case, fluctuations should depend directly on N, that is, in proportion to the first power of the light intensity.

But Campbell's aims were obstructed again. When he included the anticipated variation in ω by putting back the $\overline{\eta^2}$ term in the expression for $\overline{x^2}$, both cases were reduced to simple proportionality of fluctuation with light intensity. Having considerably extended the analysis, he was left with no more than a plausibility test of his method. He derived an order-of-magnitude estimate for the elusive quantity ω, and compared his estimate to what he called *Planck's theory,* but which was actually derived from Einstein.[35] Einstein predicted that $\omega = 1$ under all circumstances. Abram Ioffe's reworking of Ladenburg's experimental photoeffect data for the energy of electrons released from zinc suggested that $\omega = 3$.[36] Campbell's independent data on fluctuations in the current from zinc-amalgam photocathodes indicated that $\omega = 3.1$, a factor 100 times lower than that expected from any spherical wave theory of light.

ANISOTROPIC γ-RAYS

At the time that Campbell completed his analysis of light in early 1910, Egon von Schweidler applied the same method in a proposed test of the nature of γ-rays. The discussion between Stark and Sommerfeld had raised Schweidler's hopes of resolving the issue of

---

[34] Campbell, *PCPS, 15* (1910), 521.
[35] Einstein, *AP, 20* (1906), 199–206.
[36] Ioffe, *AP, 24* (1907), 939–40.

energy localization. Einstein's statement at the eighty-first *Natur-forscherversammlung* that statistical fluctuations might lie at the heart of a reconciliation of wave behavior with lightquanta also influenced Schweidler.[37] After the talk, Heinrich Rubens suggested a possible test of the idea. Like α- and β-particles, x-rays and γ-rays might produce scintillations on a fluorescent screen.[38] Schweidler quickly published his own proposed test.[39] If the effective solid angle of γ-rays allowed into an ionization chamber is reduced, either by moving the radioactive source away or by closing a diaphragm, the net ionization current will decrease. But if γ-rays are spherical impulses, the *number* of pulses entering the chamber remains constant. Therefore, their effective ionizing power must be reduced. Conversely, if the energy is localized, the ionizing power of each unit must remain constant, and only the number of units that pass into the chamber decreases with solid angle.

The analysis was in principle the same as Campbell's, giving Campbell an excellent opportunity to criticize the proposal. He pointed out that any valid analysis must include η, the fluctuation in the number of electrons released in each ionization event. And Campbell was not alone in his criticism. Sommerfeld wrote to object to the supposition that impulses can exist without an aether. Planck suggested that discontinuities in fluorescence might be due as much to differing thresholds for emission as to spatial localization of radiant energy. Bragg insisted that gas ionization must be due in either case to intermediary particles, and Schweidler specifically introduced this possibility into his analysis.[40]

Schweidler's proposal was tested by Edgar Meyer, then a *Privatdozent* under Stark at Aachen.[41] The hemispherical ionization chamber between A and B in Figure 6.3 received γ-rays from the radium–bromide source at Ra. In trial 1 the entire hemisphere was open to the rays, whereas in trial 2 the lead ring P limited the effective solid angle as shown. The ratio of saturation currents, i_1/i_2, is related differently to the ratio of the mean fluctuations of these quantities depending on whether the γ-ray energy is localized or not.

[37] Einstein, *PZ, 10* (1909), 817–25.
[38] *Ibid.*, 826.
[39] von Schweidler, *PZ, 11* (1910), 225–7.
[40] von Schweidler, *PZ, 11* (1910), 614–19.
[41] E. Meyer, Berlin *Sb, 32* (1910), 624–62. For related studies by Meyer, see *JRE, 5* (1908), 423–50; *6* (1909), 242–5; *VDpG, 12* (1910), 252–74.

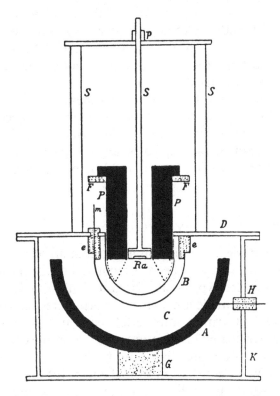

Figure 6.3. Meyer's apparatus to test fluctuations in the ionization current induced by γ-rays. [*JRE, 7* (1910), 283.]

Reporting to the Berlin Academy in June 1910, Meyer distinguished between two possibilities, calling one *isotropic* and the other *anisotropic*. The current i is always proportional to $N\omega e$ (in Campbell's notation), where e is the electronic charge. For the isotropic hypothesis, where ω varies, the ratio of fluctuations in the current for cases 1 and 2 should be

$$\frac{\bar{\epsilon i_1}}{\bar{\epsilon i_2}} = \frac{N\omega_1 e}{\sqrt{N}} \frac{\sqrt{N}}{N\omega_2 e} = \frac{i_1}{i_2} \tag{6.8}$$

that is, directly proportional to the ratio of the currents themselves. But for the anisotropic case, N varies while ω remains constant. Thus, the ratio of fluctuations is

$$\frac{\bar{\epsilon i_1}}{\bar{\epsilon i_2}} = \frac{N_1\omega e}{\sqrt{N_1}} \frac{\sqrt{N_2}}{N_2\omega e} = \sqrt{N_1/N_2} \simeq \sqrt{i_1/i_2} \tag{6.9}$$

that is, proportional to the square root of the ratio of currents. Meyer's data weakly indicated that the anisotropy was the more likely result, the mean ratio of fluctuations in each of two cases being at least 50 percent closer to the square root of the ratio of currents than to the ratio itself.

Meyer did not deny that the impulse hypothesis might explain his result. He did wonder aloud whether Sommerfeld's success in finding asymmetry in x-rays to answer his Aachen colleague, Stark, could also explain the degree of anisotropy he had now found in γ-rays. Early in 1911, Sommerfeld demonstrated that it could.

SOMMERFELD'S γ-RAY THEORY

Arnold Sommerfeld was not fully convinced that all of the issues raised in the Stark affair had been, or could be, satisfactorily answered by the impulse hypothesis. William Henry Bragg, whose discussions with Barkla made him extraordinarily interested in the Stark controversy, had written Sommerfeld early in 1910 to point out that the degree of spatial localization required in the energy of x-rays was considerably more than Sommerfeld's model could achieve. He cautioned against "dimming the light [of the neutral-pair hypothesis] which is guiding so many useful results." Although Sommerfeld had little faith in the kinds of mechanical considerations that moved Bragg, he nevertheless appealed to the triggering hypothesis rather than to any transformation of radiant energy to explain the extraordinarily high secondary electron velocities produced by γ-rays. "The production of secondary [electrons] is an intra-atomic process," he replied, one for which present knowledge is limited.[42] But in the theory of x-rays "we are on familiar ground." Sommerfeld wanted to see how much localization of energy an impulse *could* possess classically; the obvious case to try was the γ-rays.

Early in 1911 he completed the first detailed analysis of impulse γ-rays.[43] It was a pulse theory pushed to extremes to describe the

---

[42] W. H. Bragg to Sommerfeld, 7 February 1910, and Sommerfeld to W. H. Bragg, 8 February 1910. Stuewer, *BJHS, 5* (1971), 269–70. Bragg's most complete summary of the data in support of neutral-pair x-rays and γ-rays had just appered in German in Stark's *JRE, 7* (1910), 348–86.

[43] Sommerfeld, München *Sb, 41* (1911), 1–60.

radiation produced by β-electrons accelerated out of an atom at a speed close to that of light. Sommerfeld had a variety of reasons to undertake the study. He was influenced by Meyer's experiments on the fluctuations in ionization current produced by γ-rays. Sommerfeld's analysis was explicitly intended to defuse the potential problem raised by Meyer's conclusion.

Another motive for Sommerfeld was the revival of doubts about his analysis of diffraction of x-rays. Two young Berlin experimentalists began to claim that the evidence for x-ray diffraction and for Marx's x-ray velocity determination was faulty. These were revivals of concerns first voiced by one Bernhard Walter, who as early as 1898 had thought that x-rays were cathode-ray electrons that had lost their electric charge upon colliding with the anticathode.[44] Walter did competent experiments on x-rays and functioned as the chief physicist – advisor on the medical uses of x-rays in Germany.[45] Although he published a series of articles in physics journals critical of both the Haga – Wind photograph analysis and Marx's velocity determination, his views had not been taken very seriously by physicists.[46] In 1908 that situation changed. Robert Pohl and James Franck joined in Walter's arguments and forced Marx to defend his x-ray velocity experiment again.[47] Pohl and Walter revived the diffraction criticism.[48] All three insisted that they were concerned only about systematic errors in experimental technique. They did not object to the wave or impulse hypothesis of x-rays but rather to the evidence on which the impulse hypothesis had been established. But it is likely that their criticism of the

---

[44] Walter, *AP, 66* (1898), 74–81.
[45] He published extensively in the *Fortschritte auf dem Gebiete der Röntgenstrahlen*, a journal serving those with medical interest in x-rays. For a discussion of Walter's role in setting early radiological standards, see Serwer, *Radiation protection* (1976).
[46] Walter, *VDNA* (1901), part 2, p. 43; *PZ, 3* (1902), 137–40. Haga and Wind, *AP, 10* (1903), 305–12. Walter, *Fortschritte auf dem Gebiete der Röntgenstrahlen, 9* (1906), 223–5.
[47] Franck, *VDpG, 10* (1908), 117–36. Marx, *VDpG, 10* (1908), 137–56, 157–201. Franck and Pohl, *VDpG, 10* (1908), 489–94. Marx, *AP, 28* (1909), 37–56, 153–74; *33* (1910), 1305–91. Franck and Pohl, *AP, 34* (1911), 936–40. Marx stopped the exchange with the question, "Are my experiments on the velocity of x-rays by interference of electrical waves explainable?" *AP, 35* (1911), 397–400.
[48] Walter and Pohl, *AP, 25* (1908), 715–24; *29* (1909), 331–54. Pohl, *Physik der Röntgenstrahlen* (1912), 144.

experiments had been encouraged by the new theoretical objections raised by Stark in 1908 against an aether model of x-rays.

The chief motive behind Sommerfeld's analysis of γ-rays was his growing interest in the quantum transformation relation (*not* the lightquantum). Late in 1910, H. A. Lorentz had questioned whether the quantum hypothesis could explain absorption of radiation at all without some additional hypothesis to limit the spatial extent of radiant energy.[49] In 1909 Sommerfeld had denied that the quantum relation had anything to offer to the analysis of impulse radiation; in 1911 he explicitly reversed that statement.[50] The γ-ray study was intended to answer Lorentz and was Sommerfeld's first use of the quantum transformation relation. It was intended to show how the quantum relation, formulated in terms of frequency, could be extended to nonperiodic events such as the ejection of an electron by a disintegrating atom.

The same asymmetry that had answered Stark's objections regarding x-rays also served to define the localization of γ-ray impulses. The two cases are remarkably similar even though in one case the electron is stopping and in the other it is starting. The electron accelerates from rest at point *i* toward the right in Figure 6.4 and attains its final velocity at point *f*. The resulting field exhibits a variation in pulse width around the azimuth similar to that of the decelerating electron in Figure 5.5. One qualitative difference is that the quiescent and dynamic fields are interchanged; here the former, which Sommerfeld called the *electromagnetic atmosphere*, is the outer field.[51] Another difference is that once the electron reaches relativistic velocities, the dynamic field within the inner sphere deviates from the schematic symmetry shown in Figure 6.4. The field of a relativistic electron becomes restricted to the region between two cones facing each other, each with its vertex on the electron and symmetrically oriented about the axis of acceleration.[52] As the velocity approaches *c* the angle

[49] Lorentz, *PZ, 11* (1910), 1234–57. These matters will be discussed at length in Chapter 7.

[50] Sommerfeld, München *Sb, 41* (1911), 43n.

[51] *Ibid.*, 4.

[52] Heaviside, *PM, 27* (1889), 324–39. J. J. Thomson had concluded in 1898 that the electrodynamic effect of stopping an electron abruptly from a speed close to that of light is twofold. The roughly planar impulse of the surrounding dynamic field would continue as though the electron had not stopped. At the same time, a spherical impulse would arise, centered on the arrested electron and of essentially infinitesimal thickness. Thomson, *PM, 45* (1898), 172–83.

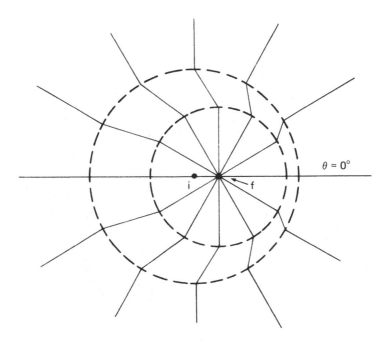

Figure 6.4. The electric field in the vicinity of a β-particle rapidly accelerated to the right.

subtended by each cone approaches 2π steradians, so that the field is finally restricted to the plane that is perpendicular to the axis and that passes through the moving electron. A schematic approximation to the actual field for an electron accelerated to $0.9c$ is presented in Figure 6.5. The electric force in the inner sphere acts only within the trailing conical "bow wave" indicated by the vertical lines; at each instant it is directed along these lines, always in, or close to, the plane normal to the velocity vector. The broken arrows indicate the direction of maximum energy flux.

Because the acceleration of the β-electron is so much greater in magnitude than even the deceleration of a cathode-ray electron, the radiated energy is more localized in the forward direction for the γ-pulse than for the x-ray impulse. Cathode ray electrons move at some 10^9 cm/sec; β-electrons travel at 0.9 to 0.99 times the velocity of light, some thirtyfold faster. To assess the difference in effect, Sommerfeld started, as he had for x-rays, with Max Abraham's reformulation of the Poynting equation.[53] Energy passing

[53] M. Abraham, *Elektrizität, 2* (1905), section 14, equations 74, 75, and 76a.

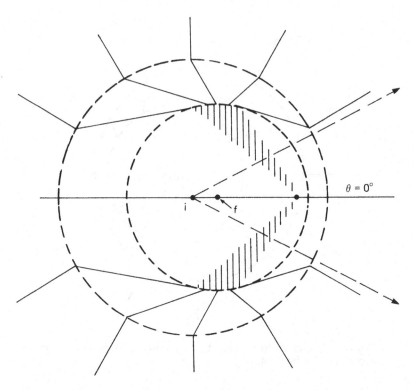

Figure 6.5. Schematic representation of the electric field in the vicinity of an electron accelerated to relativistic velocity.

per unit time at a distance r from the point i, and at an angle θ from the acceleration axis as seen from point i, is given by

$$\frac{e^2 a^2}{16\pi^2 c^3 r^2} \frac{\sin^2 \theta}{(1 - (v/c)\cos \theta)^6} \tag{6.10}$$

Here a is the magnitude of the acceleration and v is the final velocity of the electron. The factor $\sin^2 \theta$ arises from the transverse nature of the impulse; its amplitude falls to zero in the forward and backward directions and presents a maximum at 90° to the axis. This is true regardless of the time duration of the acceleration. The modulating factor $(1 - (v/c)\cos \theta)^6$ produces the asymmetry in the radiated energy. It arises from the classical interpretation of wave intensity and explains the localization of γ-ray energy in the forward direction.

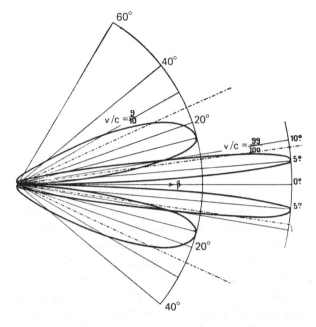

Figure 6.6. Sommerfeld's calculated distribution of γ-ray intensity as a function of azimuthal angle. [München *Sb, 41* (1911), 11.] (The two cases for v/c are not to the same scale.)

EXPANDING-RING γ-RAYS

The energy is radiated into a hollow cone that is symmetric about the axis of electron acceleration. Integrating equation 6.10 over the acceleration time produces the total energy radiated into the cone defined by the angle θ:

$$S(\theta) = \frac{\sin^2 \theta}{\cos \theta} \left[\frac{1}{(1 - (v/c) \cos \theta)^4} - 1 \right] \qquad (6.11)$$

For values of v close to c, and for small values of θ,

$$S(\theta) = \frac{\theta^2}{(1 - v/c + \theta^2/2)^4} \simeq 16 \; \theta^2/\theta^8 = 16/\theta^6 \qquad (6.12)$$

Figure 6.6 is a plot of $S(\theta)$ for two cases $v = 0.9c$ and $v = 0.99c$. In the former case S_{max} lies at about 15°, and in the latter about 5°, from the axis of acceleration. The energy never quite rises above zero in the forward direction but is the more tightly constrained

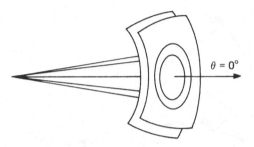

Figure 6.7. The ring-pulse γ-ray.

around the axis the closer the final velocity is to c.[54] Corrections for nonlinearity in the acceleration and for the relativistic increase in electron mass cause only a small broadening of the cone.

The net result is a ring-shaped concentration of energy. The impulse is still a temporally discontinuous event, so that the energy always lies at the intersection of the hollow cone with the expanding spherical shell. Thus it forms a ring, as in Figure 6.7, that expands in proportion to distance as it travels in the same direction as its parent β-electron. "One can see from this," Sommerfeld continued hopefully, "that for very great velocities the energy emission is almost corpuscular."[55]

"It is quite remarkable," Sommerfeld said, "that on the basis of the Maxwell theory, that is, pure undulatory theory, one is led under certain conditions to consequences that approximate the Newtonian emission theory, and that appear to have more similarity to Newtonian light projectiles than to Huygens' spherical waves."[56] In fact, $S_{\max}$ grows in proportion to $(1 - v/c)^{-3}$; thus, energy radiated into the cone of maximum effect resulting from acceleration to $0.99c$ is 10^3 times greater, as well as being more localized, than that due to the slower $0.9c$ electron.

Thus far, the calculation was entirely classical. But to compare results to experiment, Sommerfeld needed data, as yet unavailable, on the precise nature of the acceleration. He wanted to obtain an expression for E_β/E_γ, the ratio of total β-particle energy to that which appears in the resulting γ-pulse. Sommerfeld expressed the

[54] Sommerfeld calculated the extremum values of θ where S dropped to $0.1 S_{\max}$. For $v/c = 0.9$, these limits are 12.5 to 47°, for $v/c = 0.99$, they are 4 to 16°.
[55] Sommerfeld, München *Sb, 41* (1911), 13. A misprint on p. 10, l.12 puts β for γ.
[56] *Ibid.*, 4.

unknown acceleration in terms of the distance *l* over which it occurs and obtained the result

$$E_\beta/E_\gamma = \frac{6m_e\pi c^3 l}{ve^2}\sqrt{1 - v^2/c^2}$$ (6.13)

where m_e is the rest mass of the electron. It was to eliminate the unknown distance *l* that he turned to the quantum theory. "The preceding considerations are already so hypothetical," he explained, that it was reasonable to extend "the fundamental hypothesis of the Planck radiation theory to [include] radioactive emission."[57] But how to formulate the problem when the interaction is nonperiodic and so lacks a frequency? The answer emerged from a reconsideration of the relationship between the energy and the duration of discontinuous electromagnetic impulses.

An entire section of Sommerfeld's study was devoted to the relation between pulse width and the azimuthal angle from which the impulse is viewed. The pulse width is, he said, a "theoretical measure of the hardness of the γ-ray."[58] This followed from the analogy to the x-ray case, and, as was true for x-rays, rested on no direct experimental evidence. The hardness of x-rays depends on the potential of the discharge tube; the greater the kinetic energy dissipated in the collision of an electron, the harder the x-ray that results. For the γ-ray case, the analogous condition concerns the magnitude of the electron acceleration; the more rapidly the β-electron reaches full velocity, and the greater that final velocity, the harder the γ-ray pulse. Like the x-ray impulse radiated from a decelerated electron, the hardness of the γ-impulse also varies with azimuth. For Sommerfeld, by 1911 a direct connection had been formed between the energy that contributes to an x-ray or γ-ray impulse and the pulse width that results.

To discuss the azimuthal variation Sommerfeld used the symbol λ for the pulse width, but I shall continue to use the more neutral δ. The symbol λ is traditionally applied to periodic waves and was used by Sommerfeld "to remind us of the analogy of the [pulse width] to the wavelength of periodic light."[59] He calculated the azimuthal variation in δ for the γ-pulse of Figure 6.6. The really significant value, the *massgebende Impulsbreite,* was the one de-

[57] *Ibid.,* 24.
[58] *Ibid.,* 34.
[59] *Ibid.,* 35.

fined at the azimuthal angle θ_{max}, of maximum energy flow.[60] Its value is

$$\delta_m = \frac{lc}{v}(1 + \sqrt{1 - v^2/c^2}) - l \cos \theta_{max} \qquad (6.14)$$

Increasing the velocity of the final electron produces a smaller δ_m, and δ_m vanishes completely when $v = c$. In general, decreasing the value of δ implies greater hardness; the hardness of both x-rays and γ-rays increases with the velocity of the stimulating electron.

From this perspective, Sommerfeld made the same conceptual break with classical physics that Wien had made four years before. "If we accept the hypothesis that these energy ratios are connected to Planck's h, we can find a formula that derives [δ] simply from the velocity [of the electron] without the auxiliary hypotheses for the [experimental] energy measurements."[61] In its simplest form, he set the product of impulse duration τ and the electron's kinetic energy equal to h. Like Wien and Stark before him, Sommerfeld here formally adopted the nonclassical concept that the impulse duration alone suffices to determine the energy contributed to the pulse. Recognizing moreover that, although pulse *width* varies with azimuth, the *duration* of the event producing the impulse does not, he went on to claim that the inverse pulse duration defines a sort of frequency for the impulse. In retracting his claim of 1909 that the quantum condition does not apply to the *Bremsanteil* impulse rays, he admitted, "I have only recently come to realize that the pulse duration τ can enter in the place of Planck's $1/v$"[62]

So, Sommerfeld's very formulation of the quantum condition for nonperiodic systems was predicated on the novel realization that the energy contained in impulses may be characterized by pulse duration, or frequency, rather than pulse amplitude, the classical measure of intensity. He elected this condition: In each discontinuous emission, just one quantum of action h is expressed. In the general case he required that the integral of the energy,

[60] *Ibid.*, 36.
[61] *Ibid.*, 38–9.
[62] *Ibid.*, 43n. It was commonly accepted at the time, especially in Germany, that the pulse width varied with the quality or hardness of the x-rays. See, among others, Pohl, *Physik der Röntgenstrahlen* (1912), where the idea permeates the text. There was little experimental evidence to support this claim. The only significant question was raised by Seitz, *AP, 26* (1909), 448.

Table 6.1. *Ratio of kinetic energy of β-particles to that of γ-rays (E_β/E_γ) for three elements*

	Element		
	RaE	UrX	RaC
Experimental	120	44–9	2
Predicted by Sommerfeld	330	150	100

expressed over the time interval of the acceleration, is always equal to h:[63]

$$\int_0^\tau E \, dt = \tau E_{\text{total}} = h \qquad (6.15)$$

Thus, Sommerfeld reformulated the quantum condition in terms of action, the product of energy with time, rather than energy alone. This change altered the need to formulate the problem in terms of frequency. Sommerfeld succeeded in finding a generalized quantum condition that would apply even when no frequency could be isolated, and this study was the beginning of work that would lead him by 1915 to the phase integral conditions for atomic systems.

With the quantum condition of equation 6.15, Sommerfeld sought agreement between his theoretical formulation and experimental data. He first examined the ratio of electron kinetic energy to the resulting impulse radiant energy for both x-rays and γ-rays. Russell and Soddy, H. W. Schmidt, and William Wilson had all given values of the γ-ray ratio for various velocity β-rays.[64] Sommerfeld's quantum formula for the ratio predicted values that agreed only approximately with the experiment for two of three γ sources (see Table 6.1). Applying the same test to Wien's x-ray ratio, Sommerfeld found "surprising" agreement, within 15 percent.[65] He was also able to compare the value of expected pulse

[63] Should E be only the kinetic energy or the total energy? Should the time dependence of E be taken into account? Sommerfeld did not know, but he selected equation 6.15 as the most useful representation.

[64] Russell and F. Soddy, *PM, 21* (1911), 130–54. Schmidt, *PZ, 8* (1907), 361–73; *AP, 23* (1907), 671–97; *PZ, 10* (1909), 6–16. W. Wilson, *PRS, 85A* (1910), 141–50.

[65] Sommerfeld, München *Sb, 41* (1911), 33.

width of γ-rays from some radioactive elements. The resulting values ranged from 3×10^{-11} cm to 5×10^{-13} cm, some 1.0 to 0.01 percent those of hard x-rays. Sommerfeld's quantum regulation predicted an x-ray pulse width of 2×10^{-9} cm. This agreed with Wien's quantum value when corrected for Sommerfeld's use of an energy ratio involving only the polarizable (impulse) part of the x-rays. Wien, of course, had not known in 1907 that some of the x-ray energy he measured was due to characteristic x-rays.

But Sommerfeld's chief concern was to show that the difference between his purely electromagnetic calculation and the quantum treatment was more illusory than real. "The two theories do not exclude, but augment each other," he said.[66] Wien had been unable to show a connection between his electrodynamic treatment of x-rays in 1905 and his quantum analysis in 1907. Stark had given up the electrodynamic solution entirely. Sommerfeld hoped that his γ-ray theory showed that a reconciliation was not only possible but reasonable. His electromagnetic calculation for the emission of the γ-pulse was successful as far as it went, but it left the acceleration, or the impulse duration, unspecified. That was where the quantum hypothesis, reformulated in terms of action, offered assistance. "In optics," Sommerfeld said, "consequences of the wave theory must be retained even though they remain unexplained by Planck's discovery . . . In the same way consequences of the impulse theory of x-rays can be integrated with the quantum of action."[67]

SECOND-GENERATION TESTS OF ANISOTROPY

Encouraged by the success of Sommerfeld's γ-ray theory, Edgar Meyer continued with the fluctuation analysis in 1911. Campbell had been quick to point out that Meyer, like Schweidler, had neglected a possible variation in ω, the number of electrons produced by each impulse.[68] Precisely this variation had stood in the way of Campbell's tests. The issue had taken on new significance by 1911 because Bragg had begun to claim more forcibly that the value of ω is 1; each neutral-pair γ-particle gives rise to one and only one secondary β-electron. That β-particle is then responsible

[66] *Ibid.,* 41.
[67] *Ibid.,* 41–2.
[68] Campbell, *PZ, 11* (1910), 826–33.

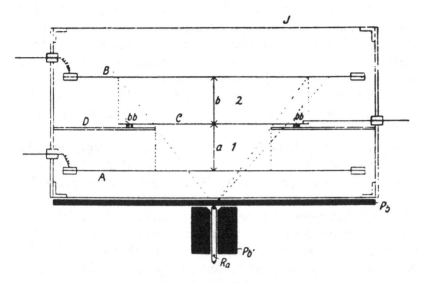

Figure 6.8. Meyer's apparatus for a second attempt to determine anisotropy of ionization by γ-rays. [*AP, 37* (1912), 705.]

for all subsequent ionization in the gas, according to Bragg.[69] If
Bragg's hypothesis was true, it seemed that the fluctuation analysis
could not distinguish dependent from independent events at all.
Meyer undertook a new series of tests that were simultaneously
designed to answer Campbell's criticism and test Bragg's hypothesis.

Meyer shot γ-rays across a double capacitor in a direction
parallel to that of the field gradient, as shown in Figure 6.8. He
compared the ionization produced in the isolated gas volumes on
both sides, reasoning that if the ionization events were fully independent, the fluctuations in saturation current in chambers 1 and 2
would remain uncorrelated.[70] The mean square fluctuation in the
differential current measured at the grounded plate C should not
change appreciably when the polarity of the field in one chamber is
changed. If plates A and B are both positively charged to 10^3 V, a
case Meyer denoted by the subscript $++$, the fluctuation ϵ_{++} will
reflect no more than the random production of electrons. If the
polarity of plate B is reversed, the positive-ion products change

[69] W. H. Bragg and Porter, *PRS, 85A* (1911), 349–65.
[70] E. Meyer, *PZ, 13* (1912), 73–81; *AP, 37* (1912), 700–20.

direction in chamber 2, but the absolute number of electrical impacts at plate C, and hence the fluctuation in the current, will be unchanged. Therefore, for uncorrelated events, $\epsilon_{++}/\epsilon_{+-} \simeq 1$. But if a given γ-ray produces more than one electron along its path, the correlated reception of electrical charges at plate C will constructively reinforce the $++$ fluctuation and cancel the $+-$; the ratio $\epsilon_{++}/\epsilon_{+-}$ will then approach some maximum value much greater than 1.

Meyer reasoned loosely that the ratio $\epsilon_{++}/\epsilon_{+-}$ represents the "coupling" between events in the two chambers; it gives the number of electrons produced, on the average, by each γ-pulse. He was apparently ignorant of the extensive British literature on differential ion mobilities; these showed that he could not count on the simultaneous arrival at C of the electron produced in chamber 1 with the ion produced in chamber 2 even if they were produced by the same γ-ray. But a typical test produced the ratio $\epsilon_{++}/\epsilon_{+-} = 2.94$; to Meyer, this high value indicated a "strong interdependence between ionization processes in [regions] 1 and 2."[71]

Then he turned the capacitor on its side and shot the γ-rays in parallel to plate C. He called this the *transverse* case and contrasted it to the earlier *longitudinal* test. The longitudinal case, he claimed, had measured the correlation between ion products produced by the same γ-ray pulse or·particle *along* the line of motion. The transverse case promised a measure of events produced along the hypothetical pulse front. Again he found a correlation; the transverse ratio $\epsilon_{++}/\epsilon_{+-}$ was also greater than 1. Meyer concluded that neither Bragg's nor Einstein's hypotheses could explain the degree of correlation he had demonstrated. Sommerfeld's electrodynamic theory of γ-rays was the "simplest explanation," even if it was not fully able to explain the high secondary β-velocities.[72] A student of Sommerfeld's, Eberhard Buchwald, analyzed Meyer's results.[73] He felt that Meyer's data allowed him to determine roughly the size of the angular region over which correlations could still be found in γ-ray-induced ionization. The half-angle he obtained was 17°, which was in reasonably good agreement with the angular width Sommerfeld assigned to γ-rays from radium C caused by electron emission at $0.95c$.

---

[71] E. Meyer, *AP, 37* (1912), 708–9.
[72] *Ibid.*, 719.
[73] Buchwald, *AP, 39* (1912), 41–52.

Campbell applied his own tests to Meyer's data.[74] He allowed for six possible outcomes, three each for the two possibilities of (A) spatially localized and (B) spreading waves. The three were (1) Planck's suggestion that atoms are predisposed by chance to ionization, (2) Bragg's suggestion that ionization is due to β-particle intermediaries, and (3) the long-honored triggering hypothesis. Campbell concluded that the evidence favored option A3, although he had known since 1907 that any triggering hypothesis was difficult to reconcile with the number of electrons in the atom. "This combination of theories," he concluded in despair, "appears to unite the disadvantages of all theories and to possess the advantages of none."[75]

Although the tests of radiation by fluctuation analysis were not over, they had run aground as of 1912. Uncertainty over the variation in the parameter ω effectively masked the significance of the results. All those who tried the experiments concluded that γ-rays are anisotropic. The question was, how far from spherical symmetry is the pattern of radiated energy? To the extent that any quantitative result was offered, the empirical data matched Sommerfeld's ring-pulse γ-rays the best.[76]

BRAGG'S APPEAL TO DUALISM

Bragg had held fast to the neutral-pair hypothesis. It was partly the completeness of Barkla's analysis for the softer x-rays that gradually pushed the domain of their discussion closer to that of the hard x-rays, where Bragg held the advantage. When Rutherford and Geiger showed in 1908 that the α-particle carries twice the charge of the electron,[77] Bragg cast aside the α-particle as the positive constituent of a pair in favor of an undefined "positive electron." In 1909 Bragg responded directly to Thomson's proposal that each electron radiates into a single narrow cone. By then, overwhelming evidence had accumulated for the forward scattering of secondary products of x-rays and γ-rays. Bragg claimed that even Thomson's "energy bundles of very small volume" must interact with individ-

[74] Campbell, *PZ, 13* (1912), 81–3.
[75] *Ibid.,* 83.
[76] Stark, *PZ, 13* (1912), 161–2. E. Meyer, *PZ, 13* (1912), 253–4. Laby and Burbidge, *PRS, 86A* (1912), 333–48. Burbidge, *PRS, 89A* (1913), 45–57.
[77] Rutherford and Geiger, *PRS, 81A* (1908), 162–73. Bragg and Madsen, *PM, 16* (1908), 938.

ual atoms to explain this behavior. "It seems hard," he sighed in this, his last paper before returning from Australia to England, "to understand the distinction between such bundles and entities generally classed as material."[78]

But in 1910, following his removal from Adelaide to Leeds, a variety of influences caused Bragg to recognize that his neutral-pair hypothesis could not be the full answer to the problems of radiation. First, he came into closer contact with the continental literature. The debate in Germany between Stark and Sommerfeld mirrored many of the issues still unresolved between himself and Barkla. A review article on his position, prepared for both British and German readers in 1910, illustrates the more reflective and less dogmatic expression of his hypothesis, which he tailored for a European audience. When a neutral pair is disrupted, the constituent electron proceeds on course; "the form of the entity may change, γ into β, X into cathode ray, and so on," he said, "but there is so little change in anything but form that practically we may assume a continuity of [particle] existence."[79]

So too did the narrowing gap between x-rays and visible light influence Bragg's position. In 1910 two independent tests, one by Bragg's compatriot R. D. Kleeman, showed that forward scattering was also found in the photoelectric effect using ultraviolet light.[80] Bragg had never presumed to extend his hypothesis to periodic light waves; finding that they too demonstrated one of the chief arguing points for neutral pairs caused him to step back for reflection. Perhaps there was some truth in the growing conviction of many physicists by 1911, British as well as German, that Barkla's characteristic x-rays bridged the gap between periodic light waves and impulse x-rays.

Bragg was also impressed by how far toward a particlelike concentration of γ-ray energy Sommerfeld had managed to come. But Bragg felt that Sommerfeld had not gone far enough. Soon after reading the γ-pulse study, he wrote to Sommerfeld to complain that the rings still dissipated their energy too rapidly.[81] How

[78] W. H. Bragg and Glasson, *PM, 17* (1909), 855–64. Bragg gave a final talk on the scattering of x-rays at the January 1909 meeting of the Australasian Association for the Advancement of Science but no text remains, *Report, 12* (1910), 113.

[79] W. H. Bragg, *PM, 20* (1910), 396.

[80] Kleeman, *Nature, 83* (1910), 339; *PRS, 84A* (1910), 92–9. Stuhlmann, *Nature, 83* (1910), 311; *PM, 20* (1910), 331–9. See Chapter 9.

[81] W. H. Bragg to Sommerfeld, 17 May 1911. Steuwer, *BJHS, 5* (1971), 271–2.

could the ring-pulse pass on virtually its entire energy to an electron after traversing macroscopic distances? But while he argued with Sommerfeld the dangers of continued attempts to make the pulse hypothesis conform with the ionization data, Bragg simultaneously hedged his bet. His basic point, he concluded, was not the physical nature of x-rays but the claim that in ionization, particlelike accumulations of energy are transferred to *single* electrons. One to one. He could avoid the quaking ground of radiation theory by separating the x-rays per se from the ionization event. With his new colleague at Leeds, H. L. Porter, Bragg proposed at the Royal Society that x-rays and γ-rays do not ionize atoms directly. One need only assume that a neutral pair disintegrates *before* ionization can occur. It is the resulting high-speed electron that strikes and ionizes the atom, they claimed.[82]

Just two weeks before, C. T. R. Wilson had reported to the Royal Society on his most recent experiments to make visible the tracks of ionizing particles by means of vapor condensation.[83] Here was the clear evidence that Bragg had long sought to demonstrate unambiguously that x-rays ionize gases in precisely the same manner as do particulate β-rays. Wilson's photograph, reproduced in Figure 6.9, shows only multiple short tracks; there is no sign of the diffuse cloud that ionization by spreading impulses would seem to require. Bragg made certain that the message got to Sommerfeld. If the energy of the rays were not concentrated in unchanging particlelike units, he claimed, exceedingly long periods would be necessary for sufficient energy to build up on a single electron from a succession of impulses.[84]

More perplexed than ever by the ambiguous action of x-rays and γ-rays, Bragg returned to the dual nature he had first ascribed to x-rays in 1907.[85] He began to call for a theory of x-rays, and to a lesser extent of γ-rays, that would embrace both particle and wave characteristics. The existence of two sorts of x-rays, the polarizable but inhomogeneous rays from the x-ray tube and the unpolarized characteristic rays from irradiated matter, reflected for him a

[82] W. H. Bragg and Porter, *PRS, 85A* (1911), 349–65. W. H. Bragg, *PM, 22* (1911), 222–3; *23* (1912), 647–50.
[83] C. T. R. Wilson, *PRS, 85A* (1911), 285–8.
[84] W. H. Bragg to Sommerfeld, 7 July 1911. Stuewer, *BJHS, 5* (1971), 273.
[85] W. H. Bragg, *TPRRSSA, 31* (1907), 94–8. He had suggested that along with the neutral pairs, impulses coexist in an x-ray beam. This was to answer Marx's velocity measurements.

Figure 6.9. C. T. R. Wilson's cloud-chamber photograph of ionization by x-rays. "The whole of the region traversed by the primary [x-ray] beam is seen to be filled with minute streaks and patches of cloud . . . mainly small thread-like objects not more than a few millimetres in length, and many of them considerably less than 1/10 mm in breadth." [*PRS, 85A* (1911), facing 286.] The x-ray beam enters from the left.

fundamental duality. Even before the riveting news that x-rays could be made to interfere, Bragg issued a clear call for a dualistic interpretation of x-rays. At the British Association meeting in 1912, he sought guidance in the inherent ambiguity of speculations on light by Newton and Huygens.[86] Earlier he had recommended forgetting "that idea of keeping touch with electromagnetic theory as we fancy it must be, which is hampering every movement."[87]

Elegant and interesting as it was, Sommerfeld's pulse analysis for γ-rays raised more problems than it solved. Grappling with some of those problems soon led Sommerfeld to significant advances in quantum theory. But in rejecting the triggering hypothesis, Sommerfeld opened himself to Bragg's criticism: No spreading-wave explanation could explain how x-rays and γ-rays retain their localized energy across macroscopic distances. Som-

[86] W. H. Bragg, BAAS *Report* (1912), 750–3. "I am very far from being averse to a reconcilement of a corpuscular and a wave theory: I think that someday it may come." Bragg to Sommerfeld, 17 May 1911. Stuewer, *BJHS, 5* (1971), 271–2. For earlier strong hints of the direction of Bragg's thoughts, see BAAS *Report* (1911), 340–1. For the influence of Bragg's dualistic claims, see Jeans, BAAS *Report* (1913), 380; *Report on radiation* (1914), 63ff.
[87] Bragg, *Studies in radioactivity* (1912), 191.

merfeld's hypothesis could not answer why, nor could Thomson's. Either the energy in the rays must be localized like actual particles, or some mechanism must be found whereby many impulses can conspire to act on a single electron before most of them have arrived at that electron.

This latter problem had first been debated, not for x-rays or γ-rays but for visible light. And in the period 1909–11, it had become clearer and clearer that whatever problems complicated the understanding of x-rays and γ-rays affected periodic light waves as well. For the latter, Sommerfeld's techniques provided no refuge, nor could one even claim the advantage of temporally discontinuous impulses. The full weight of accumulated evidence that light is a periodic, spherically propagating wave stood squarely in the way.

7

Problems with visible light

> The speaker does not wish to deny the heuristic value of
> the [lightquantum] hypothesis, only to defend the
> [classical] theory as long as possible.[1]

Between 1909 and 1912, many influential physicists first realized
that the problems preventing a consistent understanding of x-rays
applied as well to ordinary light. But unlike the relatively new
impulse theory of x-rays, the wave theory of light was exceedingly
well founded. It rested on a century's accumulation of experimental evidence. Spatial concentration in the energy of light could not
be attributed to a temporal discontinuity, as it could for x-ray
impulses. Light was known to be a repeating periodic wave. It was
not fully realized at first that hypotheses about the nature of
ordinary light required modification. For a decade after 1910, the
significance of the growing empirical evidence favoring spatial
localization of luminous energy went largely unrecognized. This
occurred because the data ran counter to the orthodox view of
visible light, the spectral region wherein the periodic properties of
radiation were most easily demonstrated and most firmly established.

 H. A. Lorentz attacked the problem in much the same spirit, but
with quite the opposite intent, as had Einstein. He showed in 1909
that the lightquantum hypothesis is incompatible with the quantum transformation relation itself, let alone with classical ideas
about radiation. In so doing, he laid the foundation for a restatement of a major difficulty with any classical explanation of the
photoelectric effect: It takes an extraordinarily long time for the
observed quantity of energy to build up from periodic waves
incident on an electron. Any nontriggering explanation of the

[1] Lorentz, *PZ, 11* (1910), 1250.

photoeffect had to clarify how this delay time, first measured in 1913, could be two orders of magnitude shorter than that predicted classically. As if to emphasize the difficulty, the triggering hypothesis itself was shown to conflict with new experiments. By 1911 the photoeffect had become a recognized problem for physicists.

This new concern over the nature of light was set against a growing awareness that x-rays and light are manifestations of the same phenomenon. A few, such as Stark, had believed this was so; between 1908 and 1912, experiments repeatedly confirmed the suspicion. The homogeneous characteristic x-rays seemed in all respects to constitute the x-ray analog to optical emission spectra. Homogeneity in the penetrating power of the secondary x-rays was increasingly attributed to uniformity of impulse width; by analogy to light, this implied that the characteristic x-rays are, in a sense, monochromatic. The existence of a well-defined threshold for the emission of characteristic x-rays suggested a mechanism similar to Stokes' law of optical fluorescence. Finally, in 1912, it was found that x-rays can interfere. These results made it imperative that a resolution to the paradoxes be found not just for x-rays but for light as well.

The sharpness of the issues was assured by the double-edged sword of analogy. In the early years of the century, impulse x-rays were thought to be sufficiently different from periodic light so that notions of discontinuity, for example in the operational definition of x-ray intensity, could be applied without requiring that similar considerations hold for light. As the data confirming the equivalence of x-rays and light increased, the need to combine understanding of both raised anew for light all of the paradoxes of x-rays. The analogy of x-rays to light, at first divergent and now convergent, began to raise serious doubts about the classical understanding of light as a periodic-wave phenomenon.

H. A. LORENTZ AND ENERGY ACCUMULATION

The highly respected Dutch theoretical physicist Hendrik Antoon Lorentz was uneasy about the lightquantum. He was universally regarded as a master mathematical physicist, and his views were taken very seriously by scientists of all nations. Although he himself had done no work on x-rays, Lorentz accepted the impulse hypothesis and thought that there was a qualitative difference between x-rays and periodic light waves.[2] He first began to take the quantum transformation relation seriously in 1908, but he raised strong arguments against the spatially localized lightquantum.[3] In a talk on the subject in April 1909 to the Dutch Congress of Scientists and Physicians, he pointed to the "striking" successes of the lightquantum: It explained Stokes' law; it seemed to be consistent with the findings of recent experiments on the photoeffect; and it appeared to explain Stark's Doppler shift in the emission spectrum from canal rays.[4] But, Lorentz concluded, "on closer examination serious difficulties arise for the hypothesis."[5]

It was primarily phenomena of interference that convinced Lorentz that light could not consist of localized lightquanta. Split beams still interfere after a separation of two million wavelengths. To achieve this, each lightquantum would have to extend almost a meter in the direction of propagation. This argument weighed heavily against the lightquantum for Planck too.[6] Lorentz pointed out that if each lightquantum spread its effect over so long a distance, it conflicted with the principle of the quantum hypothesis. To receive its full quantum of energy, a molecule would have to wait for the entire train of the lightquantum's "vibrations" to pass over it. But how can this ever occur if the minimum quantum of energy that the molecule can take up exceeds the amount contributed by each impulse? How does the molecule sense that enough vibrations are yet to come so that it can start to absorb them? How can the process of absorption even begin?

Another version of the interference argument showed that light-

---

[2] This is borne out more by what he did not say than by what he did say. In addressing the Dutch Congress of Scientists and Physicians in 1907 and 1909, talks on "Light and the structure of matter" and on the "Lightquantum hypothesis," he never mentioned x-rays. Lorentz, NNGC *Handelingen, 11* (1907), 6–21; *12* (1909), 129–39. In his famous Wolfskehl lectures in 1910 he discussed a variety of quantum issues, including the photoelectric effect. But he did not discuss x-rays. *PZ, 11* (1910), 1234–57. Nor did he include x-rays in his 1911 article on the "Nature of light" for the *Encyclopaedia Britannica,* 11th ed., *16* (1911), 617–23.

[3] For the background and story of Lorentz' conversion to the quantum transformation hypothesis, an event of major significance in the widespread adoption of the concept, see Kuhn, *Black-body theory* (1978), chap. 8.

[4] Lorentz, NNGC *Handelingen, 12* (1909), 129–39.

[5] *Ibid.,* 136.

[6] Planck to Lorentz, 7 October 1908, 16 June 1909, 7 January 1910, and draft of Lorentz to Planck, 30 July 1909. Lorentz papers. See also Planck's remarks following Einstein's lecture, *PZ, 10* (1909), 826.

quanta must also spread their effect in directions perpendicular to propagation. Imagine a telescope adjusted so that a star image is as sharply defined in the focal plane as the theory of diffraction will allow. If the objective lens or mirror is now partially blocked, so that only light from a small area of the original objective is seen, the image of the star is degraded. A *smaller* solid angle of light produces a *larger* image in the focal plane. The light must therefore spread its influence over the full objective aperture of the telescope. This is as true for the largest telescope (at the time, Hale's 150-cm telescope on Mt. Wilson) as for the pupil of the eye. But any single lightquantum that extends laterally across the Hale telescope can deliver only 10^{-4} of its energy to the eye; how then can the quantum transformation rule apply to the retinal stimulation that constitutes the "sight" of the star? "The conflict shows," Lorentz concluded, "that one cannot speak of propagating lightquanta concentrated in small regions of space that at the same time remain undivided."[7]

Lorentz was invited by the Wolfskehl commission to deliver six lectures at Göttingen in October 1910. The general theme of the lecture series – entitled "Old and new questions in physics"– was the aether: the reasons for the concept and modifications in it that had recently been suggested.[8] In the fifth lecture, Lorentz discussed the lightquantum hypothesis, with particular reference to the photoelectric effect. The triggering hypothesis he found wanting. A simple heated cathode, with no light shining on it, emits electrons with kinetic energy on the same order of magnitude as that of the cathode molecules. According to equipartition of energy, this is the expected behavior of the free electrons within the metal but not of those bound to atoms. If the light triggers the release of only unbound electrons, their velocities must also be comparable to the kinetic velocity. Experiment showed that they are fivefold greater. Moreover, it had been shown that heating the photocathode does not increase the measured velocity as it certainly should have if free electrons were being triggered.[9]

Lorentz' argument did not, in fact, address the true triggering

[7] Lorentz, NNGC *Handelingen, 12* (1909), 139.
[8] Lorentz, *PZ, 11* (1910), 1234–57.
[9] Lorentz cited only Ladenburg's studies, but there had been others beginning in 1906. Lorentz' claim in 1910 that the linear law had been confirmed was premature. Wheaton, *Photoelectric effect* (1971), chap. 13.

hypothesis because he did not consider the bound electrons. The velocity of an electron released from its atom is a function of the binding energy; depending on the details of the then unclear atomic structure, it could very easily be five times the thermal energy of a cathode molecule. The triggering hypothesis was never intended to apply to unbound electrons; the photoelectron must have a frequency, by virtue of its motion within the atom, in order to be triggered by the light.

In regard to the opposing view, that the photoeffect is a transformation of light energy to electron kinetic energy, Lorentz showed that Einstein's lightquantum explanation seemed to accord well with observations. But Lorentz judged it "impossible" if lightquanta were thought to be fully independent of one another.[10] "The speaker does not wish to deny the heuristic value of the hypothesis," he said, "only to defend the old theory as long as possible."[11]

Johannes Stark addressed the same issues at about the same time, asking rhetorically if it was really a "fantastic obstinacy," as some had suggested, that drove him to explain events "by observing their course" rather than from a "mathematical theory [such as Sommerfeld's] based on abstract concepts."[12] He felt that the very existence of the concept of an aether stood most in the way of a clear view of the problem. Like Campbell, Stark wished to do away with the aether entirely. But his hypothesis that interference effects are due to "coherent" aggregations of lightquanta that possess a periodic structure hardly convinced Lorentz. Light not only had to come in indivisible quanta, Lorentz objected, but to listen to Stark, it had to come in fixed groupings of quanta as well.

CLASSICAL ABSORPTION OF LIGHT

Lorentz analyzed classical waves to see how they might deliver quantum units of energy to bound electrons. A monochromatic light beam acts on an electron bound to an atom by a quasi-elastic force. Depending on the match between the frequency of the incident light and the oscillation frequency of the bound electron,

[10] Lorentz, *PZ*, *11* (1910), 1249.
[11] *Ibid.*, 1250.
[12] Stark, *JRE*, *7* (1910), 387.

energy is transmitted from the light to the electron. At the same time, the accelerating charged electron radiates energy – the so-called radiative damping. Lorentz used the case of maximum energy transfer, which occurs when the incident frequency is equal to the natural frequency of the electron. When the radiative loss equals the rate energy is incident from the light, the electron reaches equilibrium. The total energy available to the electron is the total contained in the periodic spherical wave that passes over the electron during the time the electron takes to reach equilibrium. The equilibrium energy of the electron is thus determined by the intensity of the incident light, although not all of the radiant energy is necessarily transferred to the electron. Lorentz showed that for visible wavelengths the minimum light intensity that could deliver one quantum of energy to an electron at equilibrium is almost 10^5 ergs/cm^2-sec. Yet the photoeffect was known to occur for visible light with an intensity at least as low as 2 ergs/cm^2-sec.

In this way, Lorentz raised a corollary to the paradox of quality: the dilemma of energy accumulation. Campbell had noted the problem first, but when Lorentz explained the issue, it could no longer be ignored.[13] Removing the damping term so that the electron could absorb an indefinite amount of energy, Lorentz calculated the number of wavelengths of light that must combine their effect before the required quantum can be absorbed. He found that for visible light the number is

$$ n = \sqrt{\frac{2hc^2}{3\pi\lambda^3 \mathcal{R} I}} \simeq \frac{2 \times 10^{10}}{\sqrt{I}} \tag{7.1} $$

$\mathcal{R}$ is the radius of the electron, λ is the wavelength of the incident light, and I is its intensity in ergs per square centimeter per second. Bright sunlight, even if its full intensity were concentrated at a single wavelength, would have to supply several million wavelengths. The low-intensity yellow light known to induce photo-emission had to supply five billion. This would take an "extraordinarily long time," he said – several millionths of a second.[14]

Lorentz did not give the derivation of equation 7.1, but he did sketch enough of the assumptions involved so that one can recon-

[13] Campbell, *Theory* (1907), 226ff.
[14] Lorentz, *PZ, 11* (1910), 1251.

struct it.[15] Despite all the simplifying assumptions – he neglected radiative losses and the intra-atomic interaction of electrons, as well as the fact that one cannot expect to have monochromatic wave trains of sufficient length – it is a revealing calculation. This is not because the result requires so many oscillations but because

[15] The equation of motion for a bound electron acted upon by a sinusoidal field and not subject to radiative losses is

$$m\ddot{x} + fx = eA_E \cos \omega t \qquad (7.1a)$$

At resonance, $\omega^2 = f/m$; hence:

$$\ddot{x} = \frac{eA_E}{m} \cos \omega t - \omega^2 x \qquad (7.1b)$$

Lorentz found that the maximum amplitude of oscillation grows in proportion to t, so we may set $x = Ct \sin \omega t + B \cos \omega t$. Then

$$\dot{x} = C\omega \sin \omega t + C\omega t \cos \omega t - B\omega \sin \omega t \qquad (7.1c)$$

and

$$\ddot{x} = \omega C \cos \omega t + C\omega \cos \omega t - (tC\omega^2 \sin \omega t + B\omega^2 \cos \omega t)$$

or

$$\ddot{x} = 2\omega C \cos \omega t - \omega^2 x \qquad (7.1d)$$

We wish to calculate the total energy of the electron $E = T + U$. To accomplish this, note that $dT/dt = d(m\dot{x}^2/2)/dt = m\ddot{x}\dot{x}$. From (7.1b):

$$\frac{dT}{dt} = \dot{x}eA_E \cos \omega t - m\omega^2 x\dot{x} = \dot{x}eA_E \cos \omega t - fx\dot{x} \qquad (7.1e)$$

But $-fx\dot{x} = d(-fx^2/2)/dt$ which is dU/dt, so

$$dE = dT + dU = \dot{x}eA_E \cos \omega t \, dt \qquad (7.1f)$$

We now set the total energy equal to $h\nu$ and solve for the time τ required to reach that threshold.

$$h\nu = \int_0^\tau eA_E \cos \omega t \, [(C - B\omega) \sin \omega t + C\omega t \cos \omega t] \, dt \qquad (7.1g)$$

The first two terms drop out when integrated over a whole number of oscillations, and because $\cos^2 \alpha = (1 + \cos 2\alpha)/2$,

$$h\nu = \int_0^\tau eA_E \, \omega t \left(\frac{1 + \cos 2\omega t}{2} \right) dt$$

Again the last term drops to zero, leaving only

$$h\nu = \frac{A_E e C \omega}{2} \frac{\tau^2}{2} \qquad (7.1h)$$

Substituting the value $C = A_E e/2\omega m$ derived from (7.1b) and (7.1d):

$$\tau^2 = 8h\nu m/e^2 A_E^2 \qquad (7.1i)$$

it requires so few. If one compares the length of a wave train needed to supply the same energy over just the cross section of the electron, or indeed over the entire volume of an atom, the result at visible wavelengths and intensity is some eight orders of magnitude greater than Lorentz' estimate![16]

Lorentz gave an analytic solution to the problem of absorption of electromagnetic energy; he did not interpret the result mechanically, as others, notably Stark and Bragg, would do. Lorentz treated the electron as a virtual point; its interaction with the field occurs only through its charge and is therefore independent of its "size." The electron radius $\mathcal{R}$ enters equation 7.1 only as a convenient combination of quantities, not as the controller of the fraction of incident radiation absorbed. The effective area of the wave front that contributes energy to the electron, according to Lorentz' calculation, is inversely proportional to the square root of the light intensity. Only for unrealistically great light intensity (on the order of 10^{18} ergs/cm^2-sec at visible frequencies) does this area become comparable to that defined by the maximum amplitude of the electron just prior to its release from the atom.[17] For lower intensities, which are known to be able to induce photoelectrons, the

---

To reconcile this result with Lorentz' formula (7.1), we need to express the field strength A_E in terms of Lorentz' I, which is an energy flux, $I = A_E^2/4\pi c$; then substitute the classical radius of the electron $\mathcal{R} = 2e^2/3m$ [M. Abraham, *AP, 10* (1902), 151]; and recognize that $\tau = n/v = n\lambda/c$:

$$n^2 = \frac{8hc^3m}{6\pi\lambda Ic\mathcal{R}\lambda^3} = \frac{4hc^2}{3\lambda^3\pi\mathcal{R}I}$$

which, except for a factor of $\sqrt{2}$, is equation 7.1.

[16] The total energy through cross-sectional area Q per second is IQ, so the time needed to attain energy hv is $\tau' = hv/IQ$. The ratio of the time calculated from energy flux to that found by Lorentz is therefore $\tau'/\tau = \sqrt{3\pi h\mathcal{R}/2\lambda I}\, c/Q$. For the cross section of an atom, $Q = 10^{-16}$ cm^2, and for a reasonable visible light intensity of $I = 500$ ergs/cm^2-sec, $\tau'/\tau \simeq 10^8$.

[17] The maximum amplitude of the harmonically bound electron is reached one-half cycle before its release when the potential enegy $fx^2/2$ is equal to hv. Thus, $x^2 = 2hc/\lambda mw^2 = h\lambda/2\pi^2mc$. For visible light, x^2 turns out to be 6×10^{-16} cm^2, the same order of magnitude as the cross-sectional area of the atom. As Lenard had noted as early as 1902, a linearly oscillating electron will have to obtain its entire T during that last half-cycle, for it will lose its potential energy as it leaves the atom unless some special mechanism is called into play to sever the binding force. The same problem afflicts an electron rotating in a central potential formed by an effective positive charge $+ e$. For a circular orbit, $U = 2T$, so the total energy is on the order of $mv^2 = m(2\pi rv)^2$. When this equals hv, r^2 for a visible frequency is also about 10^{-16} cm^2.

electromagnetic cross section is considerably greater than any mechanical interpretation would seem to allow.

Thus, when Lorentz concluded that the results predicted by equation 7.1 were "not at all favorable for the old theory," he had hardly the same discrepancy in mind that others would soon mention. But Lorentz was primarily concerned with showing that accumulation of energy over *any* finite time interval is incompatible with the quantum transformation hypothesis for absorption of light. How can an electron accept any of the energy from a classical light wave before the full quantum is presented? Since most physicists supported the classical model of light, this conflict cast doubt on the new quantum theory just as its influence was beginning to grow.[18] Lorentz concluded, despite his objections to lightquanta, that "without the assumption of lightquanta, these phenomena raise the greatest difficulties."[19]

Lorentz did not extend the argument to the more intractable case of x-ray ionization. Light possesses a repeating field oscillation, so that the accumulation of energy proceeds relatively swiftly. X-rays each contribute only a single impulse. For the case of x-rays, equation 7.1 gives a result equivalent to the total number of impulses required. Moreover, for x-rays, the advantageous resonance condition can no longer apply.

Stark liked Lorentz' argument. He devoted the second volume of his *Principles of atomic dynamics* entirely to "elementary radiation." There he marshalled all arguments at his disposal against the spreading-wave model of radiation. The first objection was that of energy accumulation; he merely reprinted Lorentz' words verbatim, judging that his opponent had made the case for lightquanta as well as he could.[20]

THE DECLINE OF THE TRIGGERING HYPOTHESIS

Philipp Lenard, the originator of the triggering hypothesis, had attained substantial recognition by 1911. He had won the Nobel

---

[18] In 1907 Einstein proposed a theory of specific heats based on the quantum postulate that soon attracted the approbation of leading physicists and physical chemists. This was the first subject, aside from the arcane problems of cavity radiation, that drew favorable reactions to the quantum. Klein, *Science*, 148 (1965), 173–80. Kuhn, XIV Congrès internationale d'histoire des sciences, *Proceedings*, 1 (1974), 183–94.

[19] Lorentz, *PZ, 11* (1910), 1251.

[20] Stark, *Elementare Strahlung* (1911), 267–70.

Prize in 1906 for his early investigation of cathode rays. The next year, he was appointed *Ordinarius* and director of the new Radiological Institute at Heidelberg. He developed a qualitative atomic model during the first decade of the century based, in part, on evidence for bound electron energies derived from experiments on the emission of photoelectrons.[21] In 1909 the triggering hypothesis was described by Rudolf Landenburg as being among "the generally accepted truths" of physics.[22] But at the same time that Lorentz raised questions about the lightquantum, Lenard discovered evidence that cast serious doubt on the triggering hypothesis.

In 1910 Lenard began an extensive series of experiments with his skilled colleague, Carl Ramsauer, to study the action of ultraviolet light on gases. They published five monographs, each directed to a different aspect of the effect.[23] Lenard was at first encouraged by the "surprising" discovery that his source of ultraviolet light released a great number of electrons from the gas, although the wavelengths he used were only minimally absorbed by the gas.[24] This seemed to be even more evidence in favor of the triggering mechanism: The incident light is not expected to lose much energy in breaking the bond that holds the electron to the atom. In solid bodies energy is lost through thermal effects, but in a gas these losses should be negligible. In 1911 Lenard and Ramsauer studied the phenomenon more closely because it "appeared to be of particular interest from the energy point-of-view."[25]

The result of their careful analysis cast strong doubt on the triggering hypothesis. They showed that most of the electrons released in the photoeffect came from a small number of highly absorbing atoms – impurities – mixed with their gas sample. The conclusion was unavoidable. "For gases, as for solids and fluids, the photoelectric effect is correlated not just with absorption [of light] but with very strong (metallic) absorption . . . Our observations leave no support for the assumption that the energy of [photoelectrons] originates in the inner atom; they indicate much more that the incident light provides the energy."[26] "A complete theory of the energy transformation in the photoelectric effect is

[21] Wheaton, *HSPS, 9* (1978), 299–323.
[22] R. Ladenburg, *JRE, 6* (1909), 427.
[23] All appeared in the Heidelberg *Sb, 1* (1910), Abh. 28, 31, 32; *2* (1911), Abh. 16, 24.
[24] Lenard and Ramsauer, Heidelberg *Sb, 1* (1910), Abh. 31, pp. 21ff.
[25] Lenard and Ramsauer, *ibid.,* Abh. 24, p. 1.
[26] *Ibid.,* 5–6.

not yet possible," they concluded.[27] At the eighty-third *Naturfor-scherversammlung* held in Karlsruhe in September 1911, Carl Ramsauer announced publicly that the photoelectric effect could no longer be thought of as exclusively a triggering action. "The energy of the ejected electron," he said, "does not come from the atom as originally assumed by Herr Lenard, but from the absorbed light."[28] If any sizable fraction of the energy of a photoelectron is transferred from the light to the electron, either the paradoxes of quantity and quality must be resolved or the mechanism by which accumulation occurs must be found. In 1913 Lenard declared the photoelectric effect a "difficulty" with no apparent solution.[29]

MAX PLANCK'S "SECOND THEORY"

Max Planck was strongly influenced by Lorentz' Wolfskehl arguments, having heard them in correspondence with Lorentz before the lectures in 1910.[30] Early in 1911, Planck offered a new approach to the problems of cavity radiation with which he had grappled since his derivation of the empirical black-body law in 1900. This, his so-called second quantum theory, appears, in light of recent research, to have been the first in which he accepted the discontinuity in energy transfer implied by the quantum transformation hypothesis.[31]

At first Planck had attempted to quantize the absorption of radiation by atoms, but the difficulties that Lorentz raised forced him to reconsider that approach.[32] The quantum hypothesis demands that a minimum quantity of energy be supplied before absorption can take place. Classical waves can deliver that energy only a bit at a time. How can the first bit be absorbed at all? And if it is absorbed, what happens if the incident radiation is cut off immediately thereafter? Either that first subquantum of energy is absorbed, violating the quantum hypothesis, or it is not. If it is not, how can light ever be absorbed?

To avoid this dilemma, Planck proposed that the absorption of radiant energy occurs continuously; only its subsequent emission

[27] *Ibid.*, 7.
[28] Ramsauer, *PZ, 12* (1911), 998.
[29] Lenard, *AP, 41* (1913), 82.
[30] See note 6 in chapter 7.
[31] Kuhn, *Black-body theory* (1978), chap. 10.
[32] Planck, *AP, 31* (1910), 758–68.

is quantized.[33] At any moment, the electrons in irradiated atoms possess among themselves a continuous distribution of energy. A small fraction have energy just less than an integral multiple n of the quantum hv (where v is the frequency of the oscillating electron). Given the small increment of energy needed to attain this threshold by the incident light, emission of the entire energy nhv can occur. The likelihood that the emission will or will not occur is determined by the character of the oscillator; for any individual oscillating electron, only a statistical probability for emission can be determined. In this last suggestion, Planck may have been influenced by the analogous recourse to a statistical theory of radioactive decay. Meyer's statistical analysis of fluctuations in the ionization produced by γ-rays, as well as the seemingly random emission of these rays, may have encouraged Planck to apply similar concepts to the emission of light.

Planck mentioned the photoelectric effect as one of the cases to which his new hypotheses might apply. It is not difficult to see how this would occur, although Planck did not provide the details. It was not until 1913 that he extended his theory of quantum emission to include the emission of electrons.[34] His new hypothesis circumvented the accumulation problem altogether. Electrons absorb energy continuously, but some of them might be ready to fly out of the atom with the stored quantity nhv in kinetic energy almost instantaneously after the illumination begins. Furthermore, the velocities of emitted electrons should be independent of both the intensity of the light and the temperature of the metal. Light energy is not transformed into electron kinetic energy; the latter already exists within the atom. Although heating the plate will increase the energy of individual electrons, photoemission occurs only at a fixed threshold, so that the resulting electron energies remain unaffected.

Planck's theory was the quantum analog of the triggering hypothesis. That was its problem. The theory suffered most of the drawbacks of the triggering hypothesis. Only a few months after Planck's announcement of it, Lenard and Ramsauer showed that photoemission occurs only where strong absorption of light is also found. According to Planck's interpretation, this should not be the case. He thought that photoemission requires only a very small

[33] Planck, *VDpG, 13* (1911), 138–48; *AP, 37* (1912), 642–56.
[34] Planck, Berlin *Sb* (1913), 350–63.

energy input from the incident light, just enough to raise the most energetic electrons to the quantum threshold. But Lenard and Ramsauer showed that strong absorption of light is correlated with strong photoemission.

There was yet another difficulty that the triggering hypothesis, whether the classical or the quantum version, had to confront. The specific form of the photorelation between incident frequency and electron velocity was still uncertain in 1911. But it was clear enough that a continuous distribution of incident light frequency sparked a continuous distribution of electron velocity. It could no longer be claimed that the data supported resonance effects at specific wavelengths. Planck's theory seemed to require that sodium atoms, for example, contain an electronic structure sufficiently complex so that a full continuum of emission threshold energies exists, with each value of the energy keyed to a particular exciting frequency.

At first, Planck denied the resonant nature of the interaction. He said, "one cannot directly determine the frequency of the excited oscillator from the wavelength of the incident ray."[35] But at the Solvay Congress later that year he retracted the claim, leaving obscure just how the immense number of required electron frequencies was to be reconciled with the limited number of electrons in the atom.[36] The difficulty would not soon go away. In 1911 it was still possible to think that the unknown structure of the atom might encompass enough electron frequencies to validate the quantum triggering hypothesis. But in 1913 Niels Bohr offered a new quantum theory of atomic structure that allowed only a small number of electron frequencies. As the acceptance of the Bohr theory grew, the unresolved problems of the photoeffect simultaneously became more intractable and less visible.

SOMMERFELD'S THEORY OF THE
PHOTOELECTRIC EFFECT

Arnold Sommerfeld was even more profoundly influenced by Lorentz' discussion of the photoeffect than was Planck. He knew that he could derive a certain degree of spatial localization for the energy of γ-rays, and even for that of x-rays. But for visible light his weapons were of no avail. A different sort of explanation had to be

---

[35] Planck, *VDpG, 13* (1911), 147.
[36] Planck, *Solvay I*, pp. 93–114, in particular 112.

found, and Lorentz had severely restricted the options. Applying the quantum of action was hindered by the fact that both absorption and emission of energy are singular, nonperiodic events. The quantum relation demands a frequency. To apply quantum theory to discontinuous events, which have no frequency, Sommerfeld had to generalize the quantum condition. When dealing with the emission of γ-rays earlier in the year, he had faced the same problem and had conceded ignorance. He had simply used Wien's experimental value for the ratio of x-ray to cathode-ray energy in order to estimate the γ-pulse width. Now, however, fresh from the realization for γ-rays that impulse duration can "take the place of Planck's $1/\nu$," Sommerfeld went considerably further. Putting the quantum transformation relation in its action form, $E/\nu = h$, and applying his newly found insight from radiation theory, he arrived at the general condition $E\tau = h$, where τ is the temporal duration of the discontinuous event.

Sommerfeld, like Planck, formulated the quantum condition in a way that sidestepped the problem of energy accumulation; he made accumulation the very basis of the new principle.[37] His general expression depended only on the time interval over which an event occurs. It could therefore be applied directly to nonperiodic events. As in Planck's theory, an atom was assumed to absorb radiant energy continuously until a specific condition was attained. Unlike Planck, Sommerfeld specified the threshold condition for emission in terms of a quantum of action, the product of energy and time, rather than of energy itself. When the time integral of the energy-determining Lagrangian taken over the duration of the process attains the value $h/2\pi$, an electron is released with the full kinetic energy T of the system. In other words, if U is the potential energy of the bound electron, the condition for emission is fixed by the relation

$$\int_0^\tau (T - U)dt = h/2\pi \qquad (7.2)$$

With his talented former assistant, Peter Debye, Sommerfeld formulated a theory of the photoelectric effect based on this quantum principle. In its final but still incomplete form, it was the most successful treatment of the photoeffect that did not require the

[37] Sommerfeld, *PZ, 12* (1911), 1057–68. See also Nisio, XIVième Congrès internationale d'histoire des sciences, *Proceedings, 2* (1975), 302–4.

assumption of the lightquantum. It was first presented at the eighty-third *Naturforscherversammlung* in 1911.[38] An electron is to be thought of as bound by an elastic force $-fx$, and loses no energy by radiation. It is acted upon by an oscillating electric field of maximum amplitude A_E and angular frequency ω. The equation of motion of the electron is therefore

$$m\ddot{x} + fx = eA_E \cos \omega t \qquad (7.3)$$

The instantaneous kinetic energy of the electron is $T = m\dot{x}^2/2$; its potential energy U is $fx^2/2$. Sommerfeld inserted these relations in the quantum condition (equation 7.2) and integrated by parts. When the equation of motion, equation 7.3, is substituted, the quantum condition becomes

$$\left.\frac{mx\dot{x}}{2}\right|_0^\tau - \frac{e}{2}\int_0^\tau xA_E \cos \omega t \, dt = h/2\pi \qquad (7.4)$$

The left-hand side of equation 7.4 describes the variation in the action of the electron as energy is pumped in by the incident wave. It is a rapidly varying sinusoid with an envelope that grows steadily but slowly from peak to peak. The quantum condition is achieved when the amplitude of the oscillation finally reaches the fixed magnitude $h/2\pi$. This can occur only very close to one of the many peaks of the growing sinusoid; otherwise the threshold would have been reached on the preceding upswing. Consequently, at the threshold the slope of the left-hand side of equation 7.4 must be close to zero. But that expression is simply $\int(T - U)dt$, so that the slope of the function is just the Lagrangian, $T - U$. The quantum condition is therefore reached when $T = U$. In other words, $m\dot{x}^2/2 = fx^2/2$. But $\sqrt{f/m}$ is ω_o, the resonant frequency of the undamped electron. Thus, when the quantum condition is fulfilled at time τ, $\dot{x} = \omega_o x$, and the first term in equation 7.4 is just T/ω_o. The second term vanishes entirely at resonance: As long as the incident frequency equals the frequency of the electron, the integral is taken over a whole number of wavelengths of a sinusoid and consequently equals zero.

Thus, for the idealized case of monochromatic light in resonance with an undamped electron, Sommerfeld's quantum condition (equation 7.2) reduces to

[38] *Ibid.*, 1064–5 and associated notes.

$$\frac{T}{\omega_0} = \frac{h}{2\pi} \tag{7.5}$$

If we express the kinetic energy of the electron in terms of the electric potential P that just suffices to arrest its motion, we obtain

$$Pe = h\nu_0 \tag{7.6}$$

This is Einstein's photoelectric effect law, equation 5.4, before the work function p is taken into consideration.

The accumulation time is an essential part of Sommerfeld's theory. The integral in equation 7.2 is always to be taken over the accumulation interval τ. Unlike Planck's theory, Sommerfeld's required the energy of the liberated electron to come from the incident light. He could not ignore the accumulation time. This was especially true if the unavoidable radiation damping was included and the incident light was more realistically granted some range of frequency. Sommerfeld had clearly known of these problems before his talk in Karlsruhe. Bragg had written in response to Sommerfeld's γ-ray theory and cited Campbell's reservations about the accumulation time for x-ray ionization. "The x-ray tube must be at work for several days," he said, "before an electron in an atom can acquire as much energy as a [released] electron is said to possess."[39] The problem is naturally more severe for aperiodic x-ray pulses than it is for the rapid and regular succession of impulses that constitute periodic light. But Lorentz had made it quite clear that a difficulty lurked even for light. His relation expressed in equation 7.1 can easily be modified to predict a delay time for visible light equal to

$$\tau = \sqrt{\frac{2h}{3\pi\lambda\mathcal{R}I}} \simeq \frac{1.5 \times 10^{-5}}{\sqrt{I}} \text{ sec} \tag{7.7}$$

For yellow light with an intensity of 2 ergs/cm²-sec, which is known to stimulate photoelectrons, the accumulation time, according to equation 7.7, is 10^{-5} sec.[40]

---

[39] W. H. Bragg to Sommerfeld, 7 July 1911. Stuewer, *BJHS*, 5 (1971), 273.

[40] The simpleminded (and misleading) textbook calculation of the delay time, by assuming that all the energy incident on the cross section of the atom accumulates on an electron, produces the result $\tau \simeq 8 \times 10^4$ sec! See Chapter 7, notes 15 and 16. The analytical classical solution given by Lorentz is nowhere nearly so far out of line with observations. *Cf.* Weidner and Sells, *Elementary modern physics* (1960), 94, 123.

In presenting his theory of the photoeffect, Sommerfeld did not mention that the accumulation time constituted a potential problem. But Stark immediately objected. He asked if Sommerfeld had calculated the magnitude of the accumulation time, claiming that Lorentz' calculation showed that it "cannot be brought into agreement with the observed conditions of [light] intensity and the maximum kinetic energy of emitted electrons."[41] Perhaps Sommerfeld did not think that the problem was so great, at least not for ordinary light. After all, one knew very little about the photoelectric effect delay time experimentally; perhaps the 10^{-5} sec prediction was not so far from the truth. Sommerfeld responded that the "difficulty of the time interval" might be alleviated when the theory was extended to nonmonochromatic light.[42] Stark persisted, saying that many of the most accurate photoeffect measurements had been done with sharply defined spectral lines. Two days later, the accumulation time took on even greater significance when Carl Ramsauer delivered the news that the triggering hypothesis was no longer viable.

It was true that not much was known about the photoelectric delay time in 1911. Some knew that Alexandr Stoletov had found in 1889 that the delay was not longer than 10^{-3} sec.[43] Wien had given an estimate of 0.01 sec in 1905 for the accumulation of x-ray pulses, and that absurdly long result had induced him to accept the triggering hypothesis.[44] Einstein, recognizing that a critical point had been raised, judged that an experimental investigation of photoeffect delay times would be "of the greatest interest."[45] Erich Marx took up Einstein's challenge, but it was almost two years before his results were known. In the meantime, Sommerfeld continued making attempts to solve the photoeffect conundrum.

THE SOLVAY CONGRESS

In October 1911 the First Solvay Congress was held in Brussels. It was here that the quantum principle was recognized officially to be a significant and unavoidable aspect of physical theory. Although

[41] Stark's comment in the discussion following Sommerfeld's presentation, *PZ, 12* (1911), 1068–9.

[42] *Ibid.*

[43] Stoletov, *Physikalische Revue, 1* (1892), 723–80. See Wheaton, *Photoelectric effect* (1971), chap. 5.

[44] Wien, *AP, 18* (1905), 1005.

[45] Einstein's contribution to the discussion, *PZ, 12* (1911), 1068–9.

the conference was titled "The Theory of Radiation and Quanta," there was little serious discussion of the lightquantum hypothesis.[46] Virtually all those assembled felt that Einstein's proposal was an unwarranted rejection of the highly verified wave theory of light. Stark was not invited to participate. Einstein was, but was asked to speak on his quantum theory of specific heats.

Sommerfeld and Planck each presented his new version of the quantum regulation. After Planck spoke, Wien asked if the long accumulation time really constituted an insurmountable problem for what he called Planck's "first theory."[47] Planck answered that it did, not so much because it disagreed with experiment but because the very process of accumulation of energy seemed to be incompatible with quantum absorption.[48] Sommerfeld repeated his arguments on the localized energy of x-rays and γ-rays. The demanding problems of absorption were submerged in his action integral representation of the quantum principle.

At the Solvay Congress, Sommerfeld expanded the treatment of the photoeffect that he had given at the earlier *Naturforscherversammlung*. The spatial concentration of light energy implied by the photoeffect he called one of the "greatest difficulties" then facing electrodynamics.[49] He asserted that "radiation theory and electromagnetic theory complement, rather than exclude, one another,"[50] but despite his best efforts, the photoeffect remained a difficulty. He was able to treat off-resonance with the help of Peter Debye; the result was "an analog to Stokes' law" of fluorescence.[51] They showed that only waves with a frequency greater than the resonant frequency of the electron are effective in passing on energy to the electron. To solve equation 7.4 when $\omega \neq \omega_0$, they defined a figure of merit $\epsilon = (\omega - \omega_0)\tau$, where τ is the accumulation time for the resonance case. They then showed that the quantum condition, equation 7.4, reformulated in terms of ϵ, became

$$(1 + \epsilon) - (\cos \epsilon + \sin \epsilon) = \frac{2m\omega_0 h}{2\pi} [(\omega^2 - \omega_0^2)/\omega_0 e A_E]^2 \quad (7.8)$$

[46] *La theorie du rayonnement et les quanta*, Langevin and M. de Broglie, eds. (Paris, 1912), henceforth *Solvay I*.
[47] Wien's comments followed Planck's presentation, *Solvay I*, p. 126.
[48] Planck, *Solvay I*, pp. 126–7.
[49] Sommerfeld, *Solvay I*, p. 344.
[50] *Ibid.*, 333.
[51] Sommerfeld, *PZ, 12* (1911), 1065.

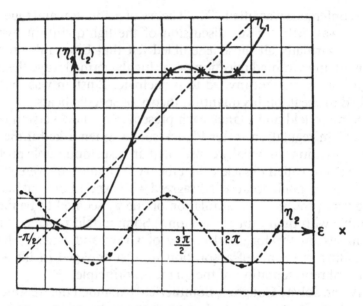

Figure 7.1. Graph used by Debye and Sommerfeld to derive Stokes' law for the photoelectric effect. [*AP, 41* (1913), 890.]

The left-hand side of equation 7.8 is the superposition of a sinusoidal oscillation, $\cos \epsilon + \sin \epsilon = 2 \sin (\epsilon + \pi/4)$, on the linear relation $1 + \epsilon$. Plotted as a function of ϵ, it is the rising sinusoid traced by the solid line in Figure 7.1. The right-hand expression of equation 7.8 is a constant for fixed ω, ω_o, and A_E. The solutions of equation 7.8 therefore lie at the intersection of the rising sinusoid with the horizontal line that represents the value of the right half of the equation. If there is more than one point of intersection, the one for the smallest value of ϵ has the shortest accumulation time τ.

The right-hand side of equation 7.8 is always positive. Thus, the only conceivable solutions for negative values of ϵ are restricted to the angular range $\epsilon = -\pi/2$ to 0. A negative value for ϵ means that the incident frequency is less than the resonant frequency. The anti-Stokes case can occur only within an extremely small fraction, some 10^{-11}, of the resonant frequency. The Stokes solution for positive values of ϵ can occur any number of times, for any ω greater than ω_o. As in any resonant interaction, the efficiency of the energy transfer drops rapidly as ϵ increases. But beyond this neat trick, Sommerfeld and Debye could not go. They used what they called their "ersatz" model to derive "a few of the most general consequences of photoelectric behavior without coming into conflict with electrodynamics."[52]

Even in their final study of the photoeffect, published in 1913, Sommerfeld and Debye could not solve half of the problems that remained.[53] They attempted to extend the analysis from mono-chromatic to "natural" radiation, and this required an appeal to statistical arguments. But despite their effort (one "somewhat tiresome calculation" of an integral extended to nine printed pages) they made no significant progress toward a resolution of the paradoxes. They did, however, calculate an expression for the expected delay time for monochromatic light. The value was "unmeasurably short" but was little different from those derived from Lorentz' equation 7.1. For the green line at 5,460 Å in the mercury spectrum, the expected delay was 8.7×10^{-5} sec, an interval too short to ensure that the light frequency remains well defined at all.[54]

One month after their prediction, the first authoritative determination of a photoeffect delay time was announced.[55] Erich Marx, who had earlier measured the velocity of x-rays, accepted the challenge proposed by Einstein in Karlsruhe. The experiments were done in Leipzig with the professor at the *Technische Staatslehranstalt,* Karl Lichtenecker.[56] They used a rotating mirror to limit the duration of the exposure of a sensitive potassium photocell to light from various sources. Photoemission could be measured for exceedingly low intensities, down to 6×10^{-8} erg/cm²-sec. For the 5,900-Å mercury line, even at the subvisible intensity of 0.5 erg/cm²-sec, they could find no delay in the onset of photo-emission right down to the limit of their equipment's temporal resolution, 1.5×10^{-7} sec. This was less than 1 percent of the value predicted by Sommerfeld and Lorentz as the minimum time needed for accumulation from classical waves. Marx pointed out that the small value could not be explained even if light energy was restricted to a diverging cone.[57] The divergence too quickly reduced the energy density of the rays. Marx referred explicitly only

[52] Debye and Sommerfeld, *AP, 41* (1913), 927, 930.
[53] *Ibid.*
[54] The data came from Buisson and Fabry, *CR, 152* (1911), 1838–41.
[55] There had been several attempts. The most significant were those by Elster and Geitel, *PZ, 13* (1912), 468–76; Ioffe, *München Sb, 43* (1913), 19–37; and E. Meyer and Gerlach, *ASPN Genève, 35* (1913), 398–400.
[56] Marx and Lichtenecker, *AP, 41* (1913), 124–60.
[57] Marx, *AP, 41* (1913), 161–90.

to Thomson's unit hypothesis, but the implication was clear that even Sommerfeld's x-ray and γ-ray models were thus rendered impotent.

Sommerfeld's theory failed in other respects too. It was necessary to leave out radiation damping entirely from the calculation. If it was included, the photoeffect could apparently not occur for any but abnormally high light intensities. Moreover, Sommerfeld and Debye had emphasized in 1911 that theirs was a true resonance model of the photoeffect, in contrast to Einstein's treatment.[58] And Sommerfeld persisted in the belief that experiments would confirm that the photoresponse shows peaks at wavelengths characteristic of the irradiated atoms. In 1913 he called "contrived" the idea that a "great number of independent resonators of all possible [frequencies] are present in the atom."[59]

A certain ambiguity surrounded the question of whether resonant peaks appear in the photoresponse from various materials. Robert Pohl and Peter Pringsheim had found in 1910 that two kinds of photoresponse are superimposed in most experiments.[60] The first is the normal effect, the second a "selective" effect controlled largely by the state of polarization of the light. Before this discovery, it was not realized that carelessness about the incident light – whether it was reflected and if so at what angle – could produce peaks in the measured photoresponse. Afterward, the evidence from the normal effect, although still uncertain, began to converge. Sommerfeld's hopes notwithstanding, it appeared that the maximum electron velocity increases monotonically in response to the frequency.[61] The peaks seemed to be an artifact of the selective photoeffect.

In the meantime, the chief impediment to accurate studies over a large wavelength range in the normal photoeffect – oxidation impurities on the photocathode – had been overcome. Karl Compton and Owen Richardson did experiments in 1912 on cathodes scraped clean and maintained in a vacuum.[62] Although the specific relation of stopping potential to light frequency was

[58] Sommerfeld, *Solvay I*, pp. 355–6.
[59] Debye and Sommerfeld, *AP, 41* (1913), 930.
[60] Pohl and Pringsheim, *VDpG, 12* (1910), 215–28, 249–60, 682–96, 697–710; *PM, 21* (1910), 155–61.
[61] Wheaton, *Photoelectric effect* (1971), chap. 13.
[62] O. Richardson and K. Compton, *PM, 24* (1912), 575–94. Hughes, *PTRS, 212A* (1913), 209–26.

still uncertain in 1913, the new evidence made it even clearer that the maximum photoelectron velocity grows monotonically with frequency.[63] If this was a resonance effect, as Sommerfeld claimed, one had to explain how an essentially infinite number of resonant frequencies are represented in the atom. Sommerfeld had no answer. He held to the faint hope that the evidence was "afflicted with great uncertainty."[64] Thereafter Sommerfeld viewed the photoeffect as a "real difficulty," but he did not publish on the subject again.[65]

THE EQUIVALENCE OF X-RAYS AND LIGHT

Two formerly separate strains of thought became firmly united in the period 1911–13: the theory of light and the theory of x-rays. The net result was to highlight the paradoxes of radiation on a general level. Before light and x-rays were recognized as common forms of radiation, localized properties could be relegated primarily to the "new" radiations: x-rays and γ-rays. About the nature of these rays there was still some uncertainty. But a constellation of events, beginning in 1911, made it clear that light is implicated too. First came Sommerfeld's γ-ray study, written in response to the criticisms of Stark and Meyer, and an attempt at a quantum theory of nonperiodic events. Then Planck offered his quantum theory of light emission, his first to employ the quantum transformation hypothesis. Soon Sommerfeld selected the photoeffect for application of his own generalized quantum principle; it provided the frequency condition that the quantum formalism seemed to require. The reaction to Sommerfeld's theory at the *Naturforscherversammlung* in September led to the first detailed experimental test of photoelectric effect delay times. And at that same meeting, Carl Ramsauer announced that the triggering hypothesis had been virtually abandoned by its creator, Philipp Lenard.

Sommerfeld had already recognized that the characteristic x-rays form a direct extension of the optical frequencies. He concluded that the *Eigenschwingungsdauern,* or oscillation periods of the characteristic x-rays, could be substituted in the quantum transformation relation "to determine [x-ray] energy, from

[63] Pohl and Pringsheim, *VDpG, 15* (1913), 637–44.
[64] Debye and Sommerfeld, *AP, 41* (1913), 928.
[65] *Ibid.,* 873.

the shortest right down to the visible and infra-red."[66] Sommerfeld provided a single quantum condition that applied equally to the periodic and aperiodic x-rays.

In England, even Barkla had begun to refer to the characteristic x-rays as fluorescent rays, implying that they were related to visible light.[67] C. T. R. Wilson presented his cloud-chamber photograph, Figure 6.9, showing dramatically that x-rays produce only a small number of electrons in their flight through a gas. And Bragg, frustrated with the inconsistencies that he had puzzled over for five years, and now nonplussed that ordinary light had to be included among the forms of radiation that act so peculiarly, began to call publicly for a dualistic conception of all forms of radiation.

On the heels of these events of 1911 came the startling news from Germany that x-rays, like light, can interfere. Our next chapter is devoted to an analysis of the impact of this influential discovery. But the finding had a discouraging effect on studies of the nature of radiation. Most physicists took the result as proof that x-rays are periodic waves like light. This was the straightforward conclusion to be drawn from seemingly uncontestable evidence. An alternative was available only to the relatively small number of physicists conversant with the evidence for paradox in the behavior of x-rays, and it was not appealing to contemplate extending those paradoxes to the venerable wave theory of light. There was very little difference in the response between British and German investigators or between mathematical and empirical physicists to the new x-ray evidence. X-ray interference did not create a crisis in radiation theory; what crisis there was existed already. For most physicists, the discovery reduced much of the uncertainty about the nature of x-radiation, and so discouraged continued research on the energy structure of free radiation.

The discovery of x-ray interference had a temporary stimulating effect on empirical studies of absorption of all forms of radiation, as shown by Figure 7.2. But that interest soon peaked and declined steadily in the middle of the decade.[68] Of course, much of the decline was due to the disruptive effect of World War I on research, especially in Britain. But in addition, the very fecundity of crystal

---

[66] Sommerfeld, München *Sb, 41* (1911), 43.
[67] Barkla, *PM, 22* (1911), 396–412. He also spoke of x-ray "line spectra."
[68] Figures are taken from the *Fortschritte der Physik,* abstracting journal of the Deutsche physikalische Gesellschaft.

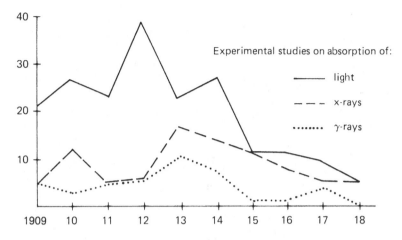

Figure 7.2. Number of experimental studies on absorption of radiation, 1909–1918.

diffraction attracted some of the research that might have been directed to the nature of x-rays. W. H. Bragg, to cite the most significant example, devoted virtually all of his research after 1914 to the analysis of crystals based on an assumption of wavelike x-rays. At the same time, the new methods of x-ray spectroscopy introduced by crystal diffraction encouraged increasing study of the x-ray emission spectra for the analysis of atomic structure. Since neither of these otherwise important research fields contributed directly to an understanding of the nature of radiation or of its quantum absorption, they do not appear in Figure 7.2.

DECLINE OF THEORETICAL INTEREST IN FREE RADIATION.

The attempts of mathematically adept physicists to grapple with the problems of radiation absorption were largely unsuccessful, and all were short-lived. Figure 7.3 traces the number of mathematical studies on the critical problems during the decade from 1909 to 1918.[69] The absolute number of studies is quite small, but

[69] The graph includes all articles concerned with the "theory" or the "nature" of one of the three forms of radiation. It also includes all articles known to treat mathematically the absorption or transformation of radiant energy. The counts come from four sections of the *Fortschritte:* Electricity and Magnetism 10B *Becquerelstrahlen,* and 10C *Röntgenstrahlen,* also 14 *Lichtelektrizität* and Optics 1 *Allgemeines.*

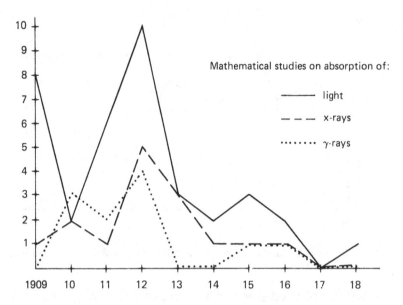

Figure 7.3. Number of mathematical studies on absorption of radiation, 1909–1918.

the marked decrease in interest, which parallels the decline in empirical studies, is clear. Note, in particular, the precipitate drop in the sum of the curves after 1913. There were a few later attempts to solve the riddle of the photoeffect, but the problems seemed to be growing and the momentum established in 1911–12 had been lost.

Even at the peak of interest in absorption phenomena, many of the studies were directed more at resolving questions of atomic structure than at resolving the paradoxes of radiation. Karl Hermann at the *Technische Hochschule* in Berlin had argued in 1912 that no resonance mechanism could explain the release of photoelectrons bound in elliptical orbits. The frequency of revolution changes as energy is added, and, according to Hermann, this would prevent any considerable accumulation of energy.[70] J. J. Thomson considered the photoelectric effect in 1913 when he proposed a contrived atom model that, like the one he had proposed in 1910, offered a continuum of electron frequencies and energies to a passing light wave.[71]

[70] K. Hermann, *VDpG, 14* (1912), 936–45.
[71] J. J. Thomson, *PM, 26* (1913), 792–9.

Interest in questions of atomic structure increased sharply after 1913 when Niels Bohr published his famous theory of atoms.[72] This was the major reason internal to physics for the concomitant decline in research on radiation. The disruption of all research in Europe during World War I played a role, but even so, consideration of radiation theory dropped far behind the treatment of atomic theory. Bohr's theory relied on the anticlassical assumption of discontinuity in both emission and absorption of radiant energy; it bypassed the problem of energy accumulation by postulating instantaneous absorption. It was also as surprisingly successful in its prediction of spectral frequencies and chemical properties as it was unorthodox. The theory was quickly taken very seriously, and it set rigid constraints on the number of possible electron frequencies in the atom as well as on the relationship between electron frequency and orbital velocity. The triggering hypothesis, indeed any resonance model of the photoelectric effect, was effectively rendered impossible.

For almost ten years after the adoption of the Bohr theory, most studies of the photoelectric effect were intended to clarify not the nature of light but the structure of atoms. In 1916, for example, Paul Epstein used the photoeffect to test his hypothesis that hyperbolic orbits followed by ejected photoelectrons are compatible with the Bohr–Sommerfeld model of the hydrogen atom.[73] He found that such orbits were possible but that the atom would have to respond selectively to particular incident frequencies, and by 1916 it was quite clear that no resonant peaks occur. Epstein did not comment on the discrepancy and did not consider the problem of how the incident light can concentrate its entire quantum effect on a single electron. Owen Richardson developed a theory of the photoelectric effect in 1914 that treated only macroscopic quantities and was entirely independent of assumptions about light. The effect was considered to be analogous to the evaporation of molecules from a liquid; the quantum condition became a sort of latent heat of vaporization.[74]

A few younger physicists outside the German and British institutes – Abram Ioffe, Jun Ishiwara, and Mieczesław Wolfke – persisted with Einstein in promoting the advantages of the lightquan-

---

[72] Bohr, *PM, 26* (1913), 1–25. See Heilbron and Kuhn, *HSPS, 1* (1969), 211–90.
[73] Epstein, *AP, 50* (1916), 815–40; *PZ, 17* (1916), 313–16.
[74] O. Richardson, *PM, 27* (1914), 476–88.

tum. But this fell on virtually deaf ears.[75] Influential mathematical physicists besides Einstein avoided the issues almost to a man. Sommerfeld turned to the elucidation of his phase-integral quantum conditions for the Bohr atom, and to detailed examination of optical and x-ray emission spectra. Lorentz undertook no study of the nature of free radiation, except for one study of x-ray impulses to which we shall return in the next chapter. Debye moved steadily toward integration of the new quantum theory of the atom with problems on the larger scale of molecular forces. Planck continued in his attempts to reconcile his thermodynamic interpretation of radiation with quantum theory. Thomson concentrated on new experimental studies of positively charged ions. He was heavily involved in war work after 1914, gave up his directorship of the Cavendish Laboratory in 1918, and effectively stopped publishing research altogether until the early 1920s.

Two who expressed the most continuing interest before 1920 in the structure of radiation were Wilhelm Wien and Max Laue. A year after receiving the Nobel Prize for his work on cavity radiation, Wien gave a series of lectures at Columbia University on "New problems of theoretical physics." He pointed again to the great difficulties that faced the quantum theory of light absorption and showed that Einstein's analysis of fluctuations did not necessarily demand recourse to the lightquantum.[76] But he soon turned almost exclusively to theoretical analysis of emission spectra from positive ions. Laue, as we shall see, was deeply involved with theoretical analysis of x-ray interference. In the second decade of the twentieth century he worked on a thermodynamic treatment of emission and absorption of light and, although unsympathetic to Einstein's lightquantum well into the 1920s, he was unusual in his belief that a statistical solution could resolve even interference phenomena.[77]

---

[75] "That [Einstein] may occasionally have missed the mark in his speculations, as for example with his hypothesis of lightquanta, ought not be held too much against him, for it is impossible to introduce new ideas, even in the exact sciences, without taking risk." These were the words of Planck, Rubens, Nernst and Warburg to the Kultusminister in their attempt to have the non-German Einstein appointed to the Kaiser Wilhelm Gesellschaft and the Prussian Academy of Sciences in 1913. Quoted in Kahan, *Archives internationales d'histoire des sciences, 15* (1962), 337–42.

[76] Wien, *Neuere Probleme* (1913); WAS *Journal, 3* (1913), 273–84; *AP, 46* (1915), 749–52.

[77] Laue, *VdPG, 19* (1917), 19–21.

Einstein himself was the most conspicuous exception among mathematical physicists; in 1916 he wrote a brilliant restatement of the problems of free radiation and offered a refined version of his statistical solution.[78] But for most physicists, the problems were too demanding and the temptation of the successes of the Bohr theory too great. No resolution to the paradoxes of radiation was found by physicists who sought a consistent mathematical representation of nature. As we shall see in the next part of this study, the important steps toward solutions in the decade from 1911 to 1921 were taken not by mathematical theorists but by empiricists who plumbed the manifold consequences of x-ray diffraction by crystals.

[78] Einstein, Zürich *Mitteilungen, 18* (1916), 47–62; *PZ, 18* (1917), 121–8.

PART IV

Interference of x-rays and the corroboration
of paradox 1912 – 1922

8

Origins of x-ray spectroscopy

[X-rays] are a kind of wave with properties no wave has
any business to have.[1]

In the spring of 1912 two research assistants in Munich directed an
x-ray beam through a crystal and found that the beam was re-
formed into a well-defined interference pattern. The property most
characteristic of periodic waves – their ability to interfere – is
shared by the x-rays. Max Laue, the man chiefly responsible for the
discovery, thought he had found proof that characteristic secondary x-rays from the crystal are periodic waves. But H. A. Lorentz
quickly pointed out that impulses should interfere too. He showed,
in a tour de force argument, that the accepted square pulse is an
impossible representation of x-rays. William Henry Bragg and his
son concluded that the interference maxima could, indeed, be due
to irregular x-ray pulses. But, as such, x-ray impulses were not
different from ordinary white light. They supported this claim with
a new technique of crystal analysis fully analogous to ordinary
optical spectroscopy.

Crystal diffraction provided a new tool for the analysis of x-rays.
Pushed furthest by Henry Moseley and Charles Darwin, the technique soon showed that some x-rays comprise periodic wave trains
of great length. The extremely sharp angular resolution of observed x-ray interference maxima indicated beyond doubt that
x-rays are no different, except in frequency, from ordinary light.
Rutherford soon extended the technique to the γ-rays. Not only
could one isolate characteristic γ-rays, he believed, one could
show, with some effort, that they interfere too.

The successful integration of the new spectroscopy with the
Bohr atom came, as had x-ray diffraction, from Sommerfeld's

[1] Moseley to his mother, 21 January 1913. Heilbron, *Moseley,* (1974), 199–200.

Munich. Walter Kossel suggested in 1915 that x-rays are the emission product of the redistribution of electrons in the close-lying orbits of the atom. This gave theoretical justification to what empiricists had known since 1913: x-rays are no more or less than high-frequency light. Rutherford analyzed the γ-ray results in the context of his recently formulated nuclear model of the atom. Certain groupings of electrons in the superstructure of the atom are responsible, he suggested incorrectly, for the characteristic high-frequency vibrations that produce monochromatic γ-rays. Although the final justification of γ-rays as the product of transitions within the nucleus did not come until 1922, the extension of the electromagnetic spectrum to include x-rays and γ-rays was essentially complete by the start of the Great War.

X-RAYS CAN INTERFERE

Max Laue, who replaced Peter Debye as Sommerfeld's theoretical assistant in 1912, thought that the periodic secondary x-rays characteristic of their scattering element might show interference effects if a suitable diffracting device could be found. It was his inspiration to suggest that the regularly spaced atoms presumed to constitute crystals would yield their characteristic x-rays from a regular array of points; this might allow observation of interference. The experimental test was performed by Walther Friedrich and Paul Knipping, both working at Sommerfeld's Institute for Theoretical Physics in Munich. The work was apparently done over Sommerfeld's objections. The resulting photographs, one of which is reproduced as Figure 8.1, vindicated Laue, and the success was announced to the world in June 1912.[2]

Sommerfeld's skepticism is easily understood. As Forman has shown, Sommerfeld was not likely concerned that thermal agitation of the lattice points would disrupt possible interference.[3] He had found long before that x-ray impulses seem not to interfere; no fringes appear at an x-ray shadow edge. Sommerfeld had other serious objections. Only a month before Laue's discovery, he had published a detailed study of x-ray diffraction.[4] Sommerfeld re-

[2] Friedrich, Knipping, and Laue, München *Sb, 42* (1912), 303–22. Reprinted with additional notes in *AP, 41* (1913), 971–88.
[3] Forman, *AHES, 6* (1969), 38–71.
[4] Sommerfeld, *AP, 38* (1912), 473–506.

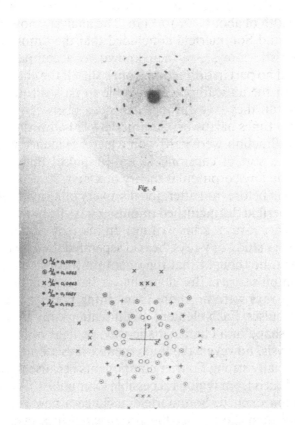

Figure 8.1. One of the original x-ray interference photographs taken in 1912 by Friedrich and Knipping at Laue's suggestion. The crystal is zinc sulfide; the x-ray beam enters parallel to the optic axis. At the bottom are Laue's predicted positions for interference spots from five homogeneous x-rays. His best estimate of the crystal spacing α was 3.38×10^{-8} cm; the λ range was $1.3-5.2 \times 10^{-9}$ cm. [Friedrich, Knipping, and Laue, München *Sb, 42* (1912), Tafel II.]

sponded shortly after the Solvay Congress to Robert Pohl and Bernhard Walter, who had been unable to find evidence of diffraction in an x-ray photograph taken under conditions similar to those originally employed by Wind and Haga.[5] Peter Koch developed an optical photometer to analyze the photograph quantitatively for Sommerfeld.[6] The resulting intensity contours allowed

[5] Walter and Pohl, *AP, 25* (1908), 715–24; *AP, 29* (1909), 331–54.
[6] Koch, *AP, 38* (1912), 507–22.

Sommerfeld to reach the gratifying conclusion that the new photograph, like the old, gave evidence of diffraction of aperiodic impulses with a width of about 4×10^{-9} cm. The analysis showed no sign of fringes, and Sommerfeld concluded that the "more periodic" characteristic x-rays, by then known to accompany the impulses, played no part in diffraction from a slit.[7] If the characteristic x-rays from the anticathode are unable to pass through the 2-μm slit, how can they pass through the spaces between crystal atoms, some 10^4 times narrower? Sommerfeld did admit that the data favoring diffraction were hard won; clearer evidence would, he said, "form a sort of capstone to the [impulse] theory and definitely exclude any corpuscular theory of x-rays."[8]

It was thus well before, not after, the discovery of x-ray interference that Sommerfeld distinguished impulse x-rays from the more periodic secondary x-rays in his writings. In his 1909 response to Stark and his 1911 study of γ-rays, he had separated the two forms of x-rays; at first, he thought that the quantum theory could not apply to the impulse part. But it might yet have been that the homogeneous x-rays were no more than strong Fourier components of the impulses. Each element might scatter x-ray impulses and alter their shape in a characteristic way. In light of his 1912 diffraction analysis, however, the fluorescent x-rays seemed to be more than especially strong Fourier components because they are apparently excluded from regions accessible to impulses. To make the emancipation explicit, Sommerfeld assigned a new name to what had, until then, been only the *Bremsanteil* in x-rays. The inhomogeneous and unpolarized *Bremsstrahlung* thus assumed a status separate from the homogeneous, partially polarized, and temporally extended fluorescent x-rays.

It was only the homogeneous x-rays that concerned Laue. Following Sommerfeld, he thought that *Bremsstrahlung* impulses of varying widths, diffracted from a confused set of crystal planes oriented in various directions, could produce only a diffuse blackening of the photo plate. Because of this, Laue expected to see only the superimposed effect of all secondary rays. The copper sulfate

---

[7] Sommerfeld, *AP, 38* (1912), 483.
[8] *Ibid.*, 474. Pohl, *Physik der Röntgenstrahlen* (1912), appeared in the summer of 1912. Before the final proofs were in, Pohl added a short section at the end on the new evidence for x-ray interference. The rest of the book is based on the impulse hypothesis.

crystal used in the first experiments was chosen carefully to maximize the homogeneity of characteristic x-rays, but it was of sufficiently low density so as not to absorb x-rays inordinately.[9] For a full year, Laue persisted in his belief that only fluorescent x-rays exhibit interference. What we now take as the distinguishing feature of x-ray diffraction – that the crystal imposes its own periodicity on the incident radiation – was, at first, unrecognized by Laue.

Problems quickly arose for his interpretation. Laue could calculate, with some ambiguity, how many specific "wavelengths" were responsible for the interference spots. He listed them beneath the photograph in Figure 8.1. When he encountered difficulty in attributing these solely to the fluorescent rays from the crystal, he raised the possibility that some of them came from the anticathode in the x-ray tube. If Sommerfeld had been wrong about the existence of interference, perhaps he was also wrong about the absence of fluorescent rays from the anticathode. Laue concluded: "that the radiation emitted by the crystal has a wave character is proven by the sharpness of the intensity maxima." But it was still unclear "whether the periodic radiation arises first in the crystal by fluorescence or whether it already is present alongside the impulses in the primary radiation and is only separated by the crystal."[10]

In his analysis of the interference photographs the following month, Laue concluded that the five responsible wavelengths ranged from 1.27 to 4.83 $\times$ 10^{-9} cm.[11] But the method used to obtain these values did not determine them uniquely. Some spots that should have appeared were conspicuously absent. Laue also had difficulty explaining why an intense interference pattern was cast by diamond, when carbon, its only elemental constituent, had been shown by Barkla to be a weak emitter of fluorescent x-rays. Even more perplexing was the evidence presented by Friedrich and Knipping that the absorption coefficient of some of the interfering rays from zinc was markedly lower than that found for the characteristic x-rays from zinc.[12] A year later, in a *Zusatz* added to the

[9] Friedrich, Knipping, and Laue, München *Sb, 42* (1912), 314. Forman, *AHES, 6* (1969), 63.

[10] Friedrich, Knipping, and Laue, München *Sb, 42* (1912), 310–11.

[11] Laue, München *Sb, 42* (1912), 363–73.

[12] Friedrich, Knipping, and Laue, München *Sb, 42* (1912), 331–2. They obtained 3.84 cm^{-1} in aluminum. Barkla and Sadler, *PM, 14* (1907), 413, had measured μ for zinc x-rays passed through aluminum as 96.0 cm^{-1}.

reprinting of the first paper, Laue still concluded that the role of the crystal is only to select certain well-defined wavelengths from the primary incident beam.[13]

The discovery of x-ray interference was not without influence on the theory of the *Bremsstrahlung* impulses. A few months after x-ray interference had been announced, H. A. Lorentz, who had never written on x-rays before, recognized that, in principle, impulses should show interference too. In December 1912 he offered an elegant argument that cast serious doubt on the accuracy of the square pulse as an approximation for the *Bremsstrahlung*.[14] The intensity of scattered impulses at point *B* due to a primary beam striking matter at point *A* consists of more than simply the secondary beam directed along line *AB*. According to Huygens' principle, the effect at *B* must take into account the elementary spherical impulses arising from every point that the scattered pulse passes. In the case of a rapidly oscillating vibration such as light, these elementary wavelets largely cancel out. But Lorentz realized that Sommerfeld's singly directed pulse shape did not allow this cancellation. A singly directed pulse is one in which the displacement of the electric field is in one direction only, transverse to the direction of propagation of the pulse, as shown in Figure 8.2. Lorentz calculated what the net effect of the scattered square impulse would be at point *B*.

He invoked the well-known rule that the sum of the distances from each focus to a point on an ellipse is constant. One must imagine space broken up into a series of nested ellipses of revolution around *A* and *B* as foci. See Figure 8.3, where one such shell is indicated in cross section. The secondary impulse from *A* will pass through different parts of a given shell at different instants, but the Huygens wavelets from each differential element on a given ellipse all converge at the same instant on point *B*. Therefore, if the impulses are singly directed, a progressively weaker constructive maximum at *B* is contributed by each of an essentially infinite number of shells. Huygens' principle applies even without

[13] Laue, *AP, 41* (1913), 989–1002.
[14] Lorentz, Amsterdam *Verslag, 21* (1913), 911–23.

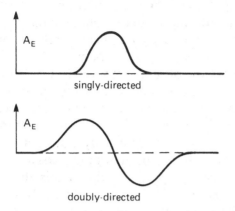

Figure 8.2. Comparison of singly and doubly directed impulses.

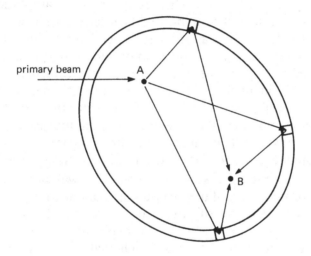

Figure 8.3. Lorentz' ellipsoid of revolution used to show that singly directed x-ray impulses cannot be valid.

material scattering centers in the surrounding space, but the actual spherical impulses that are induced at each surrounding gas molecule show the same interference effects at *B*. A conservative estimate was that the intensity due to the constructive interference of secondary impulses radiated from ambient air would exceed that amount radiated directly through *B* by a factor of 95! Barkla's repeated measurements since 1904 on the intensity of scattered x-rays showed no hint of this massive effect.

This was not the only problem. Lorentz showed that in a gas the scattered intensity at B should vary with the square of the number of scattering molecules. This directly conflicted with empirical evidence that the intensity varies as the first power. Furthermore, the temporal distinctness of the primary impulse would disappear entirely in the secondary because secondary impulses arise in each of a tremendous number of electrons spread over macroscopic distances. The net effect at B of a single primary impulse would last for an appreciable time. For a scattering body 1 cm across, the impulse seen 90° from the primary beam would persist 3×10^4 times as long as the primary impulse.[15] Add to this the rapid succession of primary impulses in any discernible x-ray beam and the net disturbance at B "would last continuously."[16]

Lorentz offered a way out of the dilemma. If the impulses are made doubly directed, so that the net excursion in amplitude is balanced on both sides of equilibrium, the interfering effect can be reduced to zero. As long as the net integral over the impulse vanishes, all is saved. But this solution brought into question the long-accepted mechanism of x-ray stimulation. How can a speeding electron, brought to rest against the anticathode, create a doubly directed impulse? "One can scarcely refrain from thinking," Lorentz said, that the "electric force in the excited impulses . . . must chiefly lie in the same direction as would be the case if the stoppage of the particle occurred in a straight line."[17]

Lorentz had touched a sensitive nerve. His was not the first published consideration of the precise shape of the x-ray impulse, but it was the first to clarify the issues. Stokes had assumed a doubly directed impulse in 1897, having explicitly rejected a singly directed one. But he published a sketch that may have misled others. Thomson adopted the square pulse later used by Sommerfeld. Stokes had merely mentioned in passing that "the dynamical theory [of diffraction] shows that [the square pulse] is not possible."[18] He very likely foresaw what Lorentz later made explicit. To

---

[15] Lorentz found that the duration of the effect at B of a single pulse is given by $2l \sin (\theta/2)/c$, where l is the linear dimension of the scattering body and θ is the angle line AB makes with the extension of the incident beam. At $\theta \simeq 90°$, and l taken to be 1 cm, the duration is about 5×10^{-11} sec. Wien had estimated that a typical beam carries 7×10^{14} pulses per second, each of them being some 1.4×10^{-15} sec at most. Each secondary is thus 3×10^4 longer than its primary.

[16] Lorentz, Amsterdam *Verslag, 21* (1913), 920.

[17] *Ibid.*, 923.

[18] Stokes, *Papers, 5* (1905), 266.

Figure 8.4. A singly directed impulse modulated by a doubly directed one.

my knowledge, the only person aware of the potential ambiguity before 1912 was Paul Ehrenfest, who wondered privately in 1905 whether the pulse might not be a step function.[19] When Wien derived a value for pulse width from energy considerations in that same year, it was only 1 percent of the value that Sommerfeld found from Wind and Haga's photograph. J. D. van der Waals, Jr., raised the possibility that Wien had measured only part of an irregular impulse produced by many zigzag motions of the electron as it stopped.[20] Wien replied accurately that his analysis was independent of a zigzag assumption.[21] In doing so, he missed the importance of the argument. For the zigzag is precisely what Lorentz used to extricate the impulse hypothesis of x-rays from the tight corner in which he found it.

Lorentz's solution depended on the ways in which impulses are scattered by matter. The singly directed impulse is preferentially scattered at long wavelengths; this is a consequence of the fact, already remarked by Sommerfeld, that its Fourier spectrum peaks at zero frequency. Physically, this implies that in passing through matter, singly directed impulses lose energy to atoms much more readily than do doubly directed impulses. A zigzag motion of the electron in slowing down will produce rapid fluctuations in the amplitude of the pulse, yet the impulse itself will remain singly directed because the net electron deceleration is in one direction only. A singly directed impulse results, modulated by in the simplest case, a doubly directed impulse, as shown in Figure 8.4. As soon as the net disturbance passes through air or the glass wall of the x-ray tube, the singly directed impulse is absorbed, leaving only the doubly directed part accessible to measurement.

[19] Ehrenfest notebook 1–06, *second* entry number 508, dated 31 December 1905. Ehrenfest archive.
[20] van der Waals, Jr., *AP, 22* (1907), 603–5.
[21] Wien, *AP, 22* (1907), 793–7.

THE BRITISH INTERPRETATION OF
X-RAY INTERFERENCE

As we noted in Chapter 6, W. H. Bragg had considerably softened his opposition to periodic x-rays by late 1912. He began to develop a dualistic concept of both x-rays and γ-rays in which a wavelike and a particlelike component exist side by side in the beam. The idea owed something to remarks he had made as early as 1907 when confronted by Marx's velocity measurement for x-rays. By 1912, Bragg was convinced that visible light as well was implicated in this duality. He expressed his dualistic speculations at the 1912 meeting of the British Association even before he assimilated Laue's interference results. He argued for a theory that would allow "a corpuscular and a wave theory of light at the same time."[22] The conceptual difficulty in this theory he thought was "one of our own making" for having in the past so rigidly demarcated the concepts of particle and wave.

Thus, even when confronted with the evidence for interference, Bragg's position could still be defended. It is quite true that he first thought of a corpuscular explanation of the x-ray interference photographs; the directions toward the spots might coincide with the directions down the "channels" between atom planes in the crystal.[23] But to emphasize this claim is to miss the dualistic stance of Bragg's revised position of 1911. In the larger context of his thought, it was hardly more than an attempt to see if the expected corpuscular interpretation was forthcoming for what clearly already had a wave interpretation. It was less particles *versus* waves as particles *and* waves: Bragg actively sought "one theory which possesses the capacities of both."[24] He was far from opposed to the wave view, but he did have doubts about attributing the entire set of interference spots to a few well-defined wavelengths in the primary x-ray beam. Even Barkla expressed doubts regarding Laue's methods.[25]

Bragg's son, William Lawrence Bragg, under the influence of his

---

[22] W. H. Bragg, *Nature, 90* (1913), 529–32, 557–60. He sought a "scheme of greater comprehensiveness, under which the light wave and corpuscular x-rays may appear as the extreme presentments of some general effect." Bragg, *Studies in radioactivity* (1912), 192–3.

[23] W. H. Bragg, *Nature, 90* (1912), 219.

[24] *Ibid.,* 361.

[25] Barkla to Rutherford, 29 October 1912. Rutherford papers. Heilbron, *Moseley* (1974), 62–3.

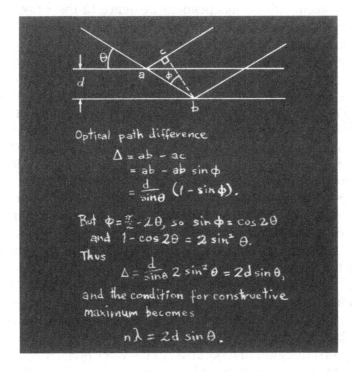

Figure 8.5. Derivation of the Bragg condition for x-ray interference in crystals.

father and of the Cambridge pulse theorists, invoked the British theory of the diffraction grating to gain an intermediate solution. Gouy, Rayleigh, and Schuster had concluded that a diffraction grating imposes its own regularity on the irregular impulses thought to constitute white light.[26] The replicated impulses then interfere with one another, reinforcing the Fourier components that make up the impulse. Each component frequency appears at its own characteristic angle relative to the incident beam. The younger Bragg recognized that Laue had found the x-ray analog to this optical process.

However irregular and aperiodic the incident impulses may be, the regularity of the crystal forces the scattered impulses to follow one another at regular intervals. The effective separation is $2d \sin \theta$ where d is the separation of the crystal planes and θ is the glancing angle of the x-ray beam against the face of the crystal (see Figure 8.5). One does not need to have Laue's few monochromatic wave-

[26] See Chapter 2.

lengths, each with an independent existence in the x-ray beam. The crystal manufactures them by the natural process analogous to Fourier analysis of the pulses.

Before the beginning of December, the younger Bragg had obtained clear interference photographs from x-rays reflected off mica at glancing angles.[27] By the time that this first demonstration was given that x-rays can reflect, it had taken on even greater value as evidence that x-rays must be periodic waves. Whenever the incident angle θ is such that the reflected impulses follow one another spaced by a whole multiple of a component wavelength, that frequency is constructively reinforced. An ideal crystal with only one effective spacing (any real crystal, of course, contains many effective spacings) should produce an accurate Fourier spectrum of the impulses thrown at it. The new approach was more than reminiscent of optical spectroscopy. The earliest attempts to chart the variation in x-ray intensity as a function of angle were done using modified optical spectroscopes: An x-ray tube and a collimating slit replaced the light source, and a crystal replaced the grating. For the most sensitive of these early surveys, an ionization chamber replaced the photographic plate.

The first question that occurred to the elder Bragg was whether the paradoxes of quality and quantity apply to the interfering x-rays. Could these monochromatic x-ray waves pass on abnormally high velocities to electrons? Or was that property limited to the nonperiodic impulse x-rays? He joined his son in the investigation, and in January 1913 wrote to Ernest Rutherford to report that indeed "the ray travels from point to point like a corpuscle [yet] the disposition of the lines of travel is governed by a wave theory. Seems pretty hard to explain, but that surely is how it stands at present."[28]

In April the Braggs were rewarded by finding distinct interference peaks at sharply defined angles.[29] It was most remarkable that

[27] W. L. Bragg, *PCPS, 17* (1912), 43–57. The younger Bragg credited C. T. R. Wilson's lectures on the analysis of white light as having influenced his reformulation of crystal diffraction analysis. *Start of x-ray analysis* (1967), 3.

[28] W. H. Bragg to Rutherford, 9 and 18 January 1913. Rutherford papers. Heilbron, *Moseley* (1974), 74.

[29] W. H. Bragg to Rutherford, 10 March 1913. Heilbron, *Moseley* (1974), 75–6. W. H. and W. L. Bragg, *PRS, 88A* (1913), 428–38. Three days before the Braggs' announcement, Julius Herweg at Greifswald reported a similar success, *PZ, 14* (1913), 417–20.

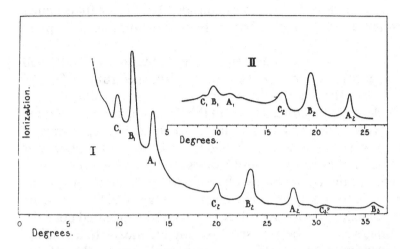

Figure 8.6. Bragg's evidence for multiple orders of x-ray spectra. [Bragg and Bragg, *PRS, 88A* (1913), 413.]

the form and spacing of the peaks did not change appreciably when different crystals were used to diffract the x-rays. The same pattern, shown in Figure 8.6, was found at just slightly different angles for the x-rays reflected from rock salt, iron pyrite, zincblende, potassium ferrocyanide, potassium bichromate, calcite, quartz, and sodium ammonium tartrate. At first, Bragg tried to discern a harmonic progression in the angle of the peaks. The distribution of the peaks was remarkably like that produced by visible light on one side of an optical grating. In each case three peaks, *A, B,* and *C,* were accompanied by three more at proportionately greater angles. The distribution looked like multiple orders generated by optical emission spectra. The new spectroscopy had been suitably rewarded with x-ray emission lines.

Bragg measured the absorption coefficient for each of the three line sets. He found that each set (e.g., C_1, C_2, C_3) has a distinct penetrability independent of the crystal. He had found "three sets of homogeneous rays," and he concluded, "since the reflection angle of each set of rays is so sharply defined, the waves must occur in trains of great length. A succession of irregularly spaced pulses could not give the observed effect."[30] Once they had estimated the spacing of atom planes in the crystals, the Braggs gave the first unambiguous estimate of the wavelength of a homogeneous x-ray.

[30] W. H. and W. L. Bragg, *PRS, 88A* (1913), 436.

The value, 1.78×10^{-8} cm, was considerably longer than Laue's hypothetical wavelengths, but in line with Sommerfeld's pulse width estimates.

By 1913, x-ray emission lines were not unexpected; Barkla had spoken in 1911 of fluorescent x-rays as a "line spectrum."[31] But the sharpness of the angular definition was unexpected. Even so, the 1/10th degree angular resolution of the Bragg apparatus prevented the full realization of homogeneity. The conflict between x-rays of definite wavelength and the paradoxes of x-ray absorption was demonstrated clearly enough to Bragg. "The problem remains," he said, "to discover how two hypotheses so different in appearance can be so closely linked together."[32] Bragg's call for a dualistic theory of light grew louder.[33] But no resolution appeared possible. The Braggs saw in the new spectroscopy the means to a detailed understanding of crystal structure. From mid-1913 on, their efforts were directed primarily to this end. They did not continue their investigation of the nature of x-rays, and for that reason their otherwise commendable work ceases to be of immediate concern to us.

HENRY MOSELEY AND THE ATOM

Corroboration of the Braggs' result was immediately forthcoming from Henry Moseley and Charles G. Darwin in Manchester. Moseley was, at the time, physics demonstrator for Rutherford, to whom he had come from Oxford to learn research methods in radioactivity.[34] His first study there was designed to corroborate Rutherford's contention that only one β-particle is emitted from each radioactive decaying atom.[35] But when Moseley heard of Laue's spots, he shifted interests. "Some Germans," he wrote his mother in October, "have recently got wonderful results by passing

[31] Barkla and Nicol, *PPSL, 14* (1911), 9–17.
[32] W. H. Bragg and W. L. Bragg, *PRS, 88A* (1913), 436.
[33] W. H. Bragg to Richardson, 20 May 1913. Richardson papers. The Braggs thought Laue's methods were "very complicated" and "unwieldy to handle." The reflection technique "made it possible to advance." W. H. and W. L. Bragg, *X rays and crystal structure* (1915), vii, 10.
[34] Heilbron, *Moseley* (1974).
[35] Moseley and Makower, *PM, 23* (1912), 203–10. Moseley, *PRS, 87A* (1912), 230–55.

X-rays through crystals and then photographing them, and I want to see if the same results are to be found with γ-rays."[36]

In preparation for his "thorough frontal attack," he first convinced Rutherford that x-rays were a worthwhile experimental subject. No previous work on x-rays had been done at Manchester. Next, he tried to clear up the more confusing aspects of Laue's discussion. Following a course parallel to that of William L. Bragg, including recourse to the same theory of diffraction of white light, Moseley arrived, step by step, at almost the same results as the Braggs. Like the Braggs, he and Darwin tested first whether the ionizing properties of the interfering x-rays were similar to those of unscattered x-rays.[37] Next, the pair retrieved a large optical spectrograph from the stores at the Manchester laboratory for a systematic survey of the intensity of scattered x-rays as a function of angle. Moseley and Darwin had already plotted $I(\theta)$ before the discovery of lines by the Braggs. Their equipment, and some inexperience, caused them to miss the emission lines entirely. The spectrograph was accurate to 1 min of arc, six times finer than that of the Braggs. In breaking up the continuum of angles for the survey, they had passed right over the extremely sharp lines.

But once informed of the peaks, they put the high resolution to good use, showing quickly that two of the three lines are doublets and that all are superimposed on a continuous background.[38] Then, realizing that the absorption coefficients were very close to those long known from Barkla's work, Moseley identified the peaks as the K and L series of secondary x-rays from the platinum anticathode of the x-ray tube. Soon Bragg was corroborating *their* results for tungsten and nickel.[39] As Moseley wrote to his mother, "there is here a whole new branch of spectroscopy, which is sure to tell one much about the nature of an atom."[40]

From the angular width of an interference peak, Moseley and Darwin tried to estimate the length of the wave train that produced it. They were unable to determine the width precisely because they

---

[36] Moseley to his mother, 10 October [1912] and 14 October [1912]. Heilbron, *Moseley* (1974), 193–4.

[37] Moseley and Darwin, *Nature, 90* (1913), 594.

[38] Moseley and Darwin, *PM, 26* (1913), 210–32.

[39] W. H. Bragg, *PRS, 89A* (1913), 246–8.

[40] Moseley to his mother, [18 May 1913]. Heilbron, *Moseley* (1974), 204–5.

did not know the effective size of the focus spot on the anticathode from which the x-rays came, but they could show that the width of a line did not exceed 5 min of arc. This upper limit ensured that the interfering rays were "very nearly monochromatic."[41] They argued from this that the wave trains must be at least fifty oscillations long, but were willing to go much further based on a purely theoretical argument. If the radiation comes from a classically oscillating electron, damped by the loss of radiated energy, its amplitude could drop to half its original value in not less than 1,000 wavelengths![42]

In no sense could it be maintained any longer that x-rays differed from ordinary light. The characteristic secondary x-rays were fully as monochromatic as visible colors. The *Bremsstrahlung* impulses constituted an exact analogy to polychromatic white light; indeed, Moseley and Darwin referred to the continuous x-ray background as *white* x-rays. They offered a qualitative analysis of the background x-ray spectrum, interpreting it as the superposition of multiple orders of Fourier spectra of the *Bremsstrahlung* impulses.

The conflict between Laue's and Moseley's views on the nature of interference was one that had arisen before. Like Lord Rayleigh before him, Moseley thought that impulses are separated into Fourier component waves by the grating; like Stokes, Laue did not. But however one chose to visualize the action of the grating, there could be no doubt that x-rays were neither more nor less than simple high-frequency light. Whatever the ambiguity that clouded the relationship of the periodic to the aperiodic x-rays, it was an ambiguity intimately shared by visible light. Whatever the resolution sought for the paradoxes of the new radiations, it was now clear that it had to apply to visible light as well.

As a test of spatial localization of the constructively interfering x-rays, Moseley and Darwin estimated how many crystal

---

[41] Moseley and Darwin, *PM*, 26 (1913), 228.

[42] They solved the equation of motion for a heavily damped electron and found that the envelope of the oscillating displacement decreases exponentially with a decay constant inversely proportional to the square of the wavelength. At an intensity maximum of the x-ray spectrum, 1.397 Å, half-amplitude is reached after 2,680 $\sqrt[3]{k}$ oscillations, where k is a constant dependent on the crystal type and ranges from 0.125 to 1.0. Thus, at least 1,000 oscillations are needed to reach half-amplitude.

"grooves" were involved in producing the beams of great angular resolution they measured. "In each reflecting plane," they reported, "an area containing at least a million atoms must have been concerned in producing so perfect a reflexion."[43] This corollary of the paradox of quantity thus added to the problem that Moseley called "this most mysterious property of energy."[44]

As is well known, Moseley went on in the next year to a systematic survey of the K and L radiations induced in a large variety of materials, and found a linear relation between the square of the x-ray frequency and the atomic number of the element. As Heilbron points out, Moseley could scarcely have avoided the interpretation of his results in light of atomic theory. While he was in Manchester, Niels Bohr was also there, bringing to fruition the first statement of his atomic theory.[45] Moseley was thereby diverted from his original interest in the interference of γ-rays, and from attempts to resolve the paradoxes of radiation. He and Darwin had promised a closer analysis of the angular resolution of x-ray lines, "as the matter is of great theoretical interest."[46] But they did not pursue the subject. For the remainder of his short life, Moseley joined the increasing number of physicists who abandoned the unrewarding landscape of radiation theory for the greener pastures of the Bohr atom.

CONTINENTAL REACTION TO
CRYSTAL INTERFERENCE

Because of the uncertainties of Laue's initial interpretation, the first reaction on the Continent to the discovery of x-ray interference was mixed. Nonetheless, by the fall of 1913, virtually all physicists believed that crystal diffraction had indisputably demonstrated that x-rays are equivalent to ordinary light. To most, this meant that x-rays, like visible light, consist of periodic electromagnetic oscillations. Only a few physicists, accustomed already to the discontinuous properties of x-rays, began to ask if light should not

---

[43] Moseley and Darwin, *PM, 26* (1913), 228.
[44] Moseley to his sister, [2 February 1913]. Heilbron, *Moseley* (1974), 200–1.
[45] Bohr, *PM, 26* (1913), 1–25.
[46] Moseley and Darwin, *PM, 26* (1913), 228.

share those unwavelike characteristics. Interest in x-ray research grew by a factor of three from 1911 to 1913.[47]

Whatever the reservations Sommerfeld held before the discovery of x-ray interference, they vanished afterward. He repeatedly spoke on x-rays before general audiences in the following years, using the Laue result as the central theme of his talk. In one of the first of these talks, delivered in April 1913 to the German Society for the Advancement of Instruction in Mathematics and Natural Science, he explained that "impulses and waves are both determined by Maxwell's equations." He saw no conflict in the existence, side by side, of both impulses and periodic waves in x-rays. It was nothing like the conflict "between the wave theory and corpuscular theory of light."[48] He also described Lorentz' concerns about the form of the *Bremsstrahlung* pulses, a rather fine point for his audience. He was intensely interested in that analysis, and had already begun an extension of Lorentz' work that he published two years later. Although a full explanation of the origin of the periodic fluorescent x-rays was still lacking, "one thing can be discerned," he said; "the theory is on the right track."[49]

Stark's response to crystal diffraction of x-rays was more guarded. He had been carrying out experiments on the mechanical effects he thought should be produced in crystals by impacts of α- and canal-ray particles.[50] He thought he had found evidence that the particles are preferentially deflected along the "avenues" of the crystal. In a paper published shortly after Laue's announcement, Stark claimed to have come already to the conclusion that the same behavior would be shown by the x-ray lightquanta. "It is remarkable," he added gratuitously, "that I had come to these views on the basis of the light-cell hypothesis even before I knew of the experiments . . . taken as evidence for 'interference' of x-rays."[51]

---

[47] The *Beiblätter zu den Annalen der Physik* lists approximately twenty-five papers on x-rays in each year from 1907 to 1910. The year 1911 has only twelve. Thereafter, the numbers rise rapidly: 33, 51, 64, 61, and 63 in 1916.

[48] Sommerfeld, *Naturwissenschaften, 1* (1913), 708. Sommerfeld spoke frequently on the new experiments on x-ray interference in the following years. *Münchener medizinische Wochenschrift, 62* (1915), 1424–30; *63* (1916), 458–60.

[49] Sommerfeld, *Naturwissenschaften, 1* (1913), 706.

[50] Stark and Wendt, *AP, 38* (1912), 941–57.

[51] Stark, *PZ, 13* (1912), 975.

Stark's inflated claims did not lead him to attempt a reconciliation of lightquanta with interference once it became clear that more was involved than channeling of particles. By mid-1913 he accepted the Braggs' explanation of crystal diffraction.[52] Subsequently, he renounced the lightquantum. Interference, he said, demands that lightquanta extend along the direction of propagation, and thus are not localized. If, on the other hand, lightquanta are localized from front to back, their spectral lines should be conspicuously broadened in an electric field.[53] Stark would win the Nobel Prize in physics in 1919 for his experimental researches on line broadening in electric fields. Yet in 1912 he could find no spectral evidence that lightquanta are localized. Soon thereafter, Stark, the first supporter of the lightquantum hypothesis after Einstein, altogether dropped the nature of radiation as a topic of research.

Laue answered Stark's early claims, and in the process showed that the Bragg reflection law, $n\lambda = 2d \sin \theta$, is a special case of his own method. It applies, he said, when the interfering rays are "so spectrally inhomogeneous as to be incapable of interfering after a difference in [optical path] of only a few wavelengths."[54] He still thought that the spots were due to a few homogeneous component x-rays. The Braggs, he felt, were pursuing only a special case of his more general analysis.

Laue concluded with the frank admission that the effect on the interference maxima of the thermal agitation of lattice points was "less than any expectation," but he was unable to explain why. The lattice point vibration was treated in detail in a pair of papers by Debye. Sommerfeld's former assistant had just finished the joint paper on the photoelectric effect, discussed in Chapter 7. Debye gave an influential talk on the quantum theory of specific heats and heat conduction at the 1913 Wolfskehl lecture series in Göttingen.[55] The theory, introduced by Einstein in 1907, became the first subject after cavity radiation to excite widespread interest in the quantum theory.[56] Debye focused on Planck's remarkable prediction of a "zero-point" energy of atoms at absolute zero

[52] Stark, *PZ, 14* (1913), 319–21.
[53] Stark, *VDpG, 16* (1914), 304–6.
[54] Laue, *PZ, 14* (1913), 422.
[55] Debye, *Kinetische Theorie der Materie* (1914), 17–60.
[56] Klein, *Science, 148* (1965), 173–80. Kuhn, *Black-body theory* (1978).

temperature. He turned to the thermal influence on crystal diffraction as a possible source of corroborating experimental data.

In July 1913 Debye concluded that thermal motion would have little or no observable effect on the definition of the Laue spots.[57] He calculated the expected decrease in intensity of the interference spots located at increasing angles from the beam direction. This relation depends on the effective wavelength of the x-rays and allowed Debye to derive values for x-ray wavelengths independently of the crystal parameters required by both Laue and Bragg.[58] The end result was a relation giving the decrease in spot intensity for increasing order of the spectrum.[59] The more compressible the crystal, that is, the weaker the interatomic binding forces, and the higher the temperature, the faster the intensity decreases. In this way, Debye provided the means to reconcile Laue's and the Braggs' interpretations of interference. If the higher orders of spectra decrease quickly enough in intensity, the continuous x-ray spectrum may be largely free of the superposed high orders. Such a spectrum would therefore be a more accurate reflection of the Fourier spectrum of the impulses that produce it than would a spectrum including the higher orders.

In the fall of 1913, Laue renounced his appeal to monochromatic primary x-rays. At the *Naturforscherversammlung* in Vienna, Walther Friedrich reported that "it is, for the first time, experimentally demonstrated [that] the absorption coefficient for x-rays is a measure of the wavelength."[60] He specifically denied the assertion that only a few wavelengths in the primary beam are responsible for the interference spots. In the discussion that followed, Laue protested that Friedrich had credited him with a viewpoint that he had "never held," that the primary x-ray beam is itself monochromatic. If he had not been entirely clear before, he claimed, it was due to his "strong reservations" about the "only possible explanation that he had seen" at the time.[61] So, Laue brought his own interpretation into line with that of the Braggs while still retaining its generality. Indeed, he now thought that the

[57] Debye, *VDpG, 15* (1913), 678–89. For the state of this issue before Debye's treatment, see Forman, *AHES, 6* (1969), 38–71
[58] Debye, *VDpG, 15* (1913), 738–52.
[59] *Ibid.*, 857–75.
[60] Friedrich, *PZ, 14* (1913), 1081.
[61] *Ibid.*, 1085.

primary beam consists of a "wide continuous spectrum." It was particularly fortunate, he said, that the Braggs, Moseley, and Darwin had found "isolated monochromatic x-rays" so that corroboration of the basic laws of interference could be found.[62]

By 1915 Sommerfeld could speak "with complete confidence of 'x-ray light' distinguished from normal light by precise determination of its wavelength."[63] As early as 1900, he had suggested the division between periodic and aperiodic waves, two ends of a continuum stretching from monochromatic light to the x-rays. This idea was now widely applied to distinguish the two types of x-rays. The characteristic x-ray lines are like monochromatic light; Sommerfeld compared them to "distinctly colored spectral lines."[64] Like white light, the *Bremsstrahlung* impulses may be thought of either as irregular pulses or as a superposition of Fourier component waves. No longer as concerned as he had been in 1900 with keeping the two concepts separate, he concluded, "it is more a question of ease and purpose which way of speaking we employ."[65]

Sommerfeld developed this idea of interchangeability between impulse and continuous spectrum representations in a careful study of the spectrum of *Bremsstrahlung* pulses in 1915. He extended Lorentz' argument against singly directed pulses using Fourier spectrum analysis to do so, and showed *inter alia* that a doubly directed square pulse has a spectrum that peaks at a frequency approximately equal to $\pi c/\delta$, rather than at zero frequency like the singly directed impulse.[66] Moreover, it was clear from the analysis that inherent in the very nature of the problem was an unavoidable tradeoff between well-defined frequency and well-defined pulse width. The more temporally restricted is the impulse, the more component frequencies have to be added to produce it, and the notion of a single frequency for the impulse becomes less well defined. Conversely, the very sharp x-ray emission lines, which define virtually pure frequencies, have to be thought of as periodic disturbances extending over macroscopic distances in space. Just where in space the effect of the x-ray is located then becomes unclear.

[62] Laue, *PZ, 14* (1913), 1077.
[63] Sommerfeld, *Naturwissenschaften, 4* (1916), 1.
[64] *Ibid.,* 5.
[65] Sommerfeld, *Naturwissenschaften, 1* (1913), 708.
[66] Sommerfeld, *AP, 46* (1915), 721–48.

It was not so much that physicists prior to this period were unable to visualize x-ray impulses as a superposition of periodic waves, but that there had been little advantage in doing so. The temporal discontinuity of impulses had attracted attention to them originally. Most physicists in the first decade of the twentieth century thought of x-rays as short temporal events rather than as the net effect of many extended processes in the aether. After 1912 that view became dated, and the new physics of x-rays exploited more and more the Fourier tradeoff between temporal discontinuity and frequency. Evidence for the interference of x-rays did not convince physicists that x-rays are periodic waves; rather, it convinced them that, whatever the true nature of radiation, x-rays act in all respects just like ordinary light.

CHARACTERISTIC γ-RAYS

Rutherford's hypothesis that γ-rays are the electromagnetic impulses produced by the expulsion of β-electrons from the atom remained purely qualitative until Sommerfeld's 1911 study. In this view, one would expect to find a correlation between β-particle velocity and γ-ray penetrating power. Although the latter property could be measured, not until 1910 were methods developed to determine with accuracy the velocities of released β-electrons. Consequently, there was little more than fragmentary evidence before 1912 that γ-rays possess properties characteristic of their parent element. McClelland had found in 1904 that the γ-rays from radium are not homogeneous in penetrating power; their absorption does not follow the exponential law.[67] A student of Rutherford's in Montreal, Tadeusz Godlewski, noted the following year that γ-rays from different sources exhibit differing penetrability.[68] Richard Kleeman, Bragg's former co-author, found at least two distinct absorption coefficients combined in the γ-rays from radium.[69] In his experiments in 1904 Paschen had found two peaks in the velocity distribution of the β-particles emitted by radium,[70] and Kleeman suggested in 1907 that these are intimately related to

[67] McClelland, *PM, 8* (1904), 67-77.
[68] Godlewski, *PM, 10* (1905), 375-9.
[69] Kleeman, *PRS, 79A* (1907), 220-33.
[70] Paschen, *AP, 14* (1904), 389-405; *PZ, 5* (1904), 502-4.

the two strong γ-ray components he had found.[71] Otto von Baeyer, Otto Hahn, and Lise Meitner found more compelling evidence of β-particle velocity peaks in 1910.[72]

There was little evidence, however, that the perceived groupings of γ-rays are characteristic of their radioactive source, even if the β-particle velocity peaks seem to be. At first, Rutherford was reluctant to believe that the patterns in γ-ray penetrating power could be, as he later put it, "ascribed to a *characteristic* radiation set up in the radiator by the primary γ-rays."[73] When D. C. H. Florance measured the intensity distribution of the γ-rays scattered from lead in 1910, Rutherford interpreted the results as proof that γ-rays cannot be stimulated by external influences on matter.[74] But the following year, J. A. Gray demonstrated that γ-rays – or what appeared to be γ-rays – could, like x-rays, be stimulated in matter by incident β-particles.[75] This set to rest several vague doubts: Some radioactive substances are known to emit a significantly lower intensity of γ-rays than would be expected were the γ-rays a product of β-particle emission. Actinium is a case in point. Rutherford thought that this might be due to "some peculiarity in the mode of emission of the β rays," and his doubts about characteristic γ-rays began to fade.[76]

First, Rutherford set Moseley to work to verify that only one β-particle is emitted by any single atom. Moseley showed that the elements called radium B and radium C emit only one β-electron when an atom decays. In the meantime, Gray found what at the time seemed convincing evidence that γ-rays can stimulate secondary γ-rays in matter,[77] although it is clear in retrospect that these were high-frequency characteristic x-rays. Initially, Rutherford doubted Gray's claim that these secondary "γ-rays" are char-

[71] Eve, *PM, 11* (1906), 586–95. Kleeman, *PM, 14* (1907), 618–44; *15* (1908), 638–63.

[72] von Baeyer and Hahn, *PZ, 11* (1910), 488–93. von Baeyer, Hahn, and Meitner, *PZ, 12* (1911), 273–9.

[73] Rutherford, *Radioactive substances* (1913), 282. Although it had very little influence on his research, Rutherford came very close to adopting Thomson's unit hypothesis for x-rays in 1913, p. 84.

[74] Florance, *PM, 20* (1910), 921–38.

[75] Gray, *PRS, 85A* (1911), 131–9. Important preparatory work was done by Starke, *Le Radium, 5* (1908), 35–41; and by Davisson, *PR, 28* (1909), 469–70.

[76] Rutherford, *Radioactive substances* (1913), 271.

[77] Gray, *PM, 26* (1913), 611–23; *PRS, 86A* (1912), 513–29; *87A* (1912), 489–501.

acteristic of the element from which they arise. But in 1912 he began to take seriously evidence that the groups of γ-rays of homogeneous penetrating power correspond to the peaks in β-electron velocity. At about the same time that the news of crystal diffraction of x-rays came from Munich, Rutherford concluded that "the transformation of energy from the β ray form to the γ ray form or vice versa takes place in definite units which are characteristic for a given ring of electrons."[78]

Rutherford now demanded the evidence for characteristic γ-rays that he had denied before. He set about to find it himself, and he soon felt he had succeeded. Frederick Soddy, with his wife and Alexander Russell, had shown in a series of experiments between 1909 and 1911 that, aside from endpoint behavior, γ-rays are absorbed exponentially along most of their path through lead.[79] This is the classical symptom of homogeneity. With a student, Rutherford measured absorption coefficients for γ-rays spontaneously emitted from several elements, finding that radium B gives three "distinct types" and radium C a single type.[80] The coefficients differed markedly; in units of inverse centimeters of aluminum, they are shown in the penultimate column of Figure 8.7, a table from a slightly later publication.[81] In a burst of overconfidence, Rutherford declared that the γ-rays of coefficient 45 from radium D were, with "little doubt," the "characteristic radiation of the L type to be expected from an element of atomic weight 210."[82]

By one of those misapprehensions that so enrich and color the history of experimental science, Rutherford was here quite misled. The electromagnetic spectrum of a radioactive element consists of more than just the γ-rays characteristic of it. The exiting γ-rays and other energetic emissions stimulate the surrounding electronic structure of the atom, releasing, in addition, the characteristic x-ray spectrum of the element. It was these high-frequency characteristic x-rays that Rutherford studied, calling them "γ-rays" all the while. We commit a historical anachronism to explain this point here, and certainly no discredit is to reflect on Rutherford. Indeed,

[78] Rutherford, *PM, 24* (1912), 453–62.
[79] F. and W. Soddy and Russell, *PM, 19* (1910), 725–57. Russell, *PRS, 86A* (1911), 240–53.
[80] Rutherford and H. Richardson, *PM, 25* (1913), 722–34.
[81] Rutherford and H. Richardson, *PM, 26* (1913), 937–44.
[82] *Ibid.,* 331. Richardson soon used the γ-ray absorption coefficients to fix the atomic weight of uranium at 230. *PM, 27* (1914), 252–6.

γ *Rays of the Thorium and Actinium Products.*

Element.	Atomic weight.	Absorption coefficient μ in aluminium.	Mass. absorption coefficient μ/d in aluminium.
Radium B	214	230 40 0·51 } $(cm.)^{-1}$ „ „	85 14·7 0·188 } $(cm.)^{-1}$ „ „
Radium C	214	0·115 „	0·0424 „
Radium D	210	45 0·99 } „ „	16·5 0·36 } „ „
Radium E	210	similar types to D but very feeble.	
Mesothorium 2	228	26 0·116 } $(cm.)^{-1}$ „	9·5 0·031 } „ „
Thorium B	212	160 32 0·36 } „ „ „	59 11·8 0·13 } „ „ „
Thorium D	208	0·096 „	0·035 „
Radioactinium	...	25 0·190 } „ „	9·2 0·070 } „ „
Actinium B	...	120 31 0·45 } „ „ „	44 11·4 0·165 } , „ „
Actinium D	...	0·198 „	0·073 „

Figure 8.7. Absorption coefficients of mixed x-rays and γ-rays from thorium and actinium products. [Rutherford, *PM, 26* (1913), 943.]

as we shall see, Rutherford's continued belief that he was studying characteristic γ-rays and thereby consolidating his hypothesis that they originate in electron groupings in the atom (where, in fact, the rays he studied *did* arise), led to much important physics and several essential discoveries. This again illustrates Francis Bacon's wise dictum that truth emerges more readily from error than from confusion. To reduce confusion for the reader, I will henceforth put crucial references to Rutherford's γ-rays *cum* x-rays in quotes thus: "γ-rays." But for the moment, we must leave Rutherford in his state of fruitful error.

By the summer of 1913, when Bohr published his theory of atomic structure, Rutherford felt he had bridged the gap in frequency that had separated x-rays from γ-rays. Never before had it been suggested that in radioactive decay one might find rays identical to those produced by electric discharges. "The radiations from the various radioactive substances can be conveniently divided into three distinct classes," Rutherford claimed. Aside from the most easily absorbed component of $\mu = 230$, which was well within the recognized domain of x-rays, the three groups are: "(1) a soft radiation, varying in different elements from $\mu = 24$ to $\mu = 45$, probably corresponding to characteristic radiations of the 'L' type excited in the radio atoms; (2) a very penetrating radiation with a value of μ in aluminum of about .1, probably corresponding to the 'K' characteristic radiation of these heavy atoms; (3) radiations of penetrating power intermediate between (1) and (2) corresponding to one or more types of characteristic radiations not so far observed with x-rays."[83] He spoke freely of x-ray "wavelength" and took the absorption data from the "γ-rays" as if they predicted new wavelengths to be found in x-ray spectra.

INTERFERENCE OF γ-RAYS

After the discovery that x-rays can interfere, many physicists thought that γ-rays ought to show interference effects too. Moseley started his x-ray investigations with this idea in mind. Walther Friedrich gave two talks to the Wednesday Munich Colloquium early in 1913 on interference of x-rays and then turned directly to a review of analogous γ-ray research.[84] It was Rutherford who actually put x-ray diffraction apparatus to work and showed that "γ-rays" can show interference effects. His results, obtained with Edward Andrade, were announced in the May 1914 issue of the *Philosophical magazine*.[85]

Exceedingly long exposure times were required to capture the penetrating "γ-rays" on photographic plates. To make the diffraction angles as large as possible, the rays from radiums B and C were reflected at glancing angles from rocksalt crystals. This decreased the effective grating spacing. The γ-ray source was placed in a strong magnetic field to deflect all β-particles and secondary electrons. As a final precaution, Rutherford used Maurice de Broglie's technique of rotating the analyzing crystal to overcome its inherent irregularities; he worried about "contorted and undulating surfaces." A very slow scan by the crystal accentuates the lines of interest by smearing out ghost lines produced by periodic irregularities in the atom planes.[86]

[83] Rutherford and Andrade, *PM, 27* (1914), 855.
[84] The Register of the *Münchener physikalisches Mittwochscolloquium*, 1 July 1912, 19 February 1913, 7 May 1913. AHQP, 20.
[85] Rutherford and Andrade, *PM, 27* (1914), 854–68. A brief first announcement was made in *Nature, 92* (1913), 267.
[86] M. de Broglie, *CR, 157* (1913), 924–6. See Heilbron, *Isis, 58* (1967), 482–4.

RADIUM B. Soft γ-ray spectrum.			PLATINUM. X-ray spectrum.
Angle of reflexion from rocksalt.	Wave-length (in cm.).	Intensity.	Angle of reflexion $\overline{1\cdot122}$.
8° 6′	$\cdot793 \times 10^{-8}$	m.	
8° 16′	·809	m.	
8° 34′	·838	m.	8° 27′
8° 43′	·853	m.	8° 43′
9° 23′	·917	f.	
9° 45′	·953	m.	
10° 3′	·982	s.	10° 2′
10° 18′	1·006	m.	10° 13′
10° 32′	1·029	m.	
10° 48′	1·055	f.	
11° 0′	1·074	f.	
11° 17′	1·100	f.	
11° 42′	1·141	m.	
12° 3′	1·175	s.	12° 3′
12° 16′	1·196	m.	
12° 31′	1·219	f.	
13° 0′	1·266	f.	
13° 14′	1·286	f.	
13° 31′	1·315	f.	
13° 52′	1·349	m.	
14° 2′	1·365	m.	

Figure 8.8. Rutherford's data for what he took to be characteristic γ-ray lines. [*PM, 27* (1914), 861.]

As early as October 1913, Rutherford and Andrade could report success. Faint lines appeared at angles of 10° and 12°. With the best current estimate of atom spacing in rocksalt (2.814 × 10^{-8} cm), they calculated the wavelengths of some twenty-one lines. The results are shown in Figure 8.8. The two strongest lines have wavelengths of 0.982 and 1.175 × 10^{-8} cm. Again, I remind the reader that what Rutherford here believed were γ-rays were actually high-frequency secondary x-rays from the source element.

Rutherford immediately tried to find room for the "γ-ray" lines in Moseley's scheme of x-rays. The wavelength of the strong line from radium B ($\mu = 40$), if inserted in the L series of x-rays, gave that element an atomic number of 82. This was "unexpected corroboration" of Kasimir Fajans' and Soddy's prediction that radium B is an isotope of lead.[87] This justification of the hypothesis of radioactive decay series encouraged Rutherford's belief that a smooth transition would be found between x-rays and γ-rays.

In August 1914 there followed studies on what Rutherford thought were the harder γ-rays from radiums B and C. This time Rutherford found it impossible to obtain the normal crystal pic-

---

[87] According to Fajans and Soddy, radium B, actinium B, and thorium B are all isotopes of lead (82).

tures. Instead, he shot the primary γ-rays directly through thin sheets of crystal and recorded interference lines on a photographic plate at a distance.[88] The results contributed two more wavelengths: 0.99 and 1.6 × 10⁻⁹ cm. These forced Rutherford to make the first of many revisions in his attempt to link the "γ-ray" data with Moseley's x-rays. Rutherford and Andrade found a line from radium B at 1°40′. This hard line ($\mu = 0.115$) and that from thorium B ($\mu = 0.36$) had therefore to be taken as evidence of a "higher mean frequency" series than the K series. "This may for convenience," Rutherford went on at a dizzying pace of speculation, "be named the 'H' series, for no doubt evidence of a similar radiation will be found in other elements when bombarded by high-speed cathode rays."[89] Rutherford was still convinced that the source of the γ-rays was intimately connected to that of the characteristic x-rays. He was therefore very much interested when, at this juncture, the source of the characteristic x-rays was identified in Germany.

A THEORY OF X-RAY EMISSION

Ernst Wagner was one of many investigators of x-rays at Röntgen's Institute for Experimental Physics in Munich; he directed his attention to the absorption of x-rays by matter. With the new spectroscopic techniques, it was possible to determine the precise wavelength at which continuous x-rays are preferentially absorbed by atoms.[90] Wagner found that the absorption is not restricted to a specific x-ray wavelength, although the low-frequency threshold for absorption is sharply defined. A typical absorption "edge" is shown schematically in Figure 8.9. Wagner began a systematic catalog of these absorption edges in metals.[91] He expected to find resonant absorption of x-rays at or near the characteristic x-ray emission frequencies of the atom; instead, he found slight differences between the onset of absorption and the slightly lower frequencies of emission. What happened to the difference? Wagner, following the Bohr model of the atom, suggested qualita-

[88] Rutherford and Andrade, *PM, 28* (1914), 263–73. See also Rutherford to Boltwood, 20 June 1914, quoted in Badash, *Rutherford and Boltwood* (1969), 293.
[89] Rutherford and Andrade, *PM, 28* (1914), 272.
[90] Heilbron, *Isis, 58* (1969), 451–85.
[91] Wagner, *AP, 46* (1915), 868–92. Kossel drew his data from Wagner, München *Sb, 44* (1914), 329–38.

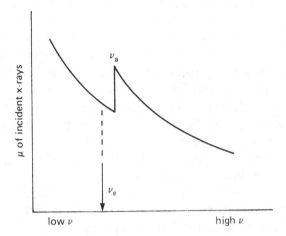

Figure 8.9. Schematic representation of the difference between x-ray emission lines and absorption edges.

tively that the discrepancies would show up as differences in electron energy levels.

Enter Walter Kossel. Educated under Philipp Lenard at Heidelberg, Kossel had come to Munich in 1913 to study with Sommerfeld. Although he never held an official position at the Institute, he kept abreast of the research work done there. He also had the benefit of close discussions with Bohr when the latter visited Munich.[92] Kossel tried to bring order to Wagner's data by arranging a table of x-ray absorption features as a function of wavelength. He soon noted that, except for the more intense of the two lines that together comprise the *K* emission feature, a strong x-ray absorption edge is *always* found at a frequency slightly greater than that of the associated emission line. As Germany was mobilizating for World War I, in October 1914 Kossel hit upon a solution to this regularity from the point of view of the Bohr atom.[93] The absorption of x-rays might be responsible for the ejection of an electron from a filled, low-lying atomic ring. Emission follows when the atom returns to equilibrium. The idea bore much similarity to Lenard's treatment of spectral radiation following photoionization. Kossel had studied the emission of electrons from metals by

[92] Bohr to Kossel, 19 July 1921. Bohr archive; AHQP BSC 4, 2.
[93] His inspiration occurred between September and October 1914. *VDpG. 16* (1914), 898–909, 953–63. Heilbron, *Isis, 58* (1969), 451–85.

incident electron beams for his dissertation under Lenard at Heidelberg.[94]

With Kossel's suggestion that characteristic x-ray lines originate in electron transitions to the innermost rings of the Bohr atom, x-rays gained theoretical credentials equivalent to those of visible light. Both were, in the context of the Bohr theory, the product of downward electron transitions: visible light to high-lying electron rings, x-rays to low-lying ones. Sommerfeld's subsequent elaboration of the Kossel scheme further established the validity of the hypothesis. Theoretical and empirical evidence had converged, justifying further the equivalence of x-rays and light.

In the first months of 1915, evidence mounted that Kossel was right. Ionization of the atom must precede the emission of characteristic x-rays. Transitions to refill the lower rings produce the characteristic lines. As Heilbron has indicated, this mechanism would not likely have suggested itself to Bohr, who had already rejected ionization as a prerequisite for optical emission. In Kossel's hypothesis, the absorption edges directly measure the energy levels in the atom. Differences in absorption edges allow calculation of the fine splitting of the main quantum levels, and this soon became the chief empirical means of refining the Bohr theory.[95] It also enabled Sommerfeld and others to make predictions about the number of electrons that shield the nucleus at each level of the atom. The main significance of the new understanding of x-rays was that it provided a means to probe more deeply into the structure of the atom, not to stimulate interest in the associated x-ray photoelectric effect. It was not until 1921 that the latter problem was seriously examined, as we shall see in a later chapter. In the meantime, the success of the Bohr theory attracted away the talents of just those physicists who were most capable of producing a theoretical solution to the problems of free radiation.

THE SOURCE OF THE γ-RAYS

The source of the x-rays was known by 1915, but the origin of the γ-rays remained obscure. Rutherford could only grope toward a qualitative γ-ray emission mechanism based on electron configu-

[94] Kossel, AP, 37 (1912), 393–424.
[95] The central document of this program is Sommerfeld, Atombau und Spektrallinien (1919), and successive editions. See also Heilbron, Atomic structure (1964), and his contribution in History of twentieth century physics (1977), 40–108.

rations in the atom. Few physicists disagreed at the time that γ-rays originate from a source in the atom equivalent to that of x-rays, but the attempt to find room in the electronic superstructure of the Bohr atom failed repeatedly. In part, the very strength of the γ-ray – x-ray analogy slowed the recognition that γ-rays might arise from a level of atomic structure that was entirely different from electron orbits.

Rutherford was convinced that a smooth transition would be found between the characteristic x-ray and γ-ray spectra. But it was very difficult to measure "γ-ray" wavelengths directly using x-ray methods. So, immediately after presenting what he believed, incorrectly, were the first few γ-ray wavelength measurements, Rutherford used them to construct a general relation between wavelength and absorption coefficient. This allowed him to derive approximate γ-ray wavelengths from simple absorption data. On the basis of this extrapolation, Rutherford submitted a hypothesis to relate the γ-rays to atomic structure.[96] The characteristic "γ-rays" were thought to be excited in special configurations of electrons within the atom. Radioactive disintegration ejects an electron "from or near the nucleus." If this β-particle happens to pass through one or more of the special regions in the electronic structure, a monochromatic γ-ray is excited. These, in turn, can reconvert their energy, either singly or in combination with other γ-rays, by stimulating a secondary β-particle. This produces the sharp peaks observed in the β-particle velocity spectrum.

There was little or no evidence for a continuous γ-radiation analogous to the x-ray *Bremsstrahlung.* But James Chadwick had found that most of the energy radiated by the β-particle lies in the continous distribution of velocity, not in the sharply defined β-particle velocity peaks.[97] Rutherford's mechanism of γ-ray emission explained why: Only γ-rays of well-defined frequency are emitted by the special electron groups. The β-rays, on the other hand, originate in the uncharted depths of the nucleus, where mechanisms may very well produce a wide distribution of energies. Moreover, β-rays that leave the nucleus without stimulating characteristic γ-rays are complemented by others that are ejected, like photoelectrons are, by combinations of γ-rays. It is the latter, relatively rare, β-particles that form the characteristic β-ray velocity peaks.

[96] Rutherford, *PM, 27* (1914), 488–98.
[97] Chadwick, *VDpG, 16* (1914), 383–91.

Rutherford could also explain why some elements, such as radium E, appear to have no β-ray lines at all. If few γ-rays are produced in the electron rings, there is no way that corresponding β-ray lines can be produced. Do atoms of radium E simply lack the necessary electron configurations? No, answered Rutherford. If the β-particles are ejected in a fixed direction relative to the electron shells, they might not pass through the special regions at all. This he thought was "not improbable" given the "remarkably definite way" that the atom disintegrates. This argument also explains how two isotopes, the atomic structure of which is the same according to the x-ray data, can differ markedly in their γ- and β-ray spectra.

Rutherford had departed from his original inspiration about the γ-rays. No longer did they accompany the emission of each β-particle; now, only certain β-rays in certain atoms were priviledged to stimulate γ-rays. He was caught, as it were, between the success of the pulse theory of γ-rays and the demands of his crystal diffraction data. He slipped into a form of expression that remarkably mirrored his own ambiguous viewpoint. Reflecting the contemporary interest in quantum energy relations, he proposed that γ-ray energy is emitted in integral multiples of $h\nu$, where ν is the frequency of the γ-ray. Clearly, this could not apply to single γ-ray impulses, each of which carries a single $h\nu$ in total energy. Rutherford was still thinking of the γ-rays more as impulses than as waves. The interference that produces the crystal pictures was due to repeating series of impulses. The emission of energy $nh\nu$ in the form of γ-rays occurs, he said, in linked "trains" of wavepulses. And the entire train of n pulses can act in concert to stimulate a single secondary β-particle; this particle obtains the entire quantity $nh\nu$ of energy. At a slightly later point, Rutherford applied the same imagery to x-rays: he spoke ambiguously of "a train of one or more waves each of energy $h\nu$."[98] He accordingly tried to demonstrate that the frequencies of the known "γ-ray" lines could be expressed as integral multiples of a unit of energy.

The war intervened and disrupted all investigations at Rutherford's Manchester laboratory. The γ-ray studies "had to be broken off."[99] Chadwick was interned in Germany as an enemy alien. In

[98] Rutherford, *PM, 28* (1914), 319.
[99] Rutherford, Chadwick, and Ellis, *Radiations from radioactive substances* (1930), 338.

fact, the demands of true γ-ray crystallography were themselves too formidable for continued studies of γ-ray wavelengths. The few high-frequency x-rays found by Rutherford and Andrade and thought to be γ-rays in 1914 were the last measured directly for over a decade. Extremely sensitive detectors were required; glancing angles as small as 10 min of arc proved necessary, as were elaborate measures to protect the photoplate from high-frequency secondary x-rays.[100] The imperfect correlation of "γ-ray" absorption with wavelength provided most of the estimates of true γ-ray wavelengths in this period. Some information about γ-ray frequencies could be extracted after 1921 by extrapolating the measured x-ray photoeffect relations to γ-rays. But as we shall see in the next chapter, the most significant of these transformation experiments on the energy balance between radiation and matter were based on principles that emphasized the particulate aspects of the radiation.

The conceptual shift needed for further understanding of the γ-ray source seems to have arisen largely by elimination of other alternatives. By 1922 it was clear that electron transitions, even to the lowest layers of the atom, are not energetic enough to produce γ-rays. There simply was no room left in the atom. When Charles Ellis showed that γ-ray frequencies, like visible and x-ray frequencies, could be reduced to differences among a set of energy values, he sought the locus of the rays in the nucleus because it was all that remained.[101] His hypothesis was soon bolstered by data on γ-rays from a number of radioactive elements; and Rutherford's few remaining doubts vanished by 1925 that true γ-rays are characteristic of the nucleus, not the electronic structure of the atom.[102]

After this reorganization of thought, it was possible to separate entirely γ-rays from x-rays: γ-rays are x-rays of nuclear origin.[103] The relatively low-frequency characteristic "γ-ray" lines that

[100] On the difficulty of γ-ray measurements, see Rutherford to Bohr, 20 May 1914. Bohr archive. Quoted in Bohr, *Collected works, 2* (1981), 593–4. Kovarik, *PR, 19* (1922), 433. Thibaud, *Thèses* (1925). Frilley, *Thèses* (1928).
[101] Ellis, *PRS, 101A* (1922), 1–17.
[102] Ellis and Skinner, *PRS, 105A* (1924), 185–98. Ellis, *PCPS, 22* (1924), 369–78. Black, *PRS, 106A* (1924), 632–40; *109A* (1925), 166–76.
[103] Rutherford and Wooster, *PCPS, 22* (1925), 834–7. Rutherford here used the best available data on his "γ-ray" frequencies, together with Moseley's relation, to show that the atomic number of radium B is 83, not 82. This result suggested to him that the "γ-ray" is emitted as a part of, or after, the β-particle emission that transforms element 82 to 83.

Rutherford had fitted so readily into Moseley's x-ray scheme are, in fact, characteristic x-rays stimulated by α- and β-particles as they disrupt the electronic structure of an atom. To be sure, γ-rays and x-rays are not very different from one another, except in energy, after each is released from the atom. But the result of believing them to be equivalent to one another was ultimately to locate the source of the γ-rays in the nucleus, a markedly different level of atomic structure than that whence come both x-rays and visible light. The γ-ray–x-ray analogy had come to an end, largely as a result of its long-lived success.

9

Quantum transformation experiments

Where does the ejected electron get its kinetic energy
when its separation from the light source becomes so
great that the light intensity almost completely vanishes?[1]

The same x-ray spectroscopic techniques that led to the identification of the atomic origin of x-ray spectra also offered means to test
the localization of energy in the radiation. In the second decade of
this century, x-ray spectrometers used as monochromatizers provided more purely defined x-rays than had been available from
natural characteristic radiation. This, coupled with techniques to
determine the kinetic energy of secondary electrons, provided
opportunities for precise testing of the quantum transformation
relation, $E = h\nu$. These experiments form the subject of this
chapter.

We are primarily concerned here with the absorption of radiation, not with its emission. To be sure, the quantum regulation of
the emission of radiant energy has no classical explanation; yet
there is no electromechanical inconsistency implied in the creation of a spherical wave containing a definite amount of energy. It
is the inverse case that causes real difficulty. How can that quantum of spherically radiating energy concentrate its full power on a
single electron? For this reason, verification of the quantum relationship for emitted radiation lies outside our direct concerns. The
Franck–Hertz experiments beginning in 1912, for example, demonstrated the quantum nature of energy transfer, but they did little
to encourage acceptance or even consideration of the lightquantum.[2] On the other hand, the experiments detailed here that
verified the quantum nature of the absorption of light and x-rays

[1] Planck, *Das Wesen des Lichts* (1920), 15.
[2] Franck and G. Hertz, *VDpG, 14* (1912), 457–67, 512–17.

gave substance to the lightquantum hypothesis because they verified the particlelike transfer of radiant energy to matter.

In none of the experiments on the energy of radiation, going
back to those by Rutherford and Wien, had it yet been possible to
make comparisons between incident radiant energy of specified
quality and the resulting velocity of released electrons. X-ray
spectroscopy made available sources of precisely defined x-ray
energy; lacking at first were means to chart with equal precision the
energies contributed to electrons by the radiation. Robert Millikan
provided unambiguous proof in 1914 that Einstein's law of the
photoelectric effect is valid over a large spectral range, but resistance to the lightquantum was strongest for visible and ultraviolet
light. Millikan himself remained unconvinced that the quantum
interpretation should replace the wave theory of light. Rutherford
and his students developed an improved β-ray velocity spectrometer on the eve of the Great War and began to make progress
toward a full understanding of the quantum absorption of x-rays.
But the exigencies of the war stopped this research, and it was not
revived systematically until the 1920s.

Thus, although these experiments set the stage for the eventual
full-scale verification of Einstein's linear photoelectric relation for
x-rays and γ-rays, and actually provided it for light, they did not
sway opinion to the lightquantum hypothesis. For that essential
step in the abandonment of deterministic physics more was required, including a fuller and clearer understanding of the energy
needed to remove a deeply bound electron from an atom. Efforts
prior to 1920 to cope with absorption of radiation effectively
mirror the dualistic interpretation that was to come. Spectroscopy,
a technique founded in principle on the wave nature of radiation,
contributed the data that most clearly demonstrated the particulate nature of light.

THE EMPIRICAL STATUS OF THE
PHOTOELECTRIC EFFECT

Experimental studies on the photoeffect after 1910 were made with
the growing realization that light and x-rays are common forms of
radiation. Even before Philipp Lenard's rejection of the triggering
hypothesis in 1911, evidence had pointed to analogies between
x-ray ionization and the photoeffect. It had been shown in 1907 by

Innes, Cooksey, and Laub that the velocity of the secondary electrons released by x-rays is independent of the intensity but increases with the quality of the x-rays.[3] Wilhelm Seitz had carried the analogy even further in 1910 with studies on the behavior of the x-ray photocurrent in the vicinity of zero external potential. The result, he said, "reminds one qualitatively of the Lenard curve for the photoelectric effect."[4] Also, in 1910 Otto Stuhlman, working with Owen Richardson at Princeton University, investigated the photoemission from a 10^{-7}-cm-thick platinum film deposited on the back side of a quartz plate that was transparent to ultraviolet light. The number of electrons released in the same direction as the transmitted light beam was some 15 to 17 percent greater than the emission in the backward direction, measured when the same film was laid on the front side of the quartz plate.[5] Within a week, R. D. Kleeman at Cambridge provided independent corroboration.[6] Electrons stimulated by light show the same asymmetrical pattern as do those released by x-rays.

The assumptions used to interpret the similarities between light and x-rays changed markedly after the discovery of x-ray interference. Neither Stuhlman nor Kleeman had felt that the asymmetry in photoemission contradicted the wave theory of light "in its ordinary electromagnetic form."[7] Charlton Cooksey, the Yale physicist who had first demonstrated the forward scattering of x-rays, summarized the data on the equivalent action in the photoeffect in 1912. He stated that periodic ultraviolet light should be evaluated on the same scale of penetrating power as x-rays.[8] This conclusion was reached even before the announcement of x-ray diffraction. After x-rays were known to interfere, the analogy between x-rays and light began to exercise its influence to alter ideas about light rather than to define x-rays in comparison to

---

[3] Innes, *PRS, 79A* (1907), 442–62. Cooksey, *AJS, 24* (1907), 285–304. Laub, *AP, 26* (1908), 712–26.

[4] Seitz, *PZ, 11* (1910), 707.

[5] Stuhlman, *Nature, 83* (1910), 311; *PM, 20* (1910), 331–9; *22* (1911), 854–64. Rubens and Ladenburg found in 1907–before the Bragg–Barkla debate clarified its importance–that a difference of 1 : 100 exists between the photocurrent stimulated in the backward and forward directions from a gold foil 10^{-5} cm thick. *VDpG, 9* (1907), 749–52.

[6] Kleeman, *Nature, 83* (1910), 339; *PRS, 84A* (1910), 92–9.

[7] Stuhlman, *PM, 20* (1910), 339.

[8] Cooksey, *PM, 24* (1912), 37–45.

light. And the number of experimental studies on the photoeffect peaked in 1912, as can be seen in Figure 7.2.

As discussed briefly in Chapter 7, the empirical understanding of the precise form of the photoelectric relation was lacking even as late as 1913. First, it was difficult to ensure that photocathode materials used in tests were free of impurities. The mose sensitive cathods are the alkali metals with loosely bound outer electrons. These substances are subject to rapid oxidation, and emission from the oxidized products clouded the results of many experiments. Second, one had to be careful to separate the effects of the normal photoeffect from the selective photoeffect discovered in 1910.

Two other influences were responsible for the strong experimental interest in the photoeffect around 1912. Of greatest significance was Lenard's renunciation of the triggering hypothesis the year before, an event that removed the basic presupposition that underlay most previous serious studies of the phenomenon. Second was the discovery of x-ray crystal diffraction, a result that fully identified x-rays with ordinary light. It consequently raised for visible light all the paradoxes of x-rays and γ-rays.

One of the most significant experiments on the photoeffect in this peak year 1912 was done by Owen Richardson and Karl Compton at Princeton. Richardson's interest in the photoeffect had been partly stimulated by his close ties to J. J. Thomson. He had been a student in Trinity College, had taken a first in the natural sciences tripos, and had done research under Thomson.[9] In 1906 he had gone to Princeton University as a faculty member, and it was there that he began research on the photoeffect with his former student, Karl Compton. Dissatisfied with previous failures to remove impurities, they directed single wavelengths of ultraviolet light onto photocathodes that had been carefully scraped clean and then kept in a vacuum.[10] Their results appeared to substantiate Einstein's relation – that light frequency is proportional to the square of electron velocity – a result that Richardson later admitted he had not expected to find.[11] Criticism by Robert Pohl and

---

[9] Swenson, *DSB, 11* (1975), 419-23.
[10] O. Richardson and K. Compton, *PM, 24* (1912), 575-94. Hughes, *PTRS, 212A* (1913), 209-26.
[11] O. Richardson, *PRS, 94A* (1918), 269.

Peter Pringsheim in 1913 brought the explicit form of the relationship into question again, but it was thereafter clear that no resonance peaks were to be found.[12] Richardson explicitly pointed out the similarities between x-rays and light, reflected on the paradoxes that stood in the way of a consistent theory of x-rays, and rehearsed the data for asymmetrical photoemission. The last, he remarked, might lead one "to the conclusion that [like x-rays, light] . . . consists of a shower of material particles."[13]

To avoid this unpalatable conclusion, Richardson offered a treatment of the photoeffect "from a rather wider standpoint" than either lightquanta or waves, "without making any definite hypothesis about the structure of the radiation."[14] Expressed in its fullest form in 1914, Richardson's derivation of the photoelectric relation rested on no assumptions about the nature of light or about the structure of the atom. It was a purely thermodynamic argument that bypassed completely the issues that led to the double paradox of quality and quantity. Photoemission of electrons was considered to be analogous to the evaporation of a gas.[15] Richardson assumed the Planck law for the energy distribution in the radiation, and also assumed that the number of electrons emitted by the light is proportional to its intensity. The result predicted a sharply defined wavelength threshold for emission, analogous to a latent heat of evaporation from the metal of the cathode. Moreover, he showed that the maximum energy of an electron has to be proportional to $hv - C$, where C, like the photoelectric work function, is a constant characteristic of the metal.

But Richardson's interest in the photoelectric effect was largely motivated by concerns other than the nature of light. An understanding of the photoeffect was important to his study of the

---

[12] Pohl and Pringsheim, *VDpG, 15* (1913), 637–44.

[13] O. Richardson, *PM, 25* (1913), 145.

[14] *Ibid.*

[15] O. Richardson, *PM, 27* (1914), 476–88. Richardson's interest in the photoeffect grew out of his research on the "glow discharge," the emission of electrons from hot metals. The earliest statement of the method was in *PM, 23* (1912), 594–627, in particular 615ff. See also *PM, 24* (1912), 570–4; a note appended in proof to the last paper appeared separately in *Science, 36* (1912), 57–8; *PR, 34* (1912), 146–9. Richardson's theory deserves closer scrutiny than it has received. For a beginning, see Stuewer, *Minnesota studies in philosophy of science, 5*, (1970), 246–63.

electron discharge from a heated cathode.[16] The "glow discharge" soon became his central research subject, and in 1916 he sought an explanation of it in terms of a photoeffect stimulated by the light emitted from the hot cathode.[17] Although Richardson's photoelectric research was restricted to purely phenomenological tests that contributed precise data on a series of metals not analyzed before, he continued to watch the developing evidence about the photoeffect carefully.[18] Having formulated his own non-lightquantum hypothesis, he was confident that a resolution would be found for the problems of the absorption of radiation that would avoid the "restricted and doubtful hypothesis used by Einstein."[19]

MILLIKAN'S DEFINITIVE EXPERIMENTS

Among those who had taken an early interest in the photoelectric effect was the talented American experimentalist Robert Millikan. Beginning his study in 1906, Millikan was unwillingly drawn ever closer to a quantum interpretation of the phenomenon. First, he found no increase in the maximum electron velocity with increasing temperature of the cathode.[20] This was one of the first recognized difficulties with the triggering hypothesis, but in the face of the then general confusion about photoeffect results, it was not fatal. Soon Millikan showed that photoelectric fatigue – the decrease in sensitivity known to affect metals after long exposure to light – is an effect of impurities created by oxidation; the fatigue disappears entirely when the metals are kept in a vacuum.[21] With this new technique, Millikan examined the response of a number of metals to ultraviolet light produced by sparks from zinc electrodes fired from a bank of twelve Leyden jars.[22] He measured greater electron velocities than had been seen before. There was no

---

[16] Richardson won the 1928 Nobel Prize in physics for his work on thermionic emission of electrons.

[17] O. Richardson, *PM, 31* (1916), 149–55.

[18] K. Compton and O. Richardson, *PM, 26* (1913), 549–67. Richardson and Rogers, *PM, 29* (1915), 618–23.

[19] O. Richardson, *PRS, 94A* (1918), 271.

[20] Millikan and Winchester, *PR, 24* (1907), 116–18.

[21] Millikan and Winchester, *PR, 29* (1909), 85. Millikan, *PR, 30* (1910), 287–8. See also von Baeyer and Gehrtz, *VDpG, 12* (1910), 870–9; Hughes, *PCPS, 16* (1911), 167–74.

[22] Millikan and Wright, *PR, 34* (1912), 68–70. See also Millikan, *PR, 35* (1912), 74–6; *VDpG, 10* (1912), 712–26.

evidence of the critical potential that Einstein had predicted should stop all emitted electrons. Moreover Millikan's student, J. R. Wright, in ignorance of the selective photoeffect, thought he had found evidence of a clear resonance maximum in aluminum at 2,166 Å.[23]

Between 1911 and 1914, Millikan's evidence against the quantum relation crumbled; again the photoelectric effect confused its students. Pohl and Pringsheim showed in 1912 that the new high-velocity electrons Millikan had found were due to electrodynamic effects of the sparks used as the light source; their experiments with arc light failed to duplicate Millikan's extreme values.[24] Wright's resonance peak could be attributed to the selective effect, the lack of a threshold to systematic errors in technique. The experiments were not easy. Carl Ramsauer, who would later distinguish himself with the most precise of all measurements of photoelectron velocities, was tricked in 1914 too. Applying magnetic deflection to sort secondary electrons by velocity, Ramsauer hinted at the conclusion, soon utterly disproven, that there is no threshold for photoemission, and that the velocity distribution is entirely independent of light frequency.[25]

Millikan tried again. This time he concentrated on the frequency–velocity relation, and provided unassailable proof that Einstein had been right all along. The experiments were done at the Ryerson Laboratory in Chicago using an ingenious and complicated apparatus – "a machine-shop in vacuo" – that allowed photocathodes to be scraped clean of impurities and tested entirely under a vacuum.[26] First, the contact potential between the photocathode and the test apparatus had to be determined; then the actual photoelectron stopping potential could be measured at various wavelengths. Millikan tested sodium and lithium and found that the relation between light frequency and the potential that just turned back the fastest electrons is a perfectly straight line, as shown by the solid line in Figure 9.1, which is based on his graphic presentation of the results. The slope of the line is equal to the best contemporary value for h/e within 0.5 percent. Millikan then corrected this best-fit line for the contact potential between

[23] Wright, *PR, 33* (1911), 43–52.
[24] Pohl and Pringsheim, *VDpG, 14* (1912), 974–82. Millikan, *PR, 1* (1913), 73–5.
[25] Ramsauer, Heidelberg *Sb, 5A* (1914), Abh. 19, 20.
[26] Millikan, *PR, 4* (1914), 73–5; *PR, 6* (1915), 55.

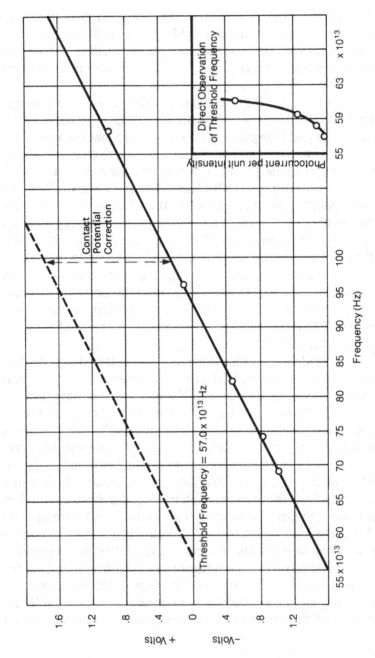

Figure 9.1. Millikan's data for lithium corroborating Einstein's photoelectric law. [Courtesy of the Exploratorium, San Francisco. From Millikan, *PR*, 7 (1916), 377.]

cathode and detector, shifting it upward to the dashed line. The intercept of the corrected relation accurately marked the threshold frequency for emission, measured separately and shown in the inset on the right side of Figure 9.1.[27] Closer corroboration of Einstein's linear law, equation 5.1, could not be desired. To my knowledge, Millikan's results were never questioned, and were quickly recognized by leading European physicists to be definitive. But acceptance of the Einstein law of the photoelectric effect did not carry with it acceptance of the hypothetical lightquantum.

MR. WHIDDINGTON'S REMARKABLE RULE

Four years before the quantitative verification of the quantum relation for the photoeffect, significant regularities had already appeared in the transformation between x-ray and electron energy. Before 1912, however, it was conceptually difficult to assign x-ray impulses a definite frequency. The corroboration of the quantum relations, for example in Wien's pioneering work in 1907, was therefore obscured. In these early years, one had to rely on classification of x-ray quality according to different penetrating powers, and in these experiments British physicists excelled. Throughout the first decade of the twentieth-century, British experimental work on x-rays outshone, and for a while outnumbered, that of any other country.[28]

The early work on the x-ray transformation relations was predominantly British. G. W. C. Kaye showed in 1908 that x-rays stimulated from different anticathodes by the same cathode ray beam show a rough proportionality between their penetrating

---

[27] Millikan, *PR, 7* (1916), 355–88. The work is described further in Wheaton, *Photoelectric effect* (1971), chap. 13, and Millikan, *Electron* (1917), 224.

[28] Numbers of research papers on x-rays listed in the *Fortschritte der Physik,* 1900–10:

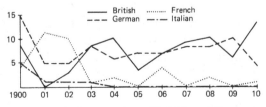

Despite the equivalent number of British and German studies, the experimental work that was cited in this period was almost entirely British.

power and the atomic weight of the anticathode.[29] The early experiments were done without the benefit of Barkla's sharply defined homogeneous x-rays. In 1910 Charles Sadler, who had collaborated with Barkla in the discovery of characteristic secondary x-rays, found that the beam of secondary electrons emitted from metals irradiated by a single fluorescent x-ray "line" is itself always absorbed exponentially.[30] This implied that the released electrons were all of similar velocity. Sadler found that the velocity was independent of both the intensity of the incident x-ray and the nature of the tertiary radiating substance. He discovered, however, that the electron velocity increases linearly with the atomic weight of the secondary radiating substance that produced the homogeneous x-rays. Beatty found at roughly the same time that the absorption of secondary electrons varies linearly with the absorption of the various homogeneous x-rays he used to stimulate release of the electrons.[31] But the most important relationship that tended to confirm the close connection between the characteristic secondary x-rays and the characteristic optical emission spectra of chemical elements was discovered by Richard Whiddington in 1911.

Whiddington studied the secondary x-rays from chromium and other elements at the Cavendish Laboratory in Cambridge. The rays obediently appeared, with their characteristic coefficient of absorption of 136 cm^{-1} in aluminum, only when the potential of the discharge tube exceeded a certain threshold. When Whiddington substituted various materials as anticathodes using Kaye's techniques, he found that there was a definite order in the penetrating power of the emitted secondary x-rays. This finding was not unexpected. Barkla had shown enough to suggest that such relationships should hold. But Whiddington thought that he had found a possible violation of Stokes' law for the fluorescent nature of secondary x-rays. "There is reason to believe," he hinted, "that this secondary radiation can be extracted by a primary radiation somewhat less penetrating than itself."[32] Thus, it was extremely important for him to refine his apparatus so that the electron energy required to stimulate a given x-ray line could be measured accurately.

[29] Kaye, *PTRS, 209A* (1908), 123–51.
[30] Sadler, *PM, 19* (1910), 337–56.
[31] Beatty, *PCPS, 15* (1910), 416–22; *PM, 20* (1910), 320–30.
[32] Whiddington, *PCPS, 15* (1910), 575; *PRS, 85A* (1911), 99–118.

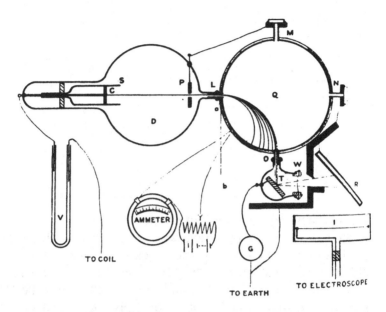

Figure 9.2. Whiddington's apparatus to show the proportionality of the atomic weight of an element to the minimum velocity of electrons capable of producing x-rays that will, in turn, stimulate characteristic x-rays in that element. [*PRS, 85A* (1911), 324.]

Six months later, Whiddington offered the results of his new apparatus.[33] It incorporated characteristics of an early cathode-ray velocity spectrometer and is shown as Figure 9.2. Bulb *D* produces the cathode-ray beam; the brass cylinder *Q* wound with wire generates a magnetic field directed out of the plane of the paper. The field selects electrons of only a single velocity; these then strike the anticathode at *T*. Whiddington substituted materials of various atomic weight *A* for the scattering plate at *R;* for each material, he determined the minimum electron velocity that would produce primary x-rays capable of stimulating characteristic rays from *R*. His results for seven elements are given in Table 9.1. They illustrate a relation approximately expressed by

$$v_{\text{critical}} \simeq A \times 10^8 \text{ cm/sec} \qquad (9.1)$$

an expression soon named *Whiddington's rule* in England.

Whiddington expected that a similar relation could be found for the x-ray photoeffect, the inverse case of that given above. He

[33] Whiddington, *PCPS, 16* (1911), 150–4; *PRS, 85A* (1911), 323–32. First announcement in *Nature, 88* (1911), 143.

Table 9.1. *Whiddington's data for seven elements*

Element	Velocity, v (10^8 cm/sec)	A	A/v
Al	20.6	27	1.3
Cr	50.9	52.5	1.03
Fe	58.3	56	0.96
Mn	61.7	58.6	0.95
Cu	62.6	63.2	1.01
Zn	63.6	65.1	1.02
Sc	73.8	78.9	1.07

promised new experiments, but problems soon arose. The velocity induced in electrons by x-rays was too great, and the electrons could not conveniently be turned back by a decelerating electric field of a practical magnitude. Moreover, the electrons released from R are not concentrated in a beam like those in tube D, and Whiddington found that he could not apply magnetic deflection to determine their maximum velocity. He could only provide a plausibility argument in 1912, based on prior data taken by Beatty and Sadler to suggest, but not to confirm, that a comparable proportionality holds for the velocities of photoelectrons created by homogeneous x-ray beams.[34]

Whiddington's attempt was premature. He was restricted to the small number of homogeneous x-ray emission lines from a few heavy atoms, and he had no quantitative means to measure specific electron velocities. He could only hope at best to determine the maximum velocity of released electrons. Even after 1912, the new specificity in x-ray quality offered by crystal diffraction had yet to be matched by techniques to measure electron energy peaks throughout the distribution of speeds. Only when these two new techniques were combined could the x-ray photoeffect be adequately investigated. We have already discussed the origins of x-ray spectroscopy. The early means developed to study the velocity of radioactively emitted β-particles now demand our attention.

β-PARTICLE VELOCITY SPECTROSCOPY

Friedrich Paschen presented a remarkable new tool to radioactive research through his defense of the claim in 1904 that γ-rays are

---

[34] Whiddington, *PRS*, *86A* (1912), 360–70, 370–8.

very fast electrons. He greatly advanced the magnetic deflection experiments done theretofore on cathode rays by Lenard, Kaufmann, and others. Paschen so strengthened the magnetic field that unwanted β-ray electrons from his γ-ray source were bent into closed circles and never left the apparatus.[35] The technique to deflect electrons in a magnetic field so that those of different velocity are separated on a photographic plate was not new; Becquerel had made such plates since 1899 before β-rays were known to be electrons.[36] But Paschen demonstrated that precise control of the electrons was possible and that correlations might be drawn between the strength of the field and the velocity of electrons bent to a predetermined radius by the magnet.

Ernest Rutherford was the first to exploit magnetic deflection systematically. In 1905 he placed a photographic plate above a wirelike source of α-particles. To arrive at the plate, the particles had to pass through a collimating slit parallel to the source and then traverse a magnetic field; this action separated electrons by velocity.[37] In this way, he showed that after passing through increasing thicknesses of absorbing matter, α-particles slow down appreciably until they are no longer able to expose the photoplate. William Wilson applied a similar technique to β-rays from radium in 1909. His apparatus was fairly sophisticated and could bend β-particle trajectories through a full 180°. The intent was to show that homogeneous β-rays are not absorbed exponentially but rather linearly in matter.[38] In these studies of α- and β-particles the γ-rays were a definite hindrance, and Wilson did not apply his methods to secondary electrons excited by the γ-rays.

The attempt to derive quantitative results for β-ray electrons was first made by Otto von Baeyer and Otto Hahn, physical chemists in Berlin. They based their work on Rutherford's magnetic deflection of the α-rays and concluded that "the line spectrum" of electron velocities produced on the photoplate offered "the possibility to determine the velocity of single β-rays."[39] They placed a sensitive photoplate above a sheet with a slit to allow through only a narrow solid angle of rays, as shown in Figure 9.3. The undeflected γ-rays

[35] See Chapter 3. A general discussion of the development is found in Jenkin, Leckey, and Liesegang, *Journal of electron spectroscopy and related phenomena, 12* (1977), 1–35.
[36] Becquerel, *CR, 130* (1900), 206–11; *PRI, 17* (1906), 85–94.
[37] Rutherford, *PM, 10* (1905), 163–76.
[38] W. Wilson, *PRS, 82A* (1909), 612–28.
[39] von Baeyer and Hahn, *PZ, 11* (1910), 493.

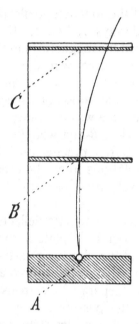

Figure 9.3. von Baeyer and Hahn's arrangement to measure β-particle velocities photographically by passing them through a magnetic field. [*PZ, 11* (1910) Tafel XII.]

mark the straight line through the slit from the radioactive wire used as the source. The curved paths followed by β-electrons in the uniform magnetic field depend on the velocity of the electrons. Since different-velocity electrons would pass through the slit on orbits of different radii of curvature, the bands in the photographic record indicated that there were distinct peaks in the velocity distribution.

Once the technique was known, modifications were instituted that led to the recognition of a remarkable focusing effect. In 1911 von Baeyer, Hahn, and Lise Meitner turned the photoplate so that it lay in the plane of β-particle motion and the entire course of the electrons was traced, in much the same manner as Villard had done in the experiments in which he discovered the γ-rays.[40] Jean Danysz, working in the laboratory of Marie Curie, placed the plate so as to take advantage of a full 180° deflection of the electrons.[41] This arrangement was soon recognized to have a distinct

---

[40] von Baeyer, Hahn, and Meitner, *PZ, 12* (1911), 273–9, 378–9; *13* (1912), 264–6.
[41] Danysz, *CR, 153* (1911), 339–41, 1066–8; *Le Radium, 9* (1912), 1–5.

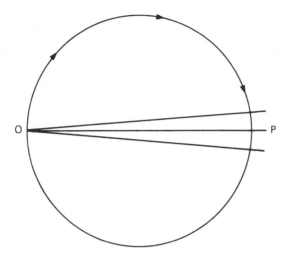

Figure 9.4. The self-focusing effect of electrons projected into a transverse magnetic field.

advantage. As Rutherford put the matter in 1913, "The β-rays of a definite velocity comprised in a comparatively wide cone of rays can be concentrated in a line of very narrow width on the photographic plate."[42]

Rutherford realized that, in a homogeneous magnetic field, electrons of a given velocity will follow circular paths with a constant radius of curvature. This is true regardless of the initial angle of projection of the electron with respect to the plane of the photographic plate. Electrons that begin their motion at any small angle from the normal to the plate will follow different orbits, but will all strike the photoplate at about the same distance from the origin. How this is done may be seen in Rutherford's Figure 9.5, but the reason that it works is more clearly rendered in Figure 9.4. The circle describes the orbit, or partial orbit, of all electrons with the same velocity. The different chords represent the relative position of the photoplate for electrons projected at various small angles from the normal. As is clear in the figure, the distance from the origin O to the darkened points P along the photoplate will be essentially constant for all electrons of the same velocity.[43] The

---

[42] Rutherford and Robinson, *PM, 26* (1913), 719.

[43] Point P is sharply defined by electrons emitted precisely normal to the plate. Electrons emitted on either side of normal will fall progressively closer to the origin. The edges are therefore smeared toward the origin but are sharply defined on the other side.

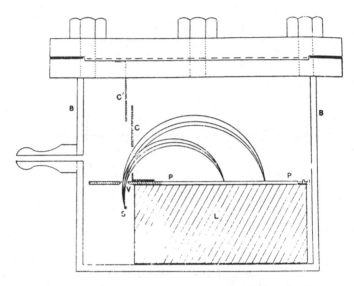

Figure 9.5. Rutherford and Robinson's β-particle velocity spectrometer.
[*PM, 26* (1913), 720.]

electrons projected in different directions are focused at point *P* in much the same way that a lens focuses on one point in the focal plane all the light that diverges from a point on the object.

COMBINING X-RAY AND β-RAY SPECTROSCOPY

Rutherford and his student, Harold Robinson, successfully followed up on Danysz's modification of the β-ray velocity analysis. Danysz had found some twenty-five distinct velocity peaks in the β-rays from radiums B and C.[44] In July 1913, Rutherford and Robinson used the refined spectrometer shown in Figure 9.5 to analyze the system of lines and derived a value of the electron velocity for each.[45] The radioactive source is placed at *S*, the magnetic field extends out of the plane of the page in the region above the photographic plate *PP*. About a year later, Rutherford realized that the method could be applied to measure the velocities of secondary electrons released from metals by γ-rays. Thus, in the same month that the first presumed γ-ray wavelengths were mea-

[44] Danysz, *Le Radium, 10* (1913), 4–6.
[45] Rutherford and Robinson, *PM, 26* (1913), 717–29.

sured in Manchester, the β-ray velocity technique was extended to the measurement of electrons produced by γ-rays.[46]

In 1914, Rutherford's students used the technique to investigate x-ray photoelectrons in a test of "Planck's theory."[47] A metal plate replaced the radioactive source at S, and it was irradiated with the x-rays produced by a nickel anticathode especially chosen for its relatively simple characteristic spectrum. Because, they said, electrons are released from various depths in the metal, the velocity edges are smeared into 2-mm bands. For an iron radiator they found three such bands, for lead two. Using the outer edge of the band for measurement, they found that the velocity of one band agreed within 2 percent with the value predicted by the quantum rule for electrons stimulated by the K x-rays from nickel.

Thus, by the beginning of World War I, the experimental techniques needed for the detailed verification of the x-ray photoeffect relations were available in Britain. The obstacles that had stopped Whiddington in 1911 were obstacles no longer. But it was not until after the war that β-ray spectroscopy was applied in a systematic way to this problem. When it did come, it came from America and France, not from Britain or Germany. Rutherford had pointed the way, and had the war not intervened, he might have followed up on the methods. But Rutherford never intended to specialize in x-ray research; his elected research domain had become the nucleus of the atom.

THE COOLIDGE X-RAY TUBE AND THE DUANE–HUNT LAW

One of the great problems that faced any attempt to quantify the characteristics of x-rays before 1913 was the inconstancy of x-ray tube output. In 1913 a great step toward the solution of that problem was made. William Coolidge at General Electric perfected a new x-ray tube that had a number of advantages over the traditional one.[48] Its cathode was made of a spiral tungsten wire; Coolidge had developed the means of forming normally brittle tungsten into a shape that could tolerate the heating needed to

---

[46] Rutherford, Robinson, and Rawlinson, *PM, 28* (1914), 281–6.
[47] Robinson and Rawlinson, *PM, 28* (1914), 277–81. Moseley to Rutherford, 27 May 1914. Rutherford papers. Quoted in Heilbron, *Moseley* (1974), 235.
[48] Coolidge, *PR, 2* (1913), 409–30.

maximize the outflow of electrons. The heated cathode allowed the tube to produce x-rays in a higher vacuum than was previously possible; harder x-rays resulted. The tube could also operate for longer periods with remarkably little variation in either the "intensity" (quantity) or "penetrating power" (quality) of the x-rays. Moreover, each parameter could be varied independently. The quality is determined by the steady potential across the electrodes, the quantity by the temperature of the tungsten filament. When combined with water cooling of the anticathode, the Coolidge tube produced x-rays of dependable properties in steady operation over extraordinarily long periods.[49] In conjunction with the versatile techniques of x-ray spectroscopy, it made possible experiments on the quantum transformation relation that were no longer limited to isolated characteristic x-ray wavelengths.

Among the first to recognize the importance to physics of the American-made Coolidge tube was William Duane, who offered strong evidence for the quantum law in x-ray emission. Duane was a relative newcomer to x-ray studies; his previous work had been almost exclusively in radioactivity.[50] When he assumed a position left vacant at Harvard after John Trowbridge left, Duane extended his research to include x-rays as being of possible therapeutic value in medicine. In 1914 he tried to improve the correlation of x-ray wavelengths with absorption, and soon turned to x-ray spectroscopy as the best means to test the quantum predictions of x-ray emission.[51] In December he showed what was no surprise to European physicists but still a novelty to Americans: that the proportionality constant relating the frequency of the x-ray to the energy of the parent cathode-ray electron is on the order of magnitude of Planck's h.[52]

Duane was fortunate in having access to two special pieces of apparatus at Harvard. The first was an accurate x-ray spectrometer that had been built by David Webster. Duane also profited from the careful work that Webster had done on the threshold voltage required for entire series of characteristic x-rays to appear.[53]

[49] See, for example, Dauvillier, *Technique des rayons x* (1924), 50ff.
[50] Bridgman, NAS *Biographical memoirs, 18* (1937), 23–41. Forman, *DSB, 4* (1971), 194–7.
[51] Duane, *PR, 4* (1914), 544.
[52] Duane, *PR, 6* (1915), 166–76.
[53] Webster, *PR, 7* (1916), 599–613.

Equally important was the large storage battery consisting of 20,000 Planté cells constructed and left by Trowbridge. It was capable of supplying extremely steady dc up to 45,000 V. In April 1915 Duane and Franklin Hunt used this equipment to test the hypothesis that a given fixed potential can excite continuous x-rays with a frequency only below a specific threshold.[54] They tried many dc potentials and found an abrupt drop in *Bremsstrahlung* x-ray intensity for spectroscopic angles less than a well-defined critical angle. That limiting angle increased as the x-ray tube potential dropped. The implication was that electrons of a given energy are able to excite x-rays of almost any frequency up to, but never greater than, a certain threshold. The fixed dc voltages were known, and the corresponding wavelengths could be calculated from the spectrometer angles. The relation followed the approximate rule used by Wien eight years before and partially verified by Whiddington in the meantime. To a few tenths of a percent:

$$V_0 = \frac{h}{e} \frac{c}{\lambda_0} \tag{9.2}$$

where V_0 is the critical potential and λ_0 the corresponding x-ray wavelength. Equation 9.2 soon became known in America as the *Duane–Hunt law,* and was recognized for many years to be the most accurate of all means to measure h.[55]

The Coolidge tube was also important to the research group at Manchester that carried on the x-ray studies begun by Henry Moseley. Charles Darwin, Moseley's collaborator, had hoped that monoenergetic electron beams might be isolated using magnetic deflection. One could then use them to produce x-rays and see if the wavelengths, measured by absorption and diffraction, correspond to the quantum law. His early experiments were prevented by the variation in x-ray output due to the unsteady vacuum in the experimental tube. The work was soon dropped altogether when he, like Moseley, left Manchester for the war.

In 1915 Rutherford resumed the work for a short while using a

---

[54] Duane and Hunt, *PR, 6* (1915), *first* 166–71. Webster argued that finding a threshold would cast doubt on any impulse representation of *Bremsstrahlung* because the Fourier spectrum of an impulse has no high-frequency limit. This is true in principle but, as Figure 2.7 illustrates, the amplitude of the high-frequency components in the pulse drop rapidly toward zero. For any practical experiment there is an effective threshold. Webster, *PR, 6* (1915), 56.

[55] Birge, *PR Supplement, 1* (1929), 57.

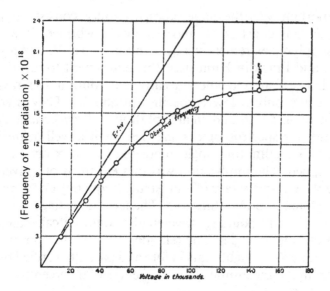

Figure 9.6. Rutherford's data showing apparent deviation from the quantum law for x-rays produced at extremely high potentials. [Rutherford, Barnes, and Robinson, *PM, 30* (1915), 353.]

new Coolidge x-ray tube. With the few students still in the laboratory, he tried to "see how far [the quantum] relation holds for the excitation of x-rays by electrons."[56] Up to 115,000 V the tube was supplied with electricity from a motor-driven Wimshurst generator; from there up to 175,000 V they used an induction coil capable of producing a 20-in. spark. Because the tube produced x-rays of unprecedented hardness, the effective wavelengths were very small and the angles required by crystal diffraction were correspondingly minute. To estimate the wavelengths, they were forced to use the imprecise relation of absorption coefficient to wavelength.

The results of the survey are shown as Figure 9.6. The calculated frequency of what Rutherford termed the *end* radiation is essentially the highest-frequency x-ray produced by the indicated voltage. They found systematic deviation from the quantum relation at high potentials; rather than increasing in proportion to the applied potential, the end frequency reached a maximum value at about 142 kV. Rather than following the Planck relation $hv = E$, the data conformed to $hv = E - kE^2$, where k is a constant. Rutherford realized that these data were only approximate. By this

[56] Rutherford, J. Barnes, and Robinson, *PM, 30* (1915), 352.

time, he was convinced that the Bohr–Kossel model of x-ray production could explain the systematic deviation as energy lost to the electron as it penetrates to the more tightly bound orbits within the atom. But in 1915, knowledge of x-ray absorption edge energies was insufficient to allow evaluation of further tests, and Rutherford, pressed by wartime work, did not pursue any.

Rutherford's primary reason for undertaking this study had little to do with x-rays; he still hoped to find evidence in support of his hypothesis that characteristic γ-rays originate in the electronic structure of the atom. He did find some x-ray frequencies at high potentials that were shorter than those estimated from absorption coefficients of what he still thought were selected γ-ray lines. But the saturation in frequency made it clear that a definite limit prevented the production of x-rays as penetrating as true hard γ-rays such as those from radium C. He wrote only one more paper on x-rays, and then after the war moved back into experiments entirely on radioactivity. The most influential voice in Britain on x-rays continued to be Charles Barkla's.

SELECTIVE X-RAY ABSORPTION

The chief means of measuring x-ray quality in Britain remained the rate at which the rays are absorbed by matter. Even after the introduction of x-ray spectroscopy, absorption coefficients continued to be the primary test. In 1912 E. A. Owen introduced a semiempirical relation to express the absorption coefficient μ for homogeneous x-rays defined for the standard absorbing material, aluminum.[57] If the rays originate in an anticathode of atomic weight A, then

$$\mu \propto A^{-5} \tag{9.3}$$

In 1914, using Moseley's data, Rutherford could reformulate Owen's relation in terms of x-ray frequency:[58]

$$\mu \propto \nu^{-5/2} \tag{9.4}$$

This was extended by Darwin and independently by Bragg and Peirce to

$$\mu \propto Z^4 \lambda^{5/2} \tag{9.5}$$

---

[57] Owen, *PRS, 86A* (1912), 426–39.
[58] Rutherford and Andrade, *PM, 28* (1914), 273.

where Z is the atomic number of the x-ray source.[59] Equation 9.5 became known in Britain as the *Bragg-Peirce law*, and it offered some convertability between spectroscopic and absorption results for x-rays. Rutherford had used it to convert his measures of highly penetrating x-rays to frequency in Figure 9.6. As Bragg said, "the older method of defining the quality of x-rays must naturally give place to the more quantitative and direct definition in terms of wavelength."[60] By this he did not mean that quality determination by absorption had outlived its usefulness, only that the results should be expressed in terms of x-ray wavelength. There were still many in Britain who questioned whether spectroscopy really offered a "more quantitative and direct" definition of x-ray quality.

There were good reasons for skepticism. The empirical relation between x-ray wavelength and the coefficient of absorption, equation 9.5, represents only the general course of the function. It was strongly suspected that resonant absorption of x-rays should occur. The analogy to visible light suggested it; the fluorescent stimulation of characteristic x-rays harmonized with it; and the Kossel model of x-ray origin fairly demanded it. But before 1914 only very imprecise data existed on resonant x-ray absorption. Barkla and Sadler had found in 1909 that the absorption coefficient for characteristic K x-rays from a given material is greatest when the absorbing substance is the same as that producing the rays.[61] Manne Siegbahn, who by 1914 was developing a first-rate laboratory for x-ray research in Sweden, pointed explicitly to the close parallel to preferential optical absorption in 1914.[62]

As discussed in Chapter 8, the first detailed study of x-ray absorption using spectroscopic methods was done by Ernst Wagner in 1914. He found that absorption increased dramatically at certain wavelengths that were characteristic of the absorbing material. But unlike the optical case, in which absorption sharply peaks at the resonant frequency, the absorption coefficient for x-rays begins to rise some distance on the high-frequency side of the actual resonant peak. Superimposed on the general equation 9.5 is a series of "edges" shown schematically in Figure 9.7. The

[59] W. H. Bragg and Peirce, *PM, 28* (1914), 626-30.
[60] *Ibid.*, 626. Albert Hull and Marion Rice showed in 1916 that the exponent 5/2 is closer to 3; *PR, 8* (1916), 326-8.
[61] Barkla and Sadler, *PM, 18* (1909), 739-60.
[62] Siegbahn, *PZ, 15* (1914), 753-56.

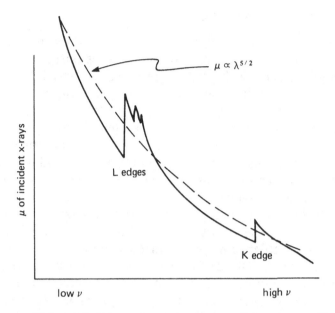

Figure 9.7. Schematic representation of the superposition of x-ray absorption edges on the law relating absorption coefficient to wavelength.

edge of each tooth always exceeds by a small fraction the frequency of the associated x-ray emission line. In their most primitive form, these were the data that Walter Kossel had used to formulate the idea that characteristic x-rays arise from electron transitions to the innermost shells of the atom.

The existence of the strong x-ray absorption features raised serious obstacles for tests of the x-ray photoeffect relation. If an incident x-ray loses energy before it stimulates an electron to leave the inner orbits, the direct quantum balance no longer holds. The "work function" in the Einstein photoeffect law is no longer negligible, nor is it constant for a given cathode. Electrons ejected from different Bohr orbits will lack differing amounts of the incident x-ray energy. This was the reason for the deviations Rutherford had found from the quantum transformation relation. To keep an accurate balance and test the quantum relations, one had to know the energy depth of the low-lying orbits. This information could be obtained only through careful study of the x-ray absorption edges, a research field that was in its infancy at the outbreak of World War I.

BARKLA'S NEGATIVE INFLUENCE

There was another way to study x-ray absorption, by the traditional means of x-ray penetrating power. Charles Barkla was the acknowledged master of the technique in Britain. Already convinced that characteristic x-rays form a line spectrum, in 1913 he briefly tried to apply crystal diffraction methods.[63] The trial reinforced his willingness to speak of true wavelengths for what had formerly been, for him, x-ray impulses, but he did not long continue the new methods.[64] After his appointment at Edinburgh in 1913, he began studies using absorption methods on the secondary electrons released by x-rays.[65] With a student, he measured the range of the electrons produced by homogeneous x-rays incident on metals, varying the pressure of the gas through which the electrons passed until the ionization current across a fixed distance reached saturation.[66] The density of the gas then gave a rough measure of the maximum energy of the electrons, and it could be reduced to actual values of electron velocity by applying a different "Whiddington's law" stating that the range of an electron is proportional to the fourth power of its velocity.[67]

The analysis convinced Barkla that, unlike the case of the optical photoeffect, the velocity of the electron does not depend at all on the material from which it is ejected; it depends, he claimed, only on the wavelength of the incident x-ray. He tested calcium, zinc, and tin using K x-rays from both silver and tin, finding that saturation occurred at the same gas pressure within 5 to 10 percent for each of the three test materials. Thus, Barkla convinced himself that no work function was subtracted from the energy donated by the x-ray. Subsequent studies, undertaken in preparation for his Bakerian Lecture to the Royal Society in 1916, led him to propose that x-ray energy is not always absorbed in whole quanta.[68] His convoluted argument depended on comparison of the energy of secondary K x-rays to the energies of secondary electrons excited at

[63] Barkla and Martyn, *PPSL, 25* (1913), 206–15.
[64] For a general account, see Stephenson, *AJP, 35* (1967), 140–52.
[65] Barkla and Philpot, *PM, 25* (1913), 832–56.
[66] Barkla and Shearer, *PM, 30* (1915), 745–53.
[67] Whiddington, *PRS, 85A* (1911), 328. See also Beatty, *PRS, 89A* (1913), 314–27.
[68] Barkla, *Nature, 94* (1915), 671–2; *95* (1915), 7.

the same time by homogeneous primary x-rays. But his experiments were able only to detect maximum energies, and rested on his assumption that no energy is lost to a work function in the ionization process.[69]

Barkla's researches were accorded more respect than they deserved. Richardson tried in 1917 to disprove his contentions, first by demonstrating that the conclusions did not follow from the data and second by an appeal to the fact that the Kossel model of x-ray emission forbids the appearance of the full x-ray quantum in the ejected electron.[70] But Barkla's reputation had been built on his solid early work and was not easily dislodged. His claims against the quantum theory and for the existence of a new high-frequency characteristic x radiation, the *J series,* muddied the waters of x-ray research between 1916 and 1923. Part of the difficulty was certainly that measurement of secondary electron energy was limited to gross methods sensitive only to the maximum electron velocity in what was actually a full distribution of velocities.

Barkla's view was remarkably influential into the 1920s.[71] Even as Owen Richardson tried to disprove Barkla's antiquantum contention, it affected research being done in Duane's laboratory at Harvard. In 1918, with his student, Kang Fuh Hu, Duane compared x-ray emission frequencies with their associated absorption edges. Despite the Kossel–Sommerfeld model, there was no clear understanding at Harvard of the differences to be expected between absorption and emission frequencies. They found that the wavelength of the emission peak was uniformly 0.3 percent greater than that for absorption. "It appears," they concluded, "that either the critical absorption wavelength is about 1/3 percent shorter than that of the [emitted] line, or else the wave lengths do not correspond to the centers of the peak in the emission spectrum."[72]

Duane's student, Hu, tried to see whether β-ray spectroscopy would contribute to an understanding of the differences. Applying

[69] Barkla, *PTRS, 217A* (1917), 315–60.
[70] O. Richardson, *PRS, 94A* (1918), 269–80.
[71] Barkla's contention that x-rays are not absorbed in entire quanta was the main point discussed by Richardson, M. de Broglie, Rutherford, Ehrenfest, Langevin, Brillouin, Lorentz, Bragg, Millikan, and Barkla at the Third Solvay Congress in 1921. The long discussion followed a presentation by Maurice de Broglie on the x-ray photoeffect. *Solvay III,* pp. 101–30.
[72] Duane and Hu, *PR, 11* (1918), 490–1.

a constant potential to his rhodium anticathode x-ray tube, Hu found that the same maximum electron velocity was elicited from both silver and lead radiators.[73] Two peaks appeared in electron velocities from silver, each within 5 percent of that predicted by the quantum relation $1/2 \ mv^2 = h\nu$ for the K_α and K_β x-ray emission lines from silver. Moreover, the x-rays from silver, incident on tin, released electrons of the velocity predicted by the quantum relation for K_α and K_β x-rays from both silver and tin. But the agreement was no closer than 5 or 6 percent; Hu reported that the electron velocity was always smaller than that expected from the quantum relation and the tube potential. "Experiments are still in progress," he said, "to see if there is a real difference from the simple equation $1/2 \ mv^2 = h\nu$, such as would be suggested by the term '$-p$' in Einstein's form of the equation."[74] Apparently the experiments, if completed at all, were inconclusive, but Hu published no more on the x-ray photorelations. Nor were the studies carried further by others at Duane's laboratory, where β-ray spectroscopy was not a standard tool for research.

ATTITUDES TOWARD THE LIGHTQUANTUM

Before 1921 the clearest evidence for spatial localization in radiation came from Millikan's results for the photoelectric effect. But this case of quantum absorption concerned visible light, in which the localized properties were not strongly pronounced and the resistance of orthodox wave theory was stiffest. Futhermore, Millikan's data concerned only one point in the spectrum of photoelectron velocities: the maximum velocity. In this period, Millikan himself opposed the lightquantum hypothesis, calling it "reckless," "untenable," and even "erroneous."[75] Despite his uncontested proof of Einstein's linear photoeffect relation, Millikan never doubted, until 1922, that an explanation of the result would be found without the suspect lightquantum. His particularly strong opposition may be attributed both to his respect for European mathematical physics – he had studied under Planck and Nernst in Berlin – and to the tradition that his teacher, Albert

[73] Hu, *PR, 11* (1918), 505–7.
[74] *Ibid.*, 506. I have corrected an obvious misprint in the equation.
[75] Millikan, *PR, 7* (1916), 355; *Electron* (1917), 230.

Michelson, had developed in American precision measurement in wave optics.[76]

Nor did William Duane adopt the lightquantum before 1922. His research had concentrated on the production, not the absorption, of x-rays. It was in the latter that the paradoxes of radiation were most strongly expressed. Duane's work contributed to the wider acceptance of the quantum relation, but it sidestepped the issues that forced consideration of the lightquantum. Duane was able to separate his empirical x-ray frequency relationship entirely from questions regarding the energy transferred in interactions; in 1916 he derived the Planck black-body law without, he claimed, any assumption that energy and frequency are related.[77] Because the entire kinetic energy of an individual electron is not necessarily radiated as an x-ray quantum, he felt that the energy in the rays could be measured only through appeal to empirical relations such as Whiddington's fourth-power law.

"We are not therefore compelled to explain the emission of radiation in quanta hv," Duane said; one had only to explain how an electron with a certain kinetic energy "can produce radiation of frequency up to, but not greater than, that given by [the quantum condition]."[78] It was a curious evasion. Following Rutherford, Duane offered a qualitative mechanism: High-frequency x-ray vibrations are associated only with the innermost electrons of an atom, to which only the highest-velocity electrons can penetrate. Thus, in 1918, Duane himself was blind to the message that emerged from his student Hu's β-ray velocity studies. The work was not continued at Harvard, and Duane nowhere, to my knowledge, made any attempt to reconcile Hu's preliminary results with his own proposed dissociation of frequency and energy.[79]

Two facts may be discerned about the experiments carried out

---

[76] Millikan had studied under Planck and Nernst in Berlin and had heard lectures by Poincaré in Paris. He sometimes published major results in German journals, including six in the period 1912–17. For the influence of Michelson, see Millikan, *Autobiography* (1950), 66.

[77] Duane, *PR, 7* (1916), 143–7.

[78] Duane, *Science, 46* (1917), 348.

[79] While Duane had worked in Paris with the Curies, he had collaborated with Jean Danysz, one of the early developers of β-electron spectroscopy. Duane may have just ended that collaboration when Danysz began the electron analysis that Duane never utilized.

on the quantum transformation relations before 1921. The first is that the most important work was done outside Britain and Germany. Although important background work had been accomplished in Rutherford's Manchester laboratory, the intervention of the war effectively stopped this research. Moreover, in Britain, Barkla refused to believe that absorption of x-ray energy occurs in quanta, and his views were influential. Richardson strongly questioned Barkla's position but was nontheless convinced that some interpretation of the photoeffect, perhaps his own, would solve the problems without the "restricted and doubtful hypothesis used by Einstein."[80] J. J. Thomson was never able to accept the quantum theory. Partly in response to his dissatisfaction with the new direction of physics, he relinquished his Cavendish Professorship to Rutherford in 1918. There were few serious empirical studies on x-ray or γ-ray absorption in Germany at all before 1921. The significant empirical work came from America and, as we shall see in the next chapter, from France.

The second thing to note is that experimentalists were no more receptive to spatially localized lightquanta than were their more mathematically inclined colleagues. It might be supposed that the evidence for quantum absorption would convince empirically minded physicists that some radical revision of the wave nature of light would have to be found. But as we have seen, this was not the case. Empiricists were no less demanding than other physicists that the well-verified evidence for interference and diffraction of radiation be granted priority. In any event, the evidence for quantum absorption of x-rays and γ-rays, where the localized properties of radiant energy are most pronounced, was sparse before 1921. But in the first months of that year, this circumstance changed abruptly, an event to which we turn in the next and the final part of this study.

[80] O. Richardson, *PRS, 94A* (1918), 269.

PART V
The conceptual origins of wave – particle
dualism
1921 – 1925

10

Synthesis of matter and light

A group of electrons that traverses a sufficiently small
aperture will exhibit diffraction effects.[1]

It is not entirely surprising that the earliest reaffirmation of the
dualism inherent in radiation came neither from Britain nor
Germany. There the issues had been recognized, discussed, and
dismissed over the preceding twenty years. And American research traditions were largely derivative of German and British
models; Duane and Millikan had both studied under Planck, and
Richardson did research with Thomson in Cambridge. It took an
outsider to these traditions to raise the issues in a way that began to
convince others.

That process began in France. Maurice and Louis de Broglie
were, by any standards, unusual physicists. Amateurs in an age of
professionals, possessing a confidence born of noble status, working on experiments that had only the meagerest precedent in their
homeland, they were not as firmly bound by the predispositions
that prevented physicists of other nations from directly assessing
the spatially localized lightquantum.

Their conversion was real. Maurice had once been convinced
that x-rays and γ-rays fit neatly into the electromagnetic spectrum.[2] In his discussion, *La nature des rayons de Röntgen* in 1915,
there was no thought that x-rays might be anything other than
periodic waves. As late as May 1920, he spoke freely of x-ray
"wavelengths" and discussed the optical fluorescence that results
from degradation of x-ray "frequency."[3] But late in 1920 he reviewed the new evidence accumulated at high x-ray frequencies.
There he hinted at unspecified difficulties in the orthodox wave

[1] L. de Broglie, *CR, 177* (1923), 549.
[2] M. de Broglie, *Revue scientifique, 53* (1915), 581.
[3] M. de Broglie, SFP *PVRC, 1920* (1921), 33.

interpretation of x-rays. In order to establish the wave nature of x-rays on as "solid" a basis as that of light, he said that one now had to ask whether "old ideas about periodic radiation propagating as waves" might not be "subject to serious revision."[4] What he had in mind became clearer just a few months later, early in 1921, when he forcefully revived the issue of the paradoxical properties of radiant energy.

CORPUSCULAR SPECTRA

Maurice de Broglie was the eldest son in a family traditionally prominent in French politics. In 1906 he assumed the hereditary title "Duc de Broglie," and at about the same time he outfitted a private laboratory in his Parisian residence.[5] Although educated in physics at the Collège de France, he was a scientific amateur, a species common in the nineteenth century but increasingly rare in the twentieth. In 1908, when he defended his theses following work with Paul Langevin, his research concerned ionization of gases and Brownian motion.[6] He soon took an interest, then uncommon for a Frenchman, in the new quantum theory. He was secretary to the First Solvay Congress in 1911 and, with Langevin, edited its proceedings.[7] With the discovery of x-ray interference in 1912, he reorganized his research plans. The following year he made a notable discovery: The slow rotation of a crystal greatly reduces the spurious x-ray interference maxima caused by periodic imperfections in the crystal lattice.[8] The rotation smooths out what Moseley had called the "horrid diffraction fringes" caused by natural ruling errors in the crystal, which greatly confused early crystallographic analysis.[9] The new technique was an important step forward and was, for example, indispensable to Rutherford in his 1914 measurement of what he thought were γ-ray wavelengths.

De Broglie did some of the most significant experimental re-

[4] M. de Broglie, *Scientia, 27* (1920), 110–11.
[5] Weill-Brunschvicg and Heilbron, *DSB, 2* (1973), 487–9. La Varende, *Les Broglie* (1950), 265–320. The laboratory was first established in de Broglie's apartments in rue Chateaubriand; later, another was set up in rue Lord Byron a block away in the 8th arrondissement.
[6] M. de Broglie, *Thèses* (1908).
[7] *Solvay I.*
[8] M. de Broglie, *CR, 157* (1913), 924–6. See also Heilbron, *Isis, 58* (1967), 482–4.
[9] Moseley to his sister, 13 August [1913]. Heilbron, *Moseley* (1974), 207–8.

search on x-ray absorption edges. In 1914, on the suggestion of Bragg and Siegbahn, he successfully interpreted anomalous absorption bands recorded on a photoplate for x-rays from silver as if they were due to preferential absorption of the x-rays by the silver halide emulsion.[10] There soon developed a lively race among Siegbahn, Wagner, and de Broglie to measure the precise frequencies of the absorption edges. Wagner found that there were two *L* edges.[11] In confirming this result in 1916, de Broglie found a third.[12] At war's end he dedicated his research to x-ray absorption; the frequencies of the edges gave the most direct measurement of the energy levels of the Bohr atom. The discovery of three *L* edges, five *M* edges, and so on directly shaped ideas about the electronic structure of the atom as interpreted by Bohr and Sommerfeld. De Broglie had entered a front-rank research field and possessed the skill and facilities to make significant contributions.

By January 1921, de Broglie realized the potential value of β-ray spectroscopy to his research. He had noted in 1920 that x-rays stimulate secondary β-electrons,[13] but now recognized that this held the key to a study of energy transformation in the absorption of x-rays. His inspiration likely had much to do with Alexandre Dauvillier, who joined de Broglie's research group late in 1920. Dauvillier's research had already moved from studies of x-ray absorption to an examination of the kinetic energies of the secondary electrons released by x-rays.[14] He began to investigate the chemical effects of electrons on photoemulsions and carefully studied the work by Barkla in which the electron range was used to fix electron energy. Early in 1921, Dauvillier completed the most thorough review in the French language on the photoelectric effect, with the explicit intention to apply that knowledge to the case of x-rays. "The fastest electron emitted by a body exposed to [x-rays] possesses precisely the same velocity as the fastest electron in the cathode beam that produces the x-rays," he claimed.[15] Dauvillier was also perplexed that the energy granted an electron by x-rays is independent of the distance over which the rays have

[10] M. de Broglie, *CR, 158* (1914), 1493–5.
[11] Wagner, *AP, 46* (1915), 868–92.
[12] M. de Broglie, *CR, 163* (1916), 352–5.
[13] M. de Broglie, SFP *PVRC, 1920* (1921), 33.
[14] Dauvillier, *Thèses* (1920).
[15] Ledoux-Lebard and Dauvillier, *Physique des rayons x* (1921), 208.

traveled.[16] "The electromagnetic and corpuscular theories seem to be very difficult to reconcile," he said. "The problem of the nature of radiation remains unresolved."[17]

In January 1921 a South African physicist, encouraged by Owen Richardson, published a study that further directed the Parisian physicists' attention to these problems.[18] Lewis Simons, like Richardson, was convinced that Barkla's antiquantum claims were simply wrong: A single homogeneous x-ray beam gives rise to an entire spectrum of electron velocities, and different amounts of energy are lost to each electron depending on its original position in the atom. Simons worked hard, but unsuccessfully, to extract convincing evidence for this claim from ionization and electron-range data. About β-ray spectroscopy, the suggestive early work by Robinson and Rawlinson, and the incomplete experiments by Hu, Simons had apparently not heard.

Maurice de Broglie had the great advantage of knowing all of these prior experiments. But he was skeptical of the claim that x-ray energy is always absorbed by electrons in integral multiples hv. When he first took up the detailed evaluation of electron energy in February 1921, he thought he had found evidence that electrons *leave* the atom with quanta of energy hv and therefore must have absorbed more than that quantized amount from the x-rays.[19] His studies of x-ray absorption edges and his understanding of the Bohr atom made it clear that some energy must be spent to remove each electron from the atom. This seemed a significant issue to clarify for presentation at the Third Solvay Congress, to which he had been invited and which would meet in two months' time. Besides, a recent disappointment in his attempt to enter the Paris Academy of Sciences made a certain broadening of his experimental inquiry attractive.[20]

And so, Maurice de Broglie, intrigued by the paradox that

[16] *Ibid.,* 399.

[17] *Ibid.,* 407–8.

[18] Simons, *PM, 41* (1921), 120–40. De Broglie knew of Simons' paper, see *Solvay III*, p. 85n.

[19] M. de Broglie, SFP *PVRC, 1921* (1922), 10.

[20] On 22 November 1920, Maurice de Broglie was nominated to replace Adolphe Carnot in the Academie des Sciences. On the 29th, after four votes, Jules Louis Breton was elected, de Broglie having received only one of about sixty votes cast in each round. *CR, 171* (1921), 1031, 1045. One of his biographers reports that Maurice de Broglie was later elected on the basis of his studies on the photoeffect. Sudre, *Revue des deux mondes* (July–Aug. 1960), 577–82.

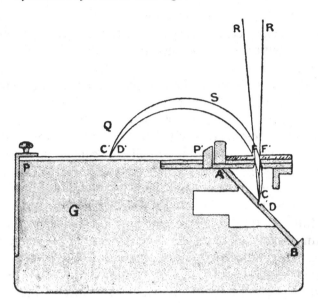

Figure 10.1. Maurice de Broglie's x-ray photoeffect velocity spectrometer. [*Rayons x* (1922), 142.]

Dauvillier had proposed, and confident he could overcome Simons' methodological difficulties, recognized that progress in quantum theory might follow from use of the electron velocity apparatus built by Rutherford, Robinson, and Rawlinson. He quickly adapted and modified the Manchester β-ray velocity spectrometer and employed it to analyze x-ray absorption using what he called *spectres corpusculaires,* or particle velocity spectra.[21] In the apparatus shown in Figure 10.1 x-rays strike the sample element placed at *C*. Secondary x-rays are unaffected by the magnetic field that is imposed in a direction normal to the plane of the page, but secondary electrons are deflected into circular paths, and thus sorted by velocity, forming characteristic patterns of bands – the corpuscular spectra – on the photoplate *PP'*.

In collaboration with his younger brother, Louis, Maurice de Broglie immediately formulated the goals of his research in terms of the Bohr model of the atom.[22] By April 1921, they agreed with Simons and Richardson that Barkla was wrong. Each electron in

[21] M. de Broglie, *CR, 172* (1921), 274–5, 527–9.
[22] M. and L. de Broglie, *CR, 172* (1921), 746–8. L. de Broglie and Dauvillier, *CR, 172* (1921), 1650–3; *CR, 173* (1921), 137–9.

an atom, granted energy $h\nu$ by the incident monochromatic x-ray beam, must spend a certain amount of that energy to escape the atom. The amount was always just the "depth" of the orbit from which the electron came, and those depths had been carefully charted in de Broglie's private laboratory in rue Chateaubriand. A monochromatic beam of frequency ν will therefore produce electrons of energy $h\nu - h\nu_K$, $h\nu - h\nu_L$, and so forth, where ν_K and ν_L are the measured frequencies of the K and L absorption edges for the cathode material. Measurement of the kinetic energy of the released electrons therefore yields directly the energy level of the vacated Bohr orbit. In this sense, the x-ray photoeffect law could not be independently verified, but the systematic deviations from the quantum relation could be explained by an appeal to the growing understanding of the electronic structure of the atom.

De Broglie began these studies in order to shed light on the atom and to verify his previous x-ray absorption results. "There is no doubt," de Broglie wrote to his friend, Frederick Lindemann, "that radiation of frequency ν expels from an element R groups of electrons with energies corresponding to

$$h\nu - \text{energy level } K$$

$$h\nu - \text{energy level } L$$

[and so on]. One can therefore find the energy levels of the secondary radiator using an [analyzing] crystal as intermediary."[23] He was impressed by the clear evidence that each feature in the x-ray spectrum is mirrored by a set of well-defined electron velocity bands in the corpuscular spectrum. "One recovers all the properties of the x-ray spectrum," he pointed out, "including lines and bands."[24] In March 1921 he had corroborated this result in ten elements, ranging from copper (29) to ytterbium (70).[25] For example, x-rays from a rhodium anticathode incident on a copper plate give rise to seven distinct electron velocity bands. Three correspond to the K_α and K_β x-rays from copper, each diminished by the absorption energies $h\nu_L$ and $h\nu_M$ for copper. What would be four lines only appears as three because the frequency difference $K_\beta - K_\alpha$ is equal to $h\nu_L - h\nu_M$, as shown in Figure 10.2. The

[23] M. de Broglie to Lindemann, 12 April 1921. Cherwell papers.
[24] M. de Broglie, *CR*, *172* (1921), 275.
[25] *Ibid.*, 806–7.

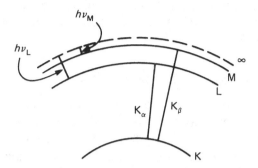

Figure 10.2. Schematic representation of the inner orbits of the copper atom to explain Maurice de Broglie's x-ray photoelectric measurements.

electron kinetic energy that remains after a K_β x-ray ionizes an L-level electron is $K_\beta - hv_L$. This is identical to that from the process $K_\alpha - hv_M$. So, only three lines appear. The other four bands correspond to the K_α and K_β x-rays from rhodium, each diminished by hv_L and hv_M for copper.[26]

By April 1921, when he reported these results to the Third Solvay Congress, de Broglie was also aware of their significance for the photoelectric relations.[27] Millikan had been able to compare optical frequencies only to the *maximum* electron velocity; de Broglie could offer corroboration at points *within* the electron distribution as well. Thus, after introducing corpuscular spectra as a means to analyze the atom, he proceeded to a general discussion of the photoelectric effect. Assuming the validity of the Bohr–Kossel model for x-ray emission, it was clear that the full energy quantum of the x-ray is transferred to individual electrons. Coupled with the evidence that each x-ray frequency gives rise to a particular set of electron velocities, it became virtually impossible for de Broglie to avoid the issue of the lightquantum. On several occasions in his presentation, he alluded to the "remarkable" behavior of the absorption process when the energy of secondary electrons is found to be the same as that of the entire spherical

[26] M. de Broglie, *JP, 2* (1921), 265–87; SFP *PVRC, 1921* (1922), 49.
[27] M. de Broglie, "La relation $hv = E$ dans les phénomènes photoélectriques; production de la lumière dans le choc des atomes par les électrons et productions des rayons de Röntgen," p. 12. Typescript draft of de Broglie's contribution to the Solvay proceedings. Solvay archives.

x-ray wave.[28] The radiation "must be corpuscular," he said, "or, if it is undulatory, its energy must be concentrated in points on the surface of the wave."[29]

C. D. ELLIS AND NUCLEAR γ-RAYS

Transformation experiments on γ-rays are limited to absorption phenomena. There was no way to stimulate the emission of γ-rays; they are a by-product of radioactive decay. The same difficulties that prevented analysis of the x-ray photorelation until the early 1920s also hampered research on the absorption of γ-rays. In addition, studies were complicated by the problems of directly measuring γ-ray wavelengths. Rutherford thought he had succeeded in 1914 in measuring wavelengths for a few homogeneous γ-ray groups, but, as we have explained, these were characteristic x-rays. The experimental difficulties of true γ-ray wavelength determination were immense. The next direct measures were attempted only in 1919 and were not successful until 1924.[30]

The work done on γ-rays in the meantime relied almost entirely on absorption coefficient measurements. A series of studies by Alexander Russell and Frederick Soddy beginning in 1909 was based on prior work by H. W. Schmidt on the absorption coefficients of γ-rays from uranium X, radium C, thorium, and actinium.[31] As discussed in Chapter 8, these experiments led to Rutherford's first interest in homogeneous groupings of γ-rays and encouraged his hypothesis that the characteristic γ-rays originate in the vibration of special groups of electrons near the center of the atom. His students, Robinson and Rawlinson, applied β-ray spectroscopy in 1914, seeking data on the electrons released by homogeneous γ-rays from matter.[32] For the same incident radiation, electrons released from gold (79) were found to be consistently 1 or 2 percent faster than those released from lead (82). To an extent not then found for the optical or x-ray case, photoelectron velocities due to γ-rays appeared to depend markedly on the material from which the electrons are emitted.

[28] *Ibid.*, 1, 12; *Solvay III*, pp. 80, 85, 89.
[29] *Ibid.*, 13; *Solvay III*, p. 89.
[30] Thibaud, *Thèses* (1925).
[31] F. and W. Soddy and Russell, *PM, 19* (1910), 725–57. Russell, *PRS, 86A* (1911), 240–53.
[32] Rutherford, Robinson, and Rawlinson, *PM, 28* (1914), 281–6.

During the war, research on these topics virtually stopped in Britain, the only country where it had been pursued at all. In 1917 Rutherford reassessed the few "γ-ray" wavelengths he had measured before the war[33] in light of new studies by Rice and Hull in America on the compass of the presumed relation between absorption coefficient and x-ray frequency.[34] Formerly, equation 9.5 or its equivalent had been shown to hold for $\lambda > 6 \times 10^{-9}$ cm; now it appeared to be valid down to 10^{-9} cm. The overlap that Rutherford had found between x-rays produced at high potentials and the secondary x-rays he believed to be γ-rays was here independently corroborated. In some cases, x-ray wavelengths derived from absorption behavior were smaller than the shortest wavelength (0.072 Å) found for "γ-rays" by Rutherford.

If the smooth transition between x-rays and γ-rays was to be believed, and Rutherford did not yet doubt that it was, the majority of γ-rays correspond to x-rays produced by potentials greater than a half million volts. "It would thus appear probable," Rutherford concluded, "that the absorbed groups of γ-rays are due to the conversion of the energy, $E = h\nu$, of a wave of frequency ν into electronic form and that consequently the energy of the β-ray groups may be utilized by the quantum relation to determine the wavelengths of the penetrating gamma rays."[35] Thus, the photoelectric relation, proven rigorously for the optical case and inferred on the basis of the Bohr atom for x-rays, was simply assumed to hold for γ-rays. By this means, wavelength measurements could be extended into the highest-frequency region despite ignorance of absorption behavior.

The man who put this program into practice, within two months of de Broglie's first papers on corpuscular spectra, was Rutherford's student at Cambridge, Charles Ellis. Even before the 1921 Solvay meeting, Rutherford had sent Ellis' unpublished results to de Broglie.[36] At the congress, still before Ellis' study had appeared in print, Rutherford announced that Ellis' results for γ-rays were in full agreement with those of de Broglie.[37] But rather than empha-

[33] Rutherford, *PM, 34* (1917), 153–62.
[34] Hull and Rice, *PR, 8* (1916), 326–8.
[35] Rutherford, *PM, 34* (1917), 161.
[36] M. de Broglie, *JP, 2* (1921), 269n.
[37] Rutherford, *Solvay III*, pp. 107–9. Ellis' first study was submitted to the Royal Society of London on 22 April 1921, sixteen days after the end of the Solvay Congress.

size, as de Broglie had, the significance of the result for energy localization in the radiation, Ellis used his result to inquire about atomic structure. By comparing the frequency of the incident γ-ray with the resulting electron velocity, "it should be possible to deduce," Ellis said, "from what part of the atom the electron originated."[38] One can here sense the growing uncertainty at Cambridge about Rutherford's model for characteristic γ-rays.

Ellis used as the source a small tube of radium emanation, which emitted the known γ-ray spectrum from radium B and radium C. The material to be tested was wrapped as a thin foil tightly around the tube, and the resulting electrons dispersed by velocity differences in a strong magnetic field. He submitted uranium (92), lead (82), platinum (78), tungsten (74), and barium (56) to the rays, and in each case found three well-defined velocity peaks in the electron velocity spectrum. He assumed that the three were due to three frequencies of γ-ray, which he denoted by energies W_1, W_2, and W_3. In each case, the observed electron energy E should be one of the unknown γ-ray energies minus an energy W_n characteristic of the atom under test.

Because the W_n values varied widely for the same presumed γ-ray energy, Ellis concluded that the released electrons had to come from the innermost rings of the atom. For example, the slowest electrons of the three groups from tungsten possess 8,000 eV more kinetic energy than the equivalent electrons stimulated in platinum. Using the most recent values of the K absorption edges for the test elements, Ellis could calculate the actual values of the γ-ray energies W_1, W_2, and W_3. The results were remarkably consistent for the five elements for which data was available, as shown in Table 10.1. The agreement across horizontal rows elicited Ellis' conclusion that the data were "in complete agreement with Rutherford's theory of the connection between the β-ray and γ-ray emission of radioactive atoms."[39]

Rutherford's theory was at the moment in a state of flux. Until the early 1920s, he had proposed special groupings of electrons in the superstructure of the atom. These were supposed to resonate to the harmonies of a relativistically speeding β-electron and then emit a γ-ray frequency characteristic of the atom. With the resumption of research in Cambridge after the war, it was clear that

[38] Ellis, *PRS, 99A* (1921), 261.
[39] *Ibid.*, 265.

Table 10.1. *Ellis' γ-ray photoelectric derivation of three γ-ray energies, W_n, incident on five elements* $(\times 10^5 \text{ eV})$

	Element				
	Ba	W	Pt	Pb	U
W_1	—	2.35	2.36	2.38	2.40
W_2	2.9	2.89	2.91	2.91	2.92
W_3	—	3.46	3.46	3.46	3.48

the Bohr theory allowed no special groupings of the sort earlier imagined by Rutherford. The book that most stimulated Ellis's analysis was Sommerfeld's influential *Atombau und Spektral linien,* the first edition of which had appeared in 1919. Using the results Sommerfeld had assembled about x-ray absorption, Ellis saw a way to justify his assumption that radium B, an isotope of lead (82), emits electrons with velocities determined photoelectrically by six well-defined frequencies of γ-ray. He subtracted, in turn, the energy of the K shell and of Sommerfeld's hypothetical Λ' shell from each of the assumed γ-ray energies, and twelve of the faster electron velocity lines emerged, accurate to less than 1 percent. "All the lines are given without any gaps and no extra ones are predicted," he remarked with evident pride.[40]

Thereafter, β-ray spectroscopic results alone could be taken to provide data on the energy of γ-rays, an extrapolation de Broglie had already made for x-rays.[41] Using the few known γ-ray wavelengths, Ellis calibrated a scale that went far below the 0.07 Å limit then set by diffraction techniques. His full report was received by the Royal Society in February 1922; in it he charted the full spectrum of true γ-ray wavelengths from radium B, radium C, thorium C, and thorium D – some twenty lines in all. "The main γ-rays from radium B correspond to energies 2.4 [to] 4.0 × 10⁵ [electron] volts," he said. If these rays "owe their origin to a transition between stationary electron states in the atom," then the final electron orbit had to be "very deep in the atom and certainly

[40] *Ibid.,* 267.
[41] M. de Broglie, *CR, 172* (1921), 806–7; *173* (1921), 1157–60. Meitner, *ZP, 9* (1922), 131–44, 145–52. M. de Broglie and Cabrera, *CR, 176* (1923), 295–6.

274   *Conceptual origins of wave–particle dualism, 1921–1925*

inside the K ring."[42] James Chadwick had just repeated Rutherford's classic α-particle scattering experiments.[43] He placed particular emphasis on the region between the nucleus and the K shell and found no evidence that any electrons were there. Ellis therefore concluded early in 1922 that "γ-rays must come from the nucleus."[44]

LES FRÈRES DE BROGLIE

Louis de Broglie developed intellectually under the influence of his elder brother, Maurice. When their father died in 1906, Maurice was thirty-one and Louis fourteen, and the elder took responsibility for the younger's education.[45] "At every stage of my life and career," Louis later told his brother, "I found you near me as guide and support."[46] After completing a *licence* in history in 1910, Louis prepared for another in physics under Léopold Brizard, the same physicist who had first taught Maurice.[47] And Maurice saw to it that Louis had an advance opportunity to study the proceedings of the First Solvay Congress in 1911, of which Maurice was a secretary.[48] Louis later recalled that exposure to these discussions encouraged him "to determine the true nature of the mysterious quanta that Max Planck had introduced into theoretical physics."[49]

After service as radio telegraph operator on the Eiffel Tower during World War I, Louis began to work in physics. He was drawn to problems that arose directly out of his brother's research, at first an interpretation of Maurice's x-ray absorption results according to the Bohr atom.[50] With Maurice and, later, others at his laboratory – Alexandre Dauvillier, Jean Thibaud, Jean Trillat – Louis published a series of papers showing the congruence between x-ray absorption and optical absorption spec-

42 Ellis, *PRS, 101A* (1922), 13.
43 Chadwick, *PM, 40* (1920), 734–46.
44 Ellis, *PRS, 101A* (1922), 14. Whiddington, *PM, 43* (1922), 1116–26.
45 M. de Broglie in *Louis de Broglie* (1953), 423–9.
46 La Varende, *Les Broglie* (1950), 306.
47 M. de Broglie, *Grandes souvenirs, belles actualités: Le recueil du jeunes, 2* (July 1948), 4–6. See also L. de Broglie *Recherches d'un demi-siecle* (1976), 7–16.
48 *Solvay I.*
49 L. de Broglie in *Louis de Broglie* (1953), 458.
50 L. de Broglie, SFP *PVRC, 1921* (1922) 15–16.

tra.[51] These studies, begun in 1920, directed Louis' attention to the structure of the atom, not to the structure of radiant energy. Like most physicists of the time, he thought x-rays to be a form of high-frequency light wave.

The central problem in the application of x-ray spectra to questions of atomic structure was then to discover the number of electrons that occupy each of the allowed Bohr orbits.[52] Arnold Sommerfeld had gone furthest with a relativistic explanation for the appearance in x-ray spectra of *doublets,* closely spaced pairs of x-ray emission lines. The separation in frequency of the pair varies in a regular manner with the atomic number of the parent atom. Sommerfeld had explained this as due to a relativistic effect of the rapidly rotating electrons closest to the atomic core.[53] Louis de Broglie therefore confronted the successful application within his own research domain of Einstein's theory of relativity, a theory that, he said, had fascinated him since his early student days.[54]

Although the younger de Broglie's interests in 1920 centered on atomic theory, within a year they had changed significantly. Maurice's experiments showed with increasing clarity in 1921 that an x-ray transfers its *entire* quantum of energy to individual electrons, and Louis began trying to find an explanation. Years later, he recalled the influence that his brother's research had had on him. It "directed my attention to the quality of waves and of x-rays and γ-rays," he said.[55] "We debated the most pressing and baffling questions of the time, in particular the interpretation of the results of your experiments on the x-ray photoeffect," he told his brother. "The insistence with which you directed my attention to the importance and the undeniable accuracy of the dual particulate and wave properties of radiation little by little redirected my thought."[56] In 1921, following his brother's report to the Solvay

[51] L. and M. de Broglie, *CR, 172* (1921), 746–8; *173* (1921), 527–9. L. de Broglie, *CR, 173* (1921), 1160–2, 1456–8. L. de Broglie and Dauvillier, *CR, 175* (1922), 685–8, 755–6. L. de Broglie, *JP, 3* (1922), 33–45.
[52] His second thesis for the doctorate in 1924 was on the closely related topic "les propriétés chimiques des éléments interprétées par leur structure électronique."
[53] Sommerfeld and Wentzel, *ZP, 7* (1922), 86–92. For de Broglie's reaction, see his contribution in *Louis de Broglie* (1953), 459–60.
[54] L. de Broglie, *Nouvelles perspectives en microphysique* (1956), 232.
[55] La Varende, *Les Broglie* (1950), 306.
[56] L. de Broglie, *Savants et decouverts* (1951), 302. Trillat, Societé Français de minéralogie et de cristallographie *Bulletin, 83* (1960), 239–41.

Congress, Louis turned his attention to the counterintuitive behavior of x-rays that, in the words of his brother, concentrates energy "in points on the surface of the wave."[57]

Louis reviewed all of Einstein's publications on the nature of light. He was now convinced, like his brother, that a synthetic theory combining wave and particulate representations was required,[58] but he confronted the formidable difficulty that Einstein had yet been unable to resolve: a physical interpretation combining contradictory wave and particle properties. In November 1921 he presented a sketch of how he expected this synthesis might be attained: A complete theory of radiation should be built in a manner parallel to that of hydrodynamics, wherein wavelike laws govern the macroscopic behavior of large numbers of water molecules.[59]

In late 1921 he applied rudimentary relativity theory to "atoms of light" in order to analyze their statistical properties "without any intervention of electromagnetism."[60] Einstein had already treated an association of lightquanta as if subject to the laws of kinetic theory. But de Broglie took lightquanta to be literally "atoms of light," and unlike Einstein he defined a mass for them using the quantum relations and relativistic dynamics. Lightquanta, "supposed [to be] of equal very small mass," move at velocities dependent on their frequency, but in all cases very near to c.[61] The effective mass m is the sum of that due to the kinetic energy and a "proper" mass m_0. Because m_0 for an atom of light must be "infinitely small," the approximation $m = h\nu_0/c^2$ will suffice, but this is only an approximation. Using it, de Broglie derived the Wien radiation law and showed, *inter alia*, that a gas of lightquanta should exert only half the radiation pressure expected using the classical corpuscular theory of light.

Several matters served in 1922 and 1923 to reinforce Louis de Broglie's interest in the lightquantum and its eventual reconciliation with the wave theory of light. Maurice completed his written contribution to the Solvay Congress proceedings; the revision, finished early in 1922, pointed up in even greater detail the di-

[57] M. de Broglie, draft of Solvay III presentation, p. 13. Solvay archives. *Solvay III*, p. 89.
[58] L. de Broglie in *Louis de Broglie* (1953), 460.
[59] L. de Broglie, *JP, 3* (1922), 33–45; *CR, 175* (1922), 811–13.
[60] L. de Broglie, *JP, 3* (1922), 422.
[61] *Ibid.*, 422n, 428.

lemma of waves and particles. At the time of Maurice's verbal presentation in April 1921, it was clear only that the incident x-ray frequencies are replicated in the electron spectra.[62] Taking up the subject again in November 1921, de Broglie established "several more recent results," among them the finding that electron velocity lines appear even for stimulation of the radiator by its own characteristic x-rays.[63] That is, not only will the incident x-rays invoke a series of scattered electron velocity lines, each decreased in energy by the internal energy losses in the atom of the scattering substance, but they will also give rise to a series of velocity lines from any of the substance's own fluorescent x-ray lines, stimulated by the incident rays, again diminished by the energy levels of the atom.[64] And de Broglie restated the evidence for the particulate nature of x-rays. "Today," he said, "our attention is again directed to a set of phenomena that suggest the new idea of [quantum] emission." These results "present facts in such a way that their quantities are sometimes described in terms of the wave theory, sometimes in terms of the emission theory [of light]."[65]

The evidence for this dualistic interpretation was in fact little different from that recognized in 1907 by Bragg, and the elder de Broglie pointed to it in some *remarques* at the end of his 1922 book on the physics of x-rays. When a spherical x-ray wave passes through a gas, two unaccountable things happen: (1) only a very small fraction of gas molecules is ionized, releasing a fast electron; (2) the kinetic energy of an electron so freed is many orders of magnitude greater than the total radiant energy in the small solid angle of the wave that passes over the electron.[66] Maurice de Broglie thus forcefully revived interest in the paradoxes of quality and quantity; now they were considerably strengthened by the compelling evidence for quantum absorption of both x-rays and γ-rays.

WAXING INTEREST IN THE LIGHTQUANTUM

Maurice de Broglie's may have been the loudest voice in Louis de Broglie's ear, but his was not the only voice. In the year following

[62] M. de Broglie, draft of Solvay III presentation. Solvay archives.
[63] M. de Broglie to Lindemann, 5 November 1921. Cherwell papers.
[64] M. de Broglie, *Solvay III*, 80–100.
[65] M. de Broglie, *Les rayons x* (1922), 17, 19.
[66] *Ibid.*, 156–9. M. and L. de Broglie, *CR, 175* (1922), 1139–41.

Maurice's presentation of evidence for the spatial concentration of radiant energy, several leading physicists in Germany and England first gave serious consideration to Einstein's claims. Just five years before, in 1916, Einstein had restated his position. In the second part of a study now remembered only for its statistical treatment of emission and absorption of light, Einstein claimed that each light-quantum must possess a linear momentum of magnitude hv/c.[67] But he did not then stimulate the interest he thought appropriate for the hypothesis.

Planck delivered a talk on the "Nature of light" to the Kaiser Wilhelm Gesellschaft in 1919 in which he surveyed the difficulties that were then facing radiation theory.[68] Planck was still far from adopting the lightquantum and did not even discuss the hypothesis directly. Instead, he suggested that the triggering hypothesis might yet explain the photoeffect: According to relativity theory, each gram of matter constitutes 2×10^{10} kilocalories of energy. The mass of even the lightest atom is enough to account for the ejection of several electrons with the observed velocities. Planck said, however, that the quantum nature of light was the "foremost and most difficult dilemma in quantum theory,"[69] and once again relativity theory was invoked as a factor in its resolution.

By late 1921, opposition to the lightquantum began to wane. W. H. Bragg pointed to the "considerable new information" that suggested localized x-ray quanta.[70] He did not say so explicitly, but the information he had in mind was certainly the results de Broglie and Ellis had announced only a few months before. Einstein proposed an experimental test that he thought might distinguish between a wave and a lightquantum interpretation of Doppler shifts in the frequency of light radiated from moving ions.[71] At first, the experimental results of Hans Geiger and Walther Bothe seemed to support the lightquantum explanation.[72] But Paul Ehrenfest soon pointed out an error in Einstein's analysis that rendered the result inconclusive.[73]

---

[67] Einstein, Zürich *Mitteilungen, 18* (1916), 47–62; *PZ, 18* (1917), 121–8.
[68] Planck, *Wesen des Lichts* (1920), given 28 October 1919.
[69] *Ibid.,* 22.
[70] W. H. Bragg, *Nature, 107* (1921), 374.
[71] Einstein, Berlin *Sb* (1921), 882–3.
[72] Geiger and Bothe's results were not published. See Einstein, Berlin *Sb* (1922), 22.
[73] See Klein, *HSPS, 2* (1970), 1–39.

Far more remarkable than the proposed experiment was the reaction to it. Unlike everything else Einstein had written on lightquanta – a hypothesis taken seven years before as an example of how an otherwise great mind can err – his work in 1921 sparked lengthy discussion by a number of leading mathematical physicists.[74] Max Laue, who had recently begun to worry about properties of radiation "not yet explained by wave theory," engaged Einstein in a spirited debate before the Berlin Academy.[75] Ehrenfest initiated a long correspondence on the significance of the proposed experiment.[76] Gregory Breit published a detailed paper on the subject.[77] C. Wilhelm Oseen, a close friend of Bohr, and several others tried to see how close special solutions of Maxwell's equations would come to the properties of what they now called Einstein's "needle radiation."[78] Erwin Schrödinger, influenced by Einstein's statistical arguments, successfully derived a form of the Doppler principle for quantum emission.[79] Large numbers of physicists did not yet take the lightquantum to heart, but compared to the lack of interest just a few years before, the increase in research studies of lightquanta by mid-1922 was significant. Finally, the paradoxes of radiation made themselves felt.

Moreover, in the period of 1914–21, there had been a remarkable change in Einstein's public image. In 1914, when he was offered a position in Berlin and the directorship of the Kaiser Wilhelm Institut für Physik, Einstein had been a demonstrably superior physicist. But by 1921, his revolutionary suggestion that light is bent by gravity had been experimentally verified amid great and growing public acclaim. It was no longer easy to dismiss an idea, however seemingly absurd, of so eminent a man. Physicists

---

[74] See Chapter 7, note 75.
[75] Laue, Berlin *Sb*, (1921), 480.
[76] Einstein to Ehrenfest, ca. 20 January 1922. AHQP 1. See Klein, *HSPS, 2* (1970), 1–39.
[77] Breit, *PM, 44* (1922), 1149–52.
[78] Oseen, *AP, 69* (1922), 202–4. The lightquantum was frequently interpreted as "needle radiation," that is, possessing longitudinal extension. This was an intermediate stage of interpretation designed to allow for the explanation of interference effects. See also Emden, *PZ, 22* (1921), 512–17; L. Brillouin, *JP, 2* (1921), 151–2; Bateman, *PM, 46* (1923), 977–91. Duane, NAS *Proceedings, 9* (1923), 158–64, should also be included, although it was published after Compton's conversion to the lightquantum. Duane adamantly opposed Compton's explanation but had adopted much of the essence of directed lightquanta.
[79] Schrödinger, *PZ, 23* (1922), 301–3.

began to take cognizance of Einstein's most revolutionary hypothesis of all, that of lightquanta.

In November 1922, while he was on a trip to the Far East, Albert Einstein was awarded the 1921 Nobel Prize in physics. The honor was not given to him for the theory of relativity, or for the introduction of statistical techniques into physical analysis, or for his quantum theory of specific heats. Any one of these contributions would have been worthy.[80] The award came instead on Oseen's nomination for Einstein's "achievements in mathematical physics, especially for his discovery of the photoelectric law."[81] In his letter of nomination, Oseen quoted with emphasis Manne Siegbahn's interpretation of *Einstein's law,* that energy transformation from radiation to matter or vice versa occurs only in multiples of Planck's constant times frequency.[82] Siegbahn's opinion was of particular significance because of his knowledge of and research on x-ray absorption features, essential preparation for interpretation of the x-ray photoeffect. Much of Einstein's theoretical work did not satisfy the criteria established in the Nobel will for an accomplishment of practical significance. But the theoretical underpinning for the photocell, already in industrial use in the postwar period, was eligible.

It was no coincidence that Einstein's award was the first Nobel Prize given in physics after de Broglie's and Ellis's detailed verification of the photoeffect law for radiations other than visible light. The prize normally to be awarded in late 1921 was reserved and then given to Einstein in late 1922, along with the 1922 Nobel Prize to Niels Bohr.[83] In his precis of Einstein's work for the 1921 prize, Oseen lauded the "original and perceptive analysis" that had led to Einstein's claim than an atom can absorb only multiples of single quanta. This hypothesis, he said, was given the "most beautiful confirmation" by Millikan's experiments on the energies of pho-

[80] See, as one of many examples, the letter of nomination from Sommerfeld, 11 January 1922, in which all of these accomplishments figure, including the quantum theory of the photoeffect. Nobel archives, 1922.
[81] *Nobel en 1921-1922* (1923), 6.
[82] Oseen to Nobel committee, 2 January 1922. Nobel archives, 1922.
[83] Friedman, *Nature, 292* (1981), 793-8. There were several nominations for the photoeffect work, among them ones from Sommerfeld (11 January), Emden (29 January), and Oseen (2 January). "Förslag till utdelning av Nobelpris i fysik år 1922," pp. 17-18, 74, and 14-15, respectively. Nobel archives. For other documents relevant to the Einstein award see Pais, *American scientist, 70* (1982), 358-65.

toelectrons,[84] but there was no explicit appeal to de Broglie's x-ray work in the justification for the Einstein award. That was reserved for the place where it would have maximum impact: the brief for Bohr's award written by Oseen at the same time.

Oseen saw the two awards as inextricably linked; in supporting Einstein's, he was furthering Bohr's. De Broglie's x-ray evidence formed the crucial empirical bridge between the two.[85] The *Einstein–Bohr hv law* – what we have called the *quantum transformation relation* – linked Bohr's atomic theory to the photoeffect. If the law of the photoeffect was valid, de Broglie's results verified Bohr's atomic structure down to the innermost shells of the atom. Thus, Oseen first used Millikan's confirmation of the photoeffect law to justify Einstein's award and then appealed to de Broglie's results in justifying the award to Bohr. The Royal Swedish Academy agreed, although only after holding off the Einstein award until both could be granted in 1922. The award of the 1923 Nobel Prize to Millikan for his precise experimental work, including his verification of the photoeffect law, was entirely consistent with the position taken by the Academy. The awards of Einstein and Bohr gave implicit sanction to the lightquantum hypothesis of radiation absorption that underlay the photoeffect law.

One measure of the increasing attention paid to the lightquantum hypothesis is the consolidation of opposition to the idea, opposition that need not have been made explicit a few years before. Bohr, for example, had never seriously concerned himself with the paradoxical problems of radiation theory. But just at the time that both he and Einstein received their Nobel Prizes, Bohr proposed a way to avoid adopting the lightquantum.[86] At the time, Bohr's atomic theory faced formidable difficulties. Two of the most pressing were the anomalous Zeeman effect – the multiplication of spectral lines when the radiating sample is placed in a magnetic field – and the dispersion of light. The first concerns the emission of light, but the second involves absorption where, as we have seen, the problems of localization of radiant energy are least avoidable.

---

[84] Oseen, "Einsteins lag för den fotoelektriska effekten." 12 pp. Bilage 2 to 1922 prize. Nobel archives.

[85] Oseen, "Den Borska atomteorien." 34 pp. Bilage 3 to 1922 prize. Nobel archives.

[86] Bohr, *ZP, 13* (1923), 117–65. See Klein, *HSPS, 2* (1970), 1–39.

When light strikes matter, some of the energy is reradiated at a different frequency. The relationship between the incident and dispersed light was then experimentally well known; classical consideration of the atom as a set of sinusoidal electron oscillators gave a good first approximation.[87] But this was accomplished by assuming that the oscillating electron is coupled to the resulting set of radiated frequencies. There can be no such coupling between the frequencies absorbed by the Bohr atom and those subsequently reradiated because radiated frequencies are differences of atomic frequencies, not electron frequencies themselves. Although some success attended attempts to treat optical dispersion as if it were due to perturbations of the electrons in their periodic Bohr orbits, a fundamental obstacle remained: The absorbed and radiated frequencies are proportional to the energy differences between states, not to the energies of the states themselves.[88]

A key question for Bohr was whether the full quantum of radiant energy is concentrated on a single atom or electron within an atom. It was, therefore, not a new dilemma, and Bohr had shown some concern over it slightly before Maurice de Broglie's presentation in 1921.[89] By late 1922, Bohr had concluded that the conservation laws governing energy transfer between light and matter may not hold on the microscopic level. He argued that the laws of energy and momentum conservation may only be approximations valid for large numbers of atoms; the laws may hold only statistically, and not apply to each microinteraction between radiation and matter. Only when the individual transfers are integrated over the large number of atoms in a testable sample does the observer find that conservation laws apply. Einstein had tried a similar approach a decade before and had eventually rejected it.[90]

Bohr's proposal illustrates the seriousness with which he now viewed the paradoxes of radiation; to arrive at a solution, he was willing to discard the microscopic validity of energy conservation. But just as Bohr voiced this desperate hypothesis, another convincing experimental demonstration was offered that further con-

---

[87] References to the literature on the classical treatment of dispersion may be found in Drude, *AP, 1* (1900), 437–40. The empirical data appeared in a series of papers beginning about 1900, mostly by R. W. Wood and published in the *PM* and *PZ*. See "Wood" in Poggendorff, *Handwörterbuch*, vols. 4 and 5.

[88] R. Ladenburg, *ZP, 4* (1921), 451–68.

[89] See, for example, Bohr's contribution to *Solvay III*, especially pp. 241ff.

[90] Einstein to J. Laub, November 1910. Einstein archive.

firmed the lightquantum interpretation; one that explicitly denied the escape from the lightquantum that Bohr had outlined. To understand better the environment within which Louis de Broglie worked in this period, we now turn to a brief discussion of this effect named after Arthur Compton.

THE COMPTON EFFECT

It had been known for several years that if monochromatic x-rays are directed against matter, something similar to optical dispersion occurs. The x-rays scattered off the target have effective frequencies less than that of the incident beam, and the decrease varies with the angle between the incident beam and the direction of the observer. Several attempts were made to find a theoretical interpretation of the phenomenon, but the most notable were those by the American experimentalist Arthur Holly Compton and, independently, by Peter Debye in Zürich. A recent book covers the technical aspects of these researches, so we need not give a detailed discussion here.[91] But we do need to place the results and influences of Compton's work in the context of the then growing recognition of the lightquantum.

Compton had attempted for several years to explain the decrease in scattered x-ray frequency by analogy to optical scattering from large molecules. When the wavelength of light is of the same order as the diameter of the scattering particles, interference of waves from opposite sides of the particle affects the angular distribution of scattered light, producing a pattern similar to that seen with x-rays. But in order to apply this argument to x-rays, the electrons that reradiate the scattered x-rays would have to be about 10^{-10} cm in diameter, a thousand times larger than generally believed.[92] In 1919–20 Compton discarded this uncomfortable notion when he found evidence to support a second hypothesis that scattered x-rays are a special kind of fluorescent x-ray.[93] They could not be ordinary fluorescent x-rays because they are not characteristic of the scattering substance, their frequency varies

---

[91] Stuewer, *Compton effect* (1975). Bartlett, *AJP, 32* (164), 120–7.

[92] A. Compton, WAS *Journal, 8* (1918), 2–11; *PR, 14* (1919), 20–43. See also Stuewer, *Compton effect* (1975), chap. 3.

[93] A. Compton, *PR, 18* (1921), 96. He was led to this interpretation by previous work on γ-rays: *PM, 41* (1921), 749–67; *PR, 17* (1921), 38–41.

with that of the incident rays, and they are polarized. Clearly, many puzzles remained.

Late in 1921, influenced but not yet convinced by the evidence that had already stimulated a new interest in the lightquantum by others, Compton tentatively applied the quantum theory to the problem for the first time.[94] He did not appeal directly to the radical lightquantum, but rather to the quantum transformation relation, using it to calculate the recoil velocity of an electron struck by the incident x-rays. He then explained the wavelength-shifted secondary x-rays as Doppler-shifted x-rays reradiated by the recoil electron. A year later, in December 1922, Compton abruptly joined those who were finally taking the lightquantum seriously. He discarded the two-step process and asserted that the incident x-rays are scattered in localized lightquantum units.[95]

Each interaction of a lightquantum with an electron may then be analyzed according to the kinematics of two-particle collisions, the crucial assumption being that the x-ray is a particle of energy $h\nu$ and momentum $h\nu/c$. Conservation of energy and linear momentum predict that a shift in wavelength follows the scattering angle θ according to the relation

$$\Delta\lambda = h/mc \, (1 - \cos\theta) \tag{10.1}$$

The experimental data that Compton had amassed corroborated the result with great accuracy.

Compton's decision to reinterpret his data with the lightquantum hypothesis was part of the reawakening of interest in the lightquantum we have seen in others; the issues were current, at least among European physicists. Maurice de Broglie's experimental evidence for lightquantum x-rays predated Compton's by eighteen months; Ellis' argument for quantum γ-rays had been in print for almost a year. Compton cited both in his report on x-rays to the National Research Council before his own conversion to the lightquantum hypothesis.[96] The spate of papers and discussions in the German literature that marked the growth of interest in the lightquantum formed the environment of (but was not initiated by) Compton's success. Einstein's Nobel Prize was awarded before word was heard of Compton's work. In short, Compton was

[94] A. Compton, *PR, 19* (1922), 267–8.
[95] A. Compton, *PR, 21* (1923), 483–502.
[96] A. Compton, NRC *Bulletin, 4* (1922), 54–5.

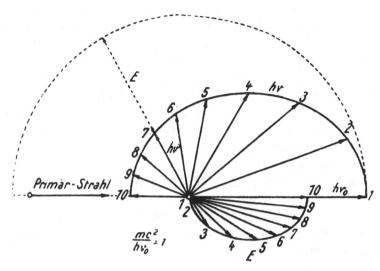

Figure 10.3. Debye's figure relating x-ray energy and direction to scattered electron energy and direction. [*PZ, 24* (1923), 164.]

reacting to a shift in opinion toward the lightquantum that had already influenced many physicists before his results contributed to that shift.

It is a measure of the increasing tolerance of the lightquantum before Compton's work that a second, virtually identical explanation of the anomalous frequency shift in scattered x-rays was given by Debye before Compton's paper was published. Debye, then *Ordinarius* at Zürich, started explicitly with Einstein's needle radiation and treated the collision of an x-ray quantum with an electron as if it were a classical interaction of two particles of known momentum and energy. He described his work as "an attempt to draw the most detailed possible conclusions from the two assumptions 'quanta' and 'needle radiation' by use of the most general laws 'energy conservation' and 'momentum conservation'." His hope was to "gain deeper insight of the quantum laws, in particular their connection with wave optics."[97] Debye's analysis was based on the experimental data that Compton himself had amassed in his report to the National Research Council just before Compton's reconciliation to the lightquantum. Debye's results were equivalent to, although somewhat more generally stated than, Compton's. Figure 10.3 summarizes his conclusions about

[97] Debye, *PZ, 24* (1923), 166.

the special case in which the incident x-ray frequency is the figure-of-merit value mc^2/h. The numbered vectors above the horizontal line of the incoming x-ray quantum express the fraction of incident radiant energy transferred to the x-ray quantum scattered in various directions; those below the horizontal express the fraction of incident energy donated to the electron in each case.

The Compton effect elicited a strong response from European physicists already primed for discussion of the unorthodox lightquantum. Sommerfeld said that it "sounded the death-knell" of the wave theory.[98] Compton's discovery had the effect of a crystal dropped into a supersaturated solution; the issues had been current for two years, and the rapid crystallization of thought was due to that advance preparation. By the time the word spread in the first months of 1923, no physicist could ignore the lightquantum any longer.

LOUIS DE BROGLIE'S QUANTUM-RELATIVISTIC PARADOX

The increasing acceptance and wider adoption of the lightquantum encouraged Louis de Broglie to intensify his search for a physical interpretation whereby the seeming incompatibility of wave and particle representations of radiation might be resolved. The theory of relativity had thus far played no essential role in the deliberations that had led to recognition of the paradoxes of radiation.[99] It had been implicated in explanation of x-ray doublets and as the source of extra energy for photoemission. In 1923 de Broglie tried to bring harmony between the two brainchildren of Einstein: the theory of relativity and the lightquantum hypothesis. He confronted in this synthesis a second dilemma, the solution to which, he quickly recognized, was closely tied to a solution of the paradoxes of radiation.

According to relativistic dynamics, precisely the sort that de Broglie had been applying to "atoms" of light, the internal energy of a particle with a rest mass of m_0 is $E = m_0 c^2$. According to the quantum theory, it possesses an intrinsic frequency ν_0, where

---

[98] Sommerfeld to Compton, about spring, 1923. Paraphrased by Compton in *JFI, 198* (1924), 70. Millikan in *Nobel en 1923* (1924).

[99] The greatest use of relativity theory had been made by Emden, *PZ, 22* (1921), 513-17.

$hv_0 = m_0c^2$. But to an observer past whom the particle travels at velocity v_p, its internal energy is $m_0c^2/\sqrt{1 - v_p^2/c^2}$, so that the observed frequency of the quantum wave is

$$v_w = \frac{m_0c^2}{h\sqrt{1 - v_p^2/c^2}} \tag{10.2}$$

On the other hand, the most directly comprehensible consequence of the theory of special relativity is that a moving clock appears to a stationary observer to be running more slowly than it does to an observer moving along with it. According to relativity theory, the observed frequency of the particle oscillator is

$$v_p = \frac{m_0c^2}{h} \sqrt{1 - v_p^2/c^2} \tag{10.3}$$

The relativistic dilation of time intervals predicts that the stationary observer measures a lower frequency than does the moving observer.

Put in its baldest terms, the theory of relativity produces an asymmetry when combined with quantum theory. A mass point's energy, and hence its intrinsic quantum frequency, must *increase* with its velocity relative to the observer; time dilation predicts that the frequency must *decrease* for the same observer. Louis de Broglie's greatest contribution to physics was to recognize that a solution to this problem would shed light on the paradox of wave and particle.

Casting about for a resolution of this new paradox for the lightquantum, de Broglie reinterpreted a study published in 1919 by Marcel Brillouin, with whom he had studied physics at the Collège de France. Brillouin had wished to explain how an electron describing a continuous periodic orbit might nonetheless be described by discontinuous quantum laws.[100] His method was based on a hypothesis originally due to Vito Volterra that, under certain circumstances, a "hereditary" field is established by a moving particle.[101] Brillouin assumed that a particle – he called it a *point mobile* – moves through a medium with a velocity greater than that of waves in the medium. The motion is assumed to be quasi-periodic and to be restricted to a circular region with a

[100] M. Brillouin, *CR, 168* (1919), 1318–20.
[101] Volterra, *Leçons sur les équations* (1913). For the fullest expression see *Journal de mathématiques pures et appliquées, 7* (1928), 249–98.

diameter less than the product of the wave velocity and the period of the particle. Brillouin had in mind the quasi-periodic motion of an electron constrained by the Coulomb force of the atomic nucleus to remain within a certain region of space. A spherical wave is to be imagined arising from the electron at the start of its periodic motion. At the end of that period, the electron must still lie within the spherical volume of space to which the wave has expanded. This condition is necessary to ensure that the particle, which moves faster than the wave, does not simply leave the wave behind. Under these special circumstances, the particle catches up at regular intervals to the waves it produces in moving through the medium.

An electron circling a nucleus creates, therefore, a series of imaginary points trailing in its wake. From each of these points a spherical wave, emitted when the electron was there, would catch up with the electron at the same instant as all the rest. The hereditary field thus established embodies, as it were, one possible past history of the particle. Brillouin showed that if the energy of the *point mobile* at an anterior position is inversely proportional to the distance to its present position, then the physical quantity that characterizes the difference between one point in the chain and the next has the dimensions of action (energy × time). Like Planck's constant of action, it changes by integral multiples from point to point.

When Louis de Broglie encountered Brillouin's suggestive study, he was deeply involved in his search for resolutions both of the wave-particle representation of light and of the relativistic-quantum discrepancy. He already had come to identify atoms of light with particles of infinitesimal, but perhaps not zero, mass. Thus, de Broglie saw that Brillouin's analysis might apply to lightquanta as well as to electrons. Viewed from this perspective, the hereditary field offered the possibility of a solution to one or both of the problems. An atom of light, like Brillouin's *point mobile,* might possess an associated wave that interacts according to some special relationship with the particle.

"Suddenly, at the end of the summer of 1923," de Broglie recalled, "all of these ideas seemed to crystallize in my mind."[102] The synthesis of particle and wave that he was developing for

[102] L. de Broglie in *Louis de Broglie* (1953), 461.

application to atoms of light applied in an equivalent form to atoms of matter as well. De Broglie's conviction that matter and light should be treated equivalently lay in the symmetry inherent in his goal: to form a synthesis between particle and wave. Much of his motivation had come from the elder de Broglie, who, in his book about x-rays in 1922, pointed out that "one finds certain kinetic aspects in undulatory radiation, and certain periodic aspects in the directedness of electrons."[103] He meant that electrons of definite velocity act in collisions just like x-rays of a precisely defined frequency.[104] His younger brother now transformed that heuristic equivalence into a physical symmetry by emphasizing the formal and physical similarity between matter and light. Louis de Broglie's initial concern about the nature of light had led to a revolutionary advance in the theory of matter.

PHASE WAVES

When de Broglie first presented his synthetic fusion of particle and wave, he did it not for the lightquantum but for a particle of matter. Consider a particle moving with velocity v_p past two stationary observers. One observer, w, watches for the quantum-wave effects of the moving point; the other, p, notices only the relativistic effects of the point oscillator. According to equations 10.2 and 10.3, the two disagree about the frequency because $v_w > v_p$. De Broglie suggested that observer w is seeing not the wave effects of the particle itself but rather those of an associated plane wave that travels in the same direction as the particle. This assumption allows a special congruence between the particle and its associated wave that seems counterintuitive at first: Even if two observers disagree about the frequencies and velocities of the internal oscillation of the particle and the wave, they nonetheless can agree that at each instant there is a particular point in space where the two oscillations remain in phase with one another. De Broglie established the criterion that this point always coincides with the changing position of the particle. That is to say, despite the fact that the quantum and relativistic frequencies of the particle differ, the two oscillations keep their phase – the fraction of a full

[103] M. de Broglie, *Les rayons x* (1922), 17.
[104] *Ibid.*, 19. He displays a table giving equivalent frequencies and the derived velocity of an electron of equivalent effect for radiations from "electric waves" through γ-rays.

wavelength that has passed – always equal to one another at the point in space occupied by the particle. Beyond establishing the connection of particle to wave, this relation defined the physical location of the particle as that where the phase wave and the internal relativistic oscillation always constructively interfere.

Imagine that the particle passes an arbitrary reference point with velocity v_p. Let us assume that it is then at the beginning of its cycle; it has phase zero at that instant. At the same instant, the phase of the plane wave is assumed to be zero at the same point. As the wave propagates, it can be imagined to leave behind in its wake an oscillation of frequency v_w. To satisfy de Broglie's criterion that the phases of particle and wave match at the particle at every instant, the wave must travel faster than the particle. Unlike Brillouin's *point mobile*, de Broglie's – which he tellingly called *le mobile* – must move more slowly than its associated wave. Indeed, one can calculate just how much faster the wave has to travel to fulfill the criterion of phase synchronization.

The amplitude of the particle's internal oscillation at time t after passing the reference point is proportional to $\sin 2\pi v_p t$. In this time, the particle travels a distance $x_p = v_p t$, so that its amplitude at x_p is

$$\Phi_p = \sin (2\pi v_p x_p/v_p) \tag{10.4}$$

In the same time interval t, the zero-phase point of the plane wave travels a distance $x_w > x_p$, so that at the same instant the amplitude of the wave at point x_p, at the particle, is

$$\Phi_w = \sin 2\pi v_w \Delta t \tag{10.5}$$

where Δt is the time interval (less than t) that the zero-phase point of the wave took to travel the distance from x_p to x_w. Equation 10.5 gives, in other words, the amplitude of the associated wave at a point $x_w - x_p$ *behind* the zero-phase point. Clearly, $\Delta t = (x_w - x_p)/v_w$. A straightforward manipulation shows that equations 10.4 and 10.5 are identically equal if and only if[105]

$$v_w = c^2/v_p \tag{10.6}$$

The velocity of the wave must exceed c by the same proportion that the velocity of the particle falls short of c.

[105] Equality of equations 10.4 and 10.5 requires $v_p x_p /v_p = v_w (x_w /v_w - x_p /v_w)$; using the relation $v_p = v_w(1 - v_p^2/c^2)$ derived from equations 10.2 and 10.3 this may be rewritten as

$$(1 - v_p^2/c^2)x_p /v_p = x_w /v_w - x_p /v_w$$

The associated wave can therefore transfer no energy. It is not a material wave; de Broglie called it an *onde fictive* when presenting this argument to the Parisian Academy of Sciences in September 1923.[106] It differs in fundamental respects from Marcel Brillouin's hypothetical wave: de Broglie's is not physically real; de Broglie's travels faster than the particle with which it is associated; finally, Brillouin's wave is spherical, and it soon became clear that de Broglie's is planar. De Broglie treated the same case, an electron traveling in a circular orbit, that Brillouin had. According to de Broglie, the fictional plane wave must precede the electron *around* the circular orbit before the electron itself has traced that path. The wave does not propagate spherically away from the coincident point in the way Brillouin imagined his wave to do.

Only in this way can one understand the means de Broglie used to calculate the time interval τ between electron–wave phase coincidences. In order to meet again at time τ after the wave front and the electron initially coincide, the wave must complete somewhat more than a single revolution in the orbit. In so doing, the wave covers a total linear distance

$$x_w = (T + \tau)v_p \tag{10.7}$$

where T is the period of the electron's revolution in its orbit. Rearranging equation 10.7 with the additional constraint imposed by equation 10.6 yields the value $\tau = T\, v_p{}^2/c^2\,(1 - v_p{}^2/c^2)$.

With this value for τ, de Broglie applied the condition of phase synchronization. The phase of the wave is, by definition, zero when it reencounters the electron. So the equality of phase demands that the phase of the relativistic oscillation of the electron also be zero at that point. Thus, in covering the distance $v_p\tau$, the electron must complete an integral number of internal oscillations. In other words:

Using the equality $x_w\,/v_w = x_p\,/v_p$, this becomes

$$1 - v_p{}^2/c^2 = 1 - x_p\,/x_w$$

And in light of the equality $x_p\,/x_w = v_p\,/v_w$, this is

$$v_p{}^2/c^2 = v_p\,/v_w$$

which leads directly to equation 10.6. This result, which follows from de Broglie's own analysis, is derived on different considerations by Kubli, *AHES*, 7 (1970), 26–68.

[106] L. de Broglie, *CR, 177* (1923), 507–10, 548–50, 630–2; *PM, 47* (1924), 446–58.

$$v_p\tau = \frac{m_0 c^2}{h}\,\frac{v_p^2/c^2}{\sqrt{1-(v_p/c)^2}}\,T = n \qquad (10.8)$$

where n is an integer. This may be rearranged to the form

$$\frac{m_0 v_p^2 T}{\sqrt{1-(v_p/c)^2}} = nh \qquad (10.9)$$

Equation 10.9 is precisely the action-integral expression for allowed electron orbits in the Bohr atom, a mathematical condition that had lacked, until now, any physical interpretation.[107]

Many years later, de Broglie described himself as always having had "a very 'realist' conception of the nature of the physical world and little given to purely abstract considerations."[108] He had found in 1923 what had eluded Bohr, a realistic conceptual explanation for the discrete orbits in the Bohr atom, and it obviously impressed him mightily. "It is almost necessary to suppose," de Broglie said at the time, "that the electron orbit is only stable if the fictional wave reencounters the electron in phase."[109] And the following year he repeated, "This beautiful result . . . is the best justification that we can give for our way of addressing the problem of quanta."[110]

EXTENDING DE BROGLIE'S HYPOTHESIS

Louis de Broglie began his studies in physics by trying to discover from x-ray analysis the principles of atomic structure. He had switched interests in 1921 to analyze the nature of radiation, and this concern had now led him directly to a successful interpreta-

---

[107] The generalized Sommerfeld condition is $\int_0^T p\,dq = nh$. In de Broglie's system this becomes

$$\int_0^T \frac{m_0(v_x + v_y + v_z)}{\sqrt{1-(v/c)^2}}\,dx\,dy\,dz = nh$$

Because of the general relation $\int p\,dq = \int p(dq/dt)dt$, this may be rewritten as

$$\int_0^T \frac{m_0}{\sqrt{1-(v/c)^2}}\,(v_x^2 + v_y^2 + v_z^2)dt = nh$$

or simply

$$\frac{m_0 v^2 T}{\sqrt{1-(v/c)^2}} = nh$$

[108] L. de Broglie in *Wave mechanics* (1973), 12.
[109] L. de Broglie, *CR*, *177* (1923), 509.
[110] L. de Broglie, *Thèses* (1924, 1963), p. 53 of reprint.

tion of the Bohr atomic theory. This mutual influence of atomic and radiation theory in his work was a product of his search for a synthetic view that would place light and matter on an equivalent footing. The revolutionary conceptual contribution of the phase wave hypothesis was that it did not distinguish between an atom of matter and an atom of light. Both consist of a *mobile* with an associated wave. In one the wave properties predominate, and in the other the localized particle properties predominate; but neither is solely particle or wave.

De Broglie still had to provide an explanation of interference and diffraction. The wave aspects of the synthetic theory of radiation allowed for both, but one had to explain why the atoms of light that carry the radiant energy behave as they do. And here he encountered formidable problems. To solve them and explain how a *mobile* can change direction, it was necessary, de Broglie said, to alter the laws of inertia. The velocity of the energy-carrying particle is equivalent to the group velocity of the phase wave, he claimed, while the wave itself, carrying no energy, travels with the phase velocity. Material points, whether lightquanta or electrons, are analogous to singularities in a group of waves; the individual waves in the group pass by, but the singularity may remain. The particle, whether lightquantum or electron, follows ray trajectories controlled by its phase wave; the trajectory is fixed by Fermat's principle of least action. In an isotropic medium, the path of the particle is always normal to the advancing wave front.

In his early studies and in his thesis of 1924, de Broglie neither used nor discussed wavepackets; he was thinking of extended monochromatic plane waves that, like Brillouin's spherical hereditary waves, determine the possible trajectories of the associated particle. Only after his success with the Bohr atom did de Broglie make a tentative identification of the velocity of his *onde fictive* with optical phase-wave velocity and that of the particle with the related group velocity. The concept of representing the particle as a superposition of phase waves that forms a localized discontinuity in the field – the now-familiar wave-packet representation – is a later invention. De Broglie's early suggestion of a parallel should not be taken as more than a suggestion of how to begin to explain interference and diffraction in his new concept of radiation.[111] The

[111] MacKinnon, *AJP*, 44 (1976), 1047–55, argues that de Broglie's identification is inconsistent with modern understanding. But the claim is ahistorical and physically misleading. See Schlegel, *AJP*, 45 (1977), 871–2.

real significance of de Broglie's inspiration lies in the influence it had on others, notably Erwin Schrödinger, whose attempts to rationalize its successes led to the full statement of wave mechanics in 1925.

De Broglie's identification of the *onde fictive* as a phase wave led to another prediction: the possibility of diffracting electron beams. Near an aperture comparable in size to the phase wavelength, the distortion produced in the wave bends the trajectory of a light-quantum or an electron. "The new dynamics of a free material point is to classical dynamics," de Broglie claimed, "like wave optics is to geometrical optics."[112] The paths of lightquanta that pass through a pair of slits are controlled by the interference pattern created in their phase waves. The energy transferred by the light, as detected for example on a photographic plate, is distributed in fringes corresponding to the well-known interference bands. This is true regardless of the intensity of the incident light. The intensity measures the number of lightquanta; it does not affect their trajectories. In an identical way, the distortion in the phase wave structure of an electron near a sufficiently small aperture could be expected to diffract the electron beam. The mechanism for the complex propagation of the phase wave that in turn directs the trajectory of an energy-carrying particle remained a desideratum of a "generalized theory of electromagnetism."

De Broglie gave the fullest expression of his hypothesis and its remarkable success for the Bohr atom as one of his theses for the doctorate under Jean Perrin and Paul Langevin in November 1924.[113] There, for the first time, he explicitly calculated the now-familiar wavelength of the phase wave that "guides the displacement of the energy" of a material particle:

$$\lambda = h/m_0 v \tag{10.10}$$

But with the exception of a refined treatment of the statistical properties of a gas of lightquantum particles, he added little in the thesis that went beyond the early papers. De Broglie had done more than simply turn Einstein's lightquantum hypothesis on its head; if there is a particlelike aspect to light, there must also be a wavelike aspect to matter. He stressed the fundamental synthesis

[112] L. de Broglie, *CR, 177* (1923), 549.
[113] L. de Broglie, *Thèses* (1924), *Annales de physique, 3* (1925), 22–128. The reprint includes a useful historical introduction by de Broglie, pp. 22–30.

that his hypothesis implied: Light and matter are equivalent. Both share in differing degrees the characteristics of localized particle and spatially extended wave. ,

Langevin was very much impressed by the work. He called it "an attempt at a solution of the most important problem of present-day physics."[114] Its most profound aspect was the general-

[114] From the draft manuscript critique by Langevin of de Broglie's thesis for-merly in the possession of Mme. Luce Langevin. I am indebted to her for permission to quote it in full here:

Rapport sur la thèse de M. L. de Broglie 1924
Le travail de M. Louis de Broglie représente un important effort vers la solution du problème le plus important de la physique actuelle, la synthèse des deux théories optiques des ondulations et des quanta, jusqu'ici contradic-toires au moins en apparence, dont chacune s'appuie sur tout un ensemble de faits et de confirmations expérimentales remarquables et dont l'opposition vient renouveler de manière imprévue l'ancien conflit de l'émission et des ondulations.

Les énoncés d'aspect paradoxal qu'a dû introduire la théorie des quanta sont eux-mêmes de deux ordres qui correspondent à ce qu'on peut appeler l'aspect dynamique d'un côté et celui des "quanta de lumière" de l'autre. Pour rendre compte de la composition du rayonnement noir et de la structure des spectres d'émission et d'absorption il a fallu, avec Planck, Bohr, Sommerfeld, admettre que parmi tous les mouvements des centres électrisés compatibles avec les lois de la dynamique classique, seuls sont possibles ceux pour lesquels les orbites électroniques satisfont à certaines relations numériques simples. D'autre part M. Einstein (et M. J. J. Thomson)* a montré que des effets dits photoélectriques où l'absorption du rayonnement s'accompagne de l'émis-sion de particules cathodiques, ainsi que les fluctuations dans le rayonnement noir semblent bien imposer l'idée que l'énergie du rayonnement lui-même n'est pas distribuée de manière continue dans l'espace mais se compose de grains d'énergie proportionnelle à la fréquence.

L'idée fondamentale de M. de Broglie consiste à unifier les deux ordres d'énoncés de quanta en associant à chaque particule électrisée considérée comme grain indépendant d'énergie un phénomène périodique de fréquence proportionnelle à l'énergie du grain exactement comme l'hypothèse des quanta de lumière associe aux ondes lumineuses périodiques des grains non électrisés où leur énergie se trouve concentrée.

La liaison générale ainsi établie entre le mouvement d'une particule électri-sée ou non et la propagation d'un phénomène périodique est rattachée par l'auteur aux principes généraux de manière très profonde et par une série de remarques telles qu'une hypothèse paradoxale au premier abord apparaît de plus en plus conforme à la nature des choses. Il montre en particulier que les relations introduites entre le mouvement d'un grain d'énergie et la propaga-tion du phénomène périodique qui l'accompagne sont exactement de même nature que celles introduites par Lord Rayleigh et Gouy entre le mouvement d'un groupe d'ondes et la propagation de la phase d'une onde exactement périodique.

D'autre part, et ceci est peut-être l'aspect le plus intéressant des idées de M. de Broglie, cette liaison semble correspondre à l'analogie, purement

*Name inserted later.

ity claimed for the relationship between particle mechanics and wave properties, between Hamiltonian dynamics of particles and Fermat's principle of least action for wave propagation. "Perhaps here is the beginning of a wave interpretation of the principles of dynamics," Langevin said, and pointed to de Broglie's method for interpreting the stationary orbits of the Bohr atom. Although

---

formelle jusqu'ici, mais peut-être très profonde, entre les énoncés d'intégrales stationnaires qui sont à la base de la dynamique d'une part avec les principes d'Hamilton et de Maupertuis et de la théorie des ondulations, d'autre part avec le principe de Fermat tel que l'introduisent et le justifient les conceptions d'Huyghens et de Fresnel.

De même que les principes de la dynamique du point matériel prennent leur expression la plus simple grâce à l'introduction d'un vecteur à quatre dimensions, l'impulsion d'Univers dont la composante d'espace est la quantité de mouvement et la composante de temps l'énergie du mobile, l'auteur montre que le principe de Fermat et toute la cinématique des ondes conduisent à introduire un "vecteur d'onde" d'Univers dont la composante d'espace est dirigée suivant le rayon et dont la composante de temps est la fréquence.

L'introduction de ces deux vecteurs conduit tout d'abord à généraliser la relation du quantum sous une forme intrinsèque valable pour tous les systèmes galiléens de référence, et à préciser la relation entre le mouvement d'une particule et le phénomène périodique associé, en affirmant l'identité des deux vecteurs [:] impulsion d'Univers de la particule et vecteur d'onde pour le phénomène périodique (ou plutôt, en raison du choix des unités, la constance de leur rapport égal à la constante h de Planck). On étend ainsi dans un sens tout à fait conforme à l'esprit de la théorie de relativité aux quatre composantes de chaque vecteur, la relation du quantum envisagée jusqu'ici pour les seules composantes de temps, énergie et fréquence.

Dans ces conditions, les principes d'Hamilton et de Fermat se confondent et représentent deux aspects d'un même principe, l'un concernant le mouvement de la particule et l'autre la propagation de l'onde associée. Peut être y a-t-il ici l'origine d'une interprétation ondulatoire des principes de la dynamique. En tout cas une conséquence immédiate de cette unification des deux vecteurs qui donnerait l'énoncé le plus général et le plus profond de l'hypothèse des quanta, est que les trajectoires dynamiquement possibles de la particule dans un champ de force se confondent avec les rayons de l'onde associée.

Un résultat particulièrement remarquable est que les conditions dynamiques de[s] quanta se déduisent immédiatement de cette conception si l'on admet qu'une trajectoire fermée ne peut être stable si l'onde associée au mobile et qui se propage le long de cette même trajectoire avec une vitesse supérieure à celle du mobile doit, en rejoignant celui-ci se retrouver en concordance de phase avec elle-même. Cette condition de résonance des trajectoires stables pour l'onde associée fournit les conditions de quanta sous la forme générale de Sommerfeld et en donne une représent[ation] concrète singulièrement suggestive malgré les obscurités qui subsiste[nt] encore quant à la nature physique du phénomène périodique associé au grain d'électricité.

En appliquant les mêmes conceptions au cas d'un gaz contenu dans un récipient à parois supposées réfléchissantes pour les ondes associées, l'auteur

others in Paris had difficulty understanding the thesis – Perrin said publicly only that it was "très intelligent"[115] – Langevin discussed the matter with Einstein and then asked de Broglie to send Einstein an advance copy of the thesis,[116] Einstein immediately recognized its value. "He has lifted a corner of the great veil," Einstein wrote to Langevin and Langevin related to de Broglie.[117] And the Parisian physicists granted the younger de Broglie his doctorate.

INTERSECTION OF RADIATION THEORY AND ATOMIC THEORY

At the time that the new synthetic understanding of matter and light evolved in France, much effort in theoretical physics lay in the attempt, largely carried on by German-speaking physicists, to extend the Bohr theory of the atom to multielectron systems. By 1923 it was widely recognized that the program was failing.[118] There were a number of difficulties, including the problem of dispersion of light, mentioned earlier, the anomalous Zeeman effect and its associated "riddles" for spectroscopy,[119] the necessity for introducing half-integral quantum numbers,[120] and problems that led eventually to the exclusion principle,[121] and to electron spin.[122] The increasing realization that semimechanical solutions to those problems were not forthcoming contributed to the environment in Germany within which Louis de Broglie's heterodoxical hypothesis was received.

The chief architect of atomic theory was Niels Bohr, who in 1923 understood that the lightquantum could be avoided no longer if he wished to bring harmony to a new nonmechanistic atomic

---

retrouve sans nouvelle hypothèse l'expression donnée par Planck pour la constante de l'entropie ou constante chimique du gaz.

[115] La Varende, *Les Broglie* (1950), 333. See also the comments of another member of de Broglie's thesis committee, Charles Mauguin, in *Louis de Broglie* (1953), 434.
[116] Langevin to L. de Broglie, 27 July 1924. "Autographes," L. de Broglie dossier.
[117] Langevin to L. de Broglie, 13 January [1925]. *Ibid.* (The original is quoted in German.)
[118] Landé, *PZ, 24* (1923), 441–4.
[119] Forman, *Isis, 59* (1968), 156–74; *HSPS, 2* (1970), 153–261.
[120] Cassidy, *HSPS, 10* (1979), 187–224.
[121] Serwer, *HSPS, 8* (1977), 189–256.
[122] Kronig in *Theoretical physics* (1960), 5–39.

theory.[123] Bohr found the Compton effect a particularly intransigent obstacle. It seemed to prove that *individual* interactions of x-rays with electrons conserve energy and momentum. Even Bohr's daring statistical interpretation of energy conservation now seemed insufficient to circumvent the lightquantum. The story need not be repeated here how the American theorist, John Slater, approached Bohr at this time with the idea that lightquanta are real and that their emission and absorption by atoms are dictated by a "virtual radiation field."[124] The resulting development of the Bohr-Kramers-Slater theory of atomic "virtual oscillators" reaffirmed in a transitional form Bohr's statistical interpretation of radiant interaction.[125] Although experimental tests of major predictions of the BKS theory by Bothe and Geiger failed to justify the idea, by the time the experimental results were known, the issues had become superfluous.[126]

The basic ideas inherent in the statistical approach of the BKS theory were used by Bohr's assistant, Hendrik Kramers, later joined by Werner Heisenberg, to treat the thorny problem of the dispersion of light by the Bohr atom.[127] This successful work strongly influenced the young Heisenberg in 1925 when he found a mathematical formalism that solved most of the pressing problems that the Bohr theory faced.[128] Heisenberg's matrix mechanics, although directed to the problems of atomic theory, was stimulated in essential ways by the renewed appreciation of Einstein's lightquantum hypothesis.

In the meantime, a separate line of development based on de Broglie's hypothesis of phase waves led to another means of analyzing the difficulties of atomic theory. This followed the route from de Broglie through Einstein to Erwin Schrödinger. The resurgence of interest since 1921 in Einstein's lightquantum had led him to try to restate its inherent flaws. He had known since 1909 that lightquanta alone were impotent to resolve the difficulties that faced radiation theory – hence his plea to Ehrenfest that

[123] Bohr, *AP, 71* (1923), 228-88.
[124] Klein, *HSPS, 2* (1970), 1-39. Slater in *Wave mechanics* (1973), 19-25.
[125] Bohr, Kramers, and Slater, *PM, 47* (1924), 785-802.
[126] Bothe and Geiger, *ZP, 26* (1924), 44; *ZP, 32* (1925), 639-63. See also A. Compton and Simon, *PR, 25* (1925), 306-13; *26* (1925), 289-99.
[127] Kramers, *Nature, 113* (1924), 673-4. Heisenberg and Kramers, *ZP, 31* (1925), 681-708.
[128] Heisenberg, *ZP, 33* (1925), 879-93.

the problem was sufficient to drive him to the madhouse.[129] Instead he developed a concept of a *Gespensterfeld,* or "ghost field," that might itself carry little or no energy, yet guide the trajectories of lightquanta in accord with laws of wave optics.[130] Little came of the concept, but it bore sufficient similarity to de Broglie's ideas so that Einstein was quick to recognize what was significant in the heterodox suggestion from France.

When Einstein encountered de Broglie's thesis, he was just developing a new statistical approach to the kinetic theory of an ideal gas using a method suggested by the Indian physicist S. N. Bose. Bose had derived the black-body law using a statistical model of interacting particles based on the lightquantum, entirely free of assumptions from classical electromagnetic theory.[131] In 1925 Einstein pointed out that de Broglie's hypothesis allowed one to treat equivalently the fluctuations in black-body radiation and the statistical fluctuations of a gas according to Bose's statistics. He then went on to propose a possible experimental test for diffraction in atomic beams.[132] The particular statistical application of de Broglie's hypothesis that Einstein made in 1925 "stuck the matter right under [Schrödinger's] nose."[133]

Among the physicists now interested in the lightquantum, Erwin Schrödinger was in a unique position to appreciate the merits of de Broglie's phase wave hypothesis. In 1922 he had called attention to a "remarkable property of the quantum orbits of an individual electron."[134] A mathematical restriction that acts on an electron in a periodic orbit suggested that with each revolution the electron "adjusts" itself to its new point in four-dimensional space–time. Thus, Schrödinger had already recognized a periodic, spatially extended effect of each electron in space. It has been suggested that Louis de Broglie's reputation among most German-

[129] Einstein to Ehrenfest, 15 March [1921]. AHQP 1, 77.
[130] Einstein, Berlin *Sb* (1923), 359–64, was as close as Einstein came to publishing his ideas. See Lorentz, *Problems* (1927), 156–7, who lectured on the idea at Caltech in 1922; Lorentz to Einstein, 13 November 1921, Einstein archive.
[131] Bose, *ZP, 26* (1924), 178–81.
[132] Einstein, Berlin *Sb* (1924), 261–7; (1925), 3–14.
[133] Schrödinger to Einstein, 23 April 1926. Einstein archive. Quoted in Przibram, *Briefe zur Wellenmechanik* (1963), 24. Schrödinger's first use of de Broglie's hypothesis was in *PZ, 27* (1926), 95–101. For the influence of Einstein's work in kinetic theory on Schrödinger's derivation of wave mechanics, see Klein, *Natural philosopher, 3* (1964), 3–49.
[134] Schrödinger, *ZP, 12* (1922), 13–23.

speaking physicists, notably those such as Bohr and Sommerfeld who were working on atomic spectroscopy, was not high.[135] But Schrödinger, like Einstein and de Broglie, was committed to relativistic and statistical formulation of theory; he was not directly concerned with problems of atomic structure.

Late in 1925, Schrödinger began a four-paper series in which he extended the formal analogy between Hamiltonian mechanics of particles and wave optics.[136] Using de Broglie's hypothesis, Schrödinger reinterpreted quantum transitions as changes in the vibrational mode of the superposition of "phase waves" that characterize the entire atomic system. He was relieved to have found an alternative interpretation of quantum theory that avoided what to him was the uncomfortable notion of mechanically indescribable electron jumps from one orbit to another. The transition of an electron could now be interpreted as a change in the vibrational mode of the entire atom. The quantum wave mechanics that resulted from Schrödinger's analysis was a direct descendant of physicists' accommodation to the lightquantum hypothesis.

In the spring of 1926, Schrödinger showed that Heisenberg's matrix mechanics, which Schrödinger said had "frightened" and "repulsed" him by its purely formal character, was mathematically equivalent to his own wave mechanics.[137] Schrödinger wrote to de Broglie in March 1926 to describe his work. "It is possible to calculate all the Born–Heisenberg matrix elements by simple differentiations and quadratures if one has the complete solution of the wave equation," he said.[138] He was especially gratified to

[135] Raman and Forman, *HSPS, 1* (1969), 291–314. Hanle, *Isis, 68* (1977), 606–9.
[136] Schrödinger, *AP, 79* (1926), 361–76. See also Wessels, *SHPS, 10* (1979), 311–40.
[137] Schrödinger, *AP, 79* (1926), 734–56. Eckart, *PR, 28* (1926), 711–26.
[138] Schrödinger to L. de Broglie, 26 March 1926. (L. de Broglie dossier.) Extracts from the original letter written in Schrödinger's French, follow [reprinted with permission of Ruth Braunizer and l'Academie des Sciences de Paris].
    "Pour des phénomènes *petites* ou *fines* il faut aussibien dans la méchanique que dans l'optique remplacer le traitement *géometrique* par un traitement *ondulatoire*–on pourrait aussi dire par un traitement *physique*,–en tenant compte de ce que la longueur d'onde est *finie* et non infiniment petite comme le suppose l'optique géometrique (et de même l'ancienne méchanique)." "Il est possible de calculer toutes les élements de matrices de Born–Heisenberg par des simples différentiations et quadratures, quand on posséde la solution complète de l'équation d'onde mentionnée plus haut, c'est à dire un système complet de 'Eigenfunktionen'."

report that the wave mechanical formulation based on de Broglie's hypothesis replaced the purely "geometrical formalism" by a wave treatment, "one could also say by a *physical* treatment."

Two formerly independent research fields – the spectroscopic analysis of atoms and the statistical analysis of radiation – were brought into contact in 1925 by the long-delayed discussion of the physical reality of lightquanta. By 1926 each had developed into equivalent formulations of the new quantum mechanics. That there were two independently formulated versions of the new quantum mechanics was in large part an effect of the virtual separation of the two parent fields – atomic theory and radiation theory – for the preceding decade. And this, in a larger sense, constituted the fullest corroboration of Louis de Broglie's remarkably fruitful but decidedly unorthodox proposal that matter and light are fundamentally one and the same thing.

Epilogue

The tiger and the shark

"We'll wait awhile," the master said, "that thus
our senses may grow used to this strong smell,
and after that, it will not trouble us."
Dante, *Inferno, XI* 10–12

A new and fundamentally indeterminate approach to physical theory had been established, and the duality of waves and particles became an inescapable aspect of the successful new physics. After Max Born proposed a probabilistic interpretation of Schrödinger's wave function in 1926, younger physicists learned that it was improper to ask what the consistent outcome would be from given initial experimental conditions.[1] One could calculate the relative likelihood of all possible results and assign a probability to each. Seeking to express the "conceptual content" of the new probabilistic mechanics – which by its very nature denied conceptualization in the ordinary sense of the word – Werner Heisenberg analyzed an issue in classical physics in which an equivalent limitation acts. The example he selected touches our study to the core, for he reverted to concepts that his teacher, Arnold Sommerfeld, had long before recognized to be true for impulse x-rays.[2]

Heisenberg's symbolic formulation of the noncommuting character of measurements to determine the position x and the momentum $p = h/\lambda$ of a particle is his famous principle of indeterminacy:

$$\Delta x \Delta p_x \sim h \tag{E.1}$$

This states that it is impossible to determine simultaneously, with arbitrary accuracy, both the position and the momentum of a

---

[1] Born, *ZP, 37* (1926), 863–7; *38* (1926), 803–27. An earlier version appeared in Göttingen *Nachrichten* (1926), 146–60.
[2] Heisenberg, *ZP, 43* (1927), 172–98. *Indeterminacy principle* is preferable to *uncertainty principle:* The former denotes the claim of *ignorabimus* essential to the Copenhagen interpretation, the latter merely *ignoramus*.

particle along a line. If the position x is known accurately, then the momentum along that axis is correspondingly indeterminate. The product of the indeterminacy in both measurements must exceed a certain small number set by Planck's constant. Momentum, according to de Broglie's relation, may be expressed in terms of wavelength. So, the wave-mechanical interpretation of the indeterminacy relation is equivalent to a prohibition of simultaneously determining the position of a particle and the wavelength of its wave function. Conceptually, it is a direct analog of the tradeoff that Sommerfeld had utilized to distinguish between localized x-ray impulses and periodic waves.

One may not speak of a piece of a monochromatic wave; its wavelength can be precisely defined only if the wave extends to infinity in space. If one interprets the momentum of the particle in terms of its de Broglie wavelength, Heisenberg's relation E.1 makes the same point about the electron. Either the velocity or the position of the particle may be defined by experiment with arbitrary accuracy, but the two cannot be determined simultaneously. If the electron is known to occupy a particular location in space, the Fourier integral of its wave function spreads over many frequencies and the momentum becomes accordingly less definite. If, on the other hand, the momentum is specified, the wave function must be monochromatic; the wave must extend to great distances in space. But then the range of possible positions that the particle may occupy with equal likelihood becomes correspondingly great.

In a later study of the continuous x-ray spectrum as an example of continuity in physics before and after quantum mechanics, Sommerfeld referred explicitly to the resuscitation of his action-integral representation of the quantum conditions in Heisenberg's principle of indeterminacy.[3] He might have gone further. In the form relating position and momentum, Heisenberg's indeterminacy relation is a conceptual extension of the Fourier interpretation of impulse and periodic waves that Sommerfeld had first raised in 1901.

ELECTRON DIFFRACTION

The wave properties of light and the localized properties of matter are familiar characteristics. The accumulated experimental evi-

---

[3] Sommerfeld, *Scientia, 51* (1932), 41–50.

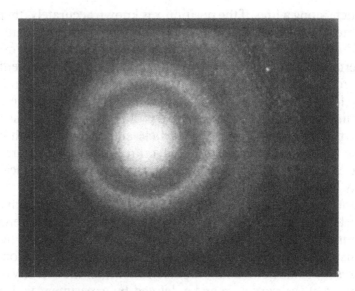

Figure E.1. Electron diffraction pattern from gold obtained in 1927 by G. P. Thomson. Compare this to the x-ray picture in Figure 8.1, imagining that photograph spinning rapidly. [*PRS, 117A* (1928), 604.]

dence that we have discussed convinced physicists in the early 1920s that radiation also demonstrates particlelike properties. Experimental evidence in corroboration of the remaining permutation of wave and particle with light and matter was soon offered.

A beam of electrons directed toward a slit of molecular dimension should show diffraction effects. "In this way," de Broglie had predicted, "one might seek experimental confirmation of the [pilot wave] hypothesis."[4] Louis de Broglie was, by his own admission, not able to undertake the tests. And he had no success in convincing the experimentalists in rue Lord Byron to assist him.[5] In mid-1925 Walter Elsasser suggested that previously unexplained results of electron-beam experiments might be interpreted as due to the wavelike behavior of electrons.[6] Anomalous peaks in the intensity of electrons scattered from matter noted in experiments two years before by Clinton Davisson and Charles Kunsman could, Elsasser argued, be evidence for diffraction of electrons.[7]

[4] L. de Broglie, *CR, 177* (1923), 549.
[5] See the interview with L. de Broglie by T. S. Kuhn in AHQP.
[6] Elsasser, *Naturwissenschaften, 13* (1925), 711.
[7] Davisson and Kunsman, *PR, 22* (1923), 242–58. See also Gehrenbeck, *Electron diffraction* (1973).

Within two years, Davisson and Lester Germer in America and George Thomson in Britain confirmed Elsasser's prediction.[8] (See Figure E.1.)

Elsasser also suggested that an anomaly in the absorption of slow electrons by noble gases might also be due to a wave effect. The absorption generally increases as the electron velocity decreases. But the Swedish physicist Nils Åkesson noted in 1916 that electrons accelerated across only a few volts seem to be *more* penetrating than faster ones.[9] The effective cross section for the transfer of electron kinetic energy to atoms of the gas drops abruptly for electrons with energy under a few electron volts; the electron mean free paths become extraordinarily long. Starting in 1921, extensive research on the phenomenon was undertaken by Herbert Mayer and Carl Ramsauer in Heidelberg.[10] Neither classical nor semiquantum treatments succeeded in explaining the "Ramsauer effect."[11] But when a full explanation using the new quantum mechanics was provided in 1929, it was, as Elsasser had suggested, of an effect similar to colloidal scattering of light. The phenomenon sets in only when the de Broglie wavelength of the incident electrons exceeds a threshold value.[12]

In 1929 Louis de Broglie received the Nobel Prize in physics for his theoretical hypothesis of matter waves.[13] The award followed results by Davisson and G. P. Thomson that verified the hypothesis experimentally.[14] It had taken seventeen years from the time

[8] Davisson and Germer, *Nature, 119* (1927), 558–60. G. P. Thomson, *PRS, 117A* (1928), 600–9. See also Russo, *HSPS, 12* (1981), 117–60.

[9] Åkesson, Lunds Universitets *Årsskrift,* 12 (1916), part 11. See also his earlier studies under Lenard, Heidelberg *Sb, 5A* (1914), Abh. 21.

[10] Ramsauer, *PZ, 21* (1920), 576–8. Mayer, *AP, 64* (1921), 451–80. Ramsauer, *PZ, 22* (1921), 613–15; *AP, 64* (1921), 513–40.

[11] The classical treatment was by Hund, *ZP, 13* (1923), 241–63. The semiquantum was presented by Zwicky, *PZ, 24* (1923), 171–83.

[12] Faxén and Holtsmark, *ZP, 45* (1927), 307–24. Wentzel, *PZ, 29* (1928), 321–37. Mott, *PCPS, 25* (1929), 304–9.

[13] The award was part of a remarkably continuous series of Nobel Prizes given for studies on the interaction of radiation with matter. The prizes in 1921–3 to Einstein, Bohr, and Millikan have already been discussed. In 1924 Manne Siegbahn's work on x-ray spectroscopy was honored, and in 1925 Franck and Hertz's work on the quantum excitation of atoms. In 1927 Compton was awarded half of the prize for his work on x-ray scattering, and the 1928 prize went to Owen Richardson for work on thermionic emission (closely related to the photoeffect). Louis de Broglie's prize in 1929 recognized the importance of the wave interpretation of matter; it was matched in 1933 by the half-prize awarded to Erwin Schrödinger.

[14] *Nobel en 1929* (1930), 16–27.

Einstein first suggested the lightquantum until he received the Nobel Prize for related work, and in virtually all of those years his revolutionary hypothesis was rejected by most physicists. It was only five years after de Broglie's equally revolutionary idea that he was awarded the prize. Both Louis and Maurice de Broglie had been nominated for the Nobel Prize as early as 1925.[15] At that time, the evaluation of their work described in detail the importance of Maurice's research on the x-ray photoeffect, but Louis's theoretical research was judged to have thus far "little significance for science."[16] Although at first even de Broglie's closest associates in France had great difficulty believing in matter waves, in Schrödinger's formulation the inherent idea was widely accepted. There was still great opposition to the conceptual demands of the new synthetic view of matter and light. But the difficulties it relieved and the symmetry inherent in the new dualistic interpretation served to recommend its attractions despite its drawbacks. Regardless of their intuitive predispositions, by 1927 most influential European physicists had learned to accept the duality of waves and particles.

Scientists raised on mechanical representations of nature found the new indeterministic physics hard to swallow. J. J. Thomson described the new dualistic basis of physics as a struggle "between a tiger and a shark, each is supreme in his own element but helpless in that of the other."[17] He persevered in attempts to find a mechanically consistent model for lightquanta. In 1925 he pictured them as disconnected closed loops or rings of electric force, their frequency determined by the time light takes to loop around the circumference.[18] William Henry Bragg joked in resignation that physicists trot out a wave description on Mondays, Wednesdays, and Fridays and a particle description on Tuesdays, Thursdays, and Saturdays.[19]

---

[15] Khvol'son to Nobel Committee, 8 January 1925. In "Förslag till utdelning av Nobelpris i fysik år 1925," p. 61. Nobel archives.

[16] Oseen, "Utredning beträffande M. Siegbahn samt M. och L. de Broglie," 16 pp. Bilage 1 to 1925 prize. Nobel archives.

[17] J. J. Thomson, *Structure of light* (1925), 15.

[18] *Ibid.,* 21ff. Thomson's proposal bears a resemblance to a casual suggestion by Philipp Lenard in 1910, Heidelberg *Sb, 1* (1910), Abh. 16, pp. 17–18.

[19] W. H. Bragg, BAAS *Report, 98* (1928), 20.

[20] Planck, *JFI, 204* (1927), 13–18. Lorentz, *Nature, 114* (1924), 608–11. For the classical portrayal of German willingness to accept dualistic theory, see Forman, *HSPS, 3* (1971), 1–115. That this sort of issue represented a crucial break with tradition is sensitively treated in McCormmach, *Night thoughts* (1981).

But many physicists in Germany, already used to a less mechanistic ontology than their British counterparts, and many younger physicists in other nations were able to adopt the new approach to physical theory with less difficulty.[20] Some, such as Schrödinger and Einstein, saw the indeterminacy relations as a statement of man's limited ability to penetrate the details of natural processes, not as a complete description of how nature acts. Others, such as Heisenberg and Bohr, felt that in the indeterminacy relations and in Bohr's associated principle of complementary physical descriptions, humans had discovered the deep truth that nature itself follows only statistical rules. In its fully developed form – the *Copenhagen interpretation* – this latter view represents the public philosophical view of most physicists to this day. But these issues are largely unrelated to doing physics. Whether nature is only statistically consistent or whether humans are simply unable to penetrate to a consistent core soon became matters of only marginal professional concern to physicists.

Physics had undergone a remarkable transformation. The inconsistencies that had plagued atom theory and confounded radiation theory were resolved in the new quantum mechanical formalism. But successful form was gained at a high conceptual price – the abandonment of deterministic theory. Problems in atomic theory were instrumental in bringing about that change in world view. But it was the attempt to formulate a consistent theory of radiation after 1921 and the influence of this program on atomic theory that sparked the most dramatic shift in understanding. And the specific problems that led to the adoption of the new dualistic interpretation of light extend back, as we have seen, to the earliest years of the century.

BIBLIOGRAPHY

All works mentioned in the notes and text are listed here by author or editor. A few liberties have been taken in the interests of consistency. Lords Cherwell, Kelvin, and Rayleigh appear here under their family names – Frederick Lindemann, William Thomson, John W. Strutt and Robert J. Strutt – even though many of their works are published under their baronial title. Similarly, English transliteration is used for Russian physicists who published papers in a German or French transliteration; in particular, Chwolson appears here as Khvol'son, Joffe as Ioffe. Names with prepositions (de Broglie, Des Coudres, van der Waals) are listed under the final part.

Abraham, H., and P. Langevin, eds. *Ions, électrons, corpuscles.* 2 vols. Paris, 1905.
Abraham, M. "Prinzipien der Dynamik des Electrons." *AP, 10* (1903), 105–79.
– *Theorie der Elektrizität,* vol. 2, *Elektromagnetische Theorie der Strahlung.* Leipzig, 1905.
Åkesson, N. "Über die Geschwindigkeitsverluste bei langsamen Kathodenstrahlen." Heidelberg *Sb, 5A* (1914), Abh. 21.
– "Über die Geschwindigkeitsverluste bei den langsamen Kathodenstrahlen und über deren selektive Absorption." Lunds Universitets *Årsskrift, 12* (1916), part 11.
Allen, S. J. "The velocity and ratio e/m for the primary and secondary rays of radium." *PR, 23* (1906), 65–94.
Andrade, E. N. da C. "William Henry Bragg." RSL *Obituary Notices of Fellows, 4* (1943), 277–300.
Ångström, K. "Beiträge zur Kenntnis der Wärmeabgabe des Radiums." *PZ, 6* (1905), 685–8.

Ashworth, J. R. "γ rays from radium." *Nature, 69* (1904), 295.

Badash, L., ed. *Rutherford and Boltwood: Letters on radioactivity.* New Haven, 1969.

Baeyer, O. von, and A. Gehrtz. "Die Anfangsgeschwindigkeiten lichtelektrisch ausgelöster Elektronen." *VDpG, 12* (1910), 870–9.

Baeyer, O. von, and O. Hahn. "Magnetische Linienspektren von β Strahlen." *PZ, 11* (1910), 488–93.

Baeyer, O. von, O. Hahn, and L. Meitner. "Über die β Strahlen des activen Niederschlags des Thorium." *PZ, 12* (1911), 273–9.

- "Nachweis von β Strahlen bei Radium D." *PZ, 12* (1911), 378–9.

- "Das magnetische Spektrum der β Strahlen des Thoriums." *PZ, 13* (1912), 264–6.

Barkla, C. G. "Secondary radiation from gases subject to x-rays." *PM, 5* (1903), 685–98.

- "Polarization in Röntgen rays." *Nature, 69* (1904), 463.

- "Polarized Röntgen radiation." *PTRS, 204A* (1905), 467–79.

- "Secondary Röntgen radiation." *PM, 11* (1906), 812–28.

- "Polarization in secondary Röntgen radiation." *PRS, 77A* (1906), 247–55.

- "The atomic weight of nickel." *Nature, 75* (1907), 368.

- "The nature of x rays." *Nature, 76* (1907), 661–62.

- "Note on x rays and scattered x rays." *PM, 15* (1908), 288–96.

- "Der Stand der Forschung über die sekundäre Röntgenstrahlung." *JRE, 5* (1908), 246–324.

- "Phenomena of x ray transmission." *PCPS, 15* (1909), 257–69. German translation in *JRE, 7* (1910), 1–15.

- "The spectra of the fluorescent Röntgen radiations." *PM, 22* (1911), 396–412. German translation in *JRE, 8* (1911), 471–88.

- "Problems of radiation." *Nature, 94* (1915), 671–2.

- "X-ray fluorescence and the quantum theory." *Nature, 95* (1915), 7.

- "Bakerian Lecture – on x-rays and the theory of radiation." *PTRS, 217A* (1917), 315–60.

Barkla, C. G., and G. H. Martyn. "Interference of Röntgen radiation (preliminary account)." *PPSL, 25* (1913), 206–15.

Barkla, C. G., and J. Nicol, "Homogeneous fluorescent x radiations of a second series." *PPSL, 14* (1911), 9–17.

- "X ray spectra." *Nature, 84* (1910), 139.

Barkla, C. G., and A. J. Philpot. "Ionization in gases and gaseous mixtures by Röntgen and corpuscular (electronic) radiations." *PM, 25* (1913), 832–56.

Barkla, C. G., and C. A. Sadler. "Secondary x rays and the atomic weight of nickel." *PM, 14* (1907), 408–22.

- "Homogeneous secondary radiations." *PM, 16* (1908), 550–84.

- "The absorption of Röntgen rays." *PM, 17* (1909), 85–94.
- "The absorption of Röntgen rays." *PM, 18* (1909), 739–60.
Barkla, C. G., and G. Shearer. "Note on the velocity of electrons expelled by x-rays." *PM, 30* (1915), 745–53.
Bartlett, A. "Compton effect: Historical background." *AJP, 32* (1964) 120–7.
Bassler, E. "Polarisation der X-Strahlen, nachgewiesen mittels Sekundärstrahlung." *AP, 28* (1909), 808–84.
Bateman, H. "On the theory of light quanta." *PM, 46* (1923), 977–91.
Beatty, R. T. "The production of cathode particles by homogeneous Röntgen radiations." *PCPS, 15* (1910), 416–22.
- "The production of cathode particles by homogeneous Röntgen radiations, and their absorption by hydrogen and air." *PM, 20* (1910), 320–30.
- "The energy of Röntgen rays." *PRS, 89A* (1913), 314–27.
Becquerel, H. "Sur les radiations invisibles émises par les corps phosphorescents." *CR, 122* (1896), 501–3.
- "Sur quelques propriétés nouvelles des radiations invisibles émises par divers corps phosphorescents." *CR, 122* (1896), 559–64.
- "Sur les radiations invisibles émises par les sels d'uranium." *CR, 122* (1896), 689–94.
- "Sur les propriétes differentes des radiations invisibles émises par les sel d'uranium, et du rayonnement de la paroi anticathode d'un tube de Crookes." *CR, 122* (1896), 762–7.
- "Influence d'un champ magnétique sur le rayonnement des corps radio-actifs." *CR, 129* (1899), 996–1001.
- "Sur le rayonnement de l'uranium et sur diverses propriétés physiques du rayonnement des corps radio-actifs." In *Rapports présentés au Congrès International de Physique,* C. Guillaume and L. Poincaré, eds., vol. 3, pp. 47–78, Paris, 1900.
- "Contribution à l'étude du rayonnement du radium." *CR, 130* (1900), 206–11.
- "Sur la dispersion du rayonnement du radium dans un champ magnétique." *CR, 130* (1900), 372–76.
- "Déviation du rayonnement du radium dans un champ électrique." *CR, 130* (1900), 809–15.
- "Sur la transparence de l'aluminum pour le rayonnement du radium." *CR, 130* (1900), 1154–57.
- "Sur la radio-activité secondaire des métaux." *CR, 132* (1901), 371–73.
- "Sur l'analyse magnétique des rayons du radium et du rayonnement secondaire provoqué par ces rayons." *CR, 132* (1901), 1286–9.
- "Sur quelques effets chimiques produits par le rayonnement du radium." *CR, 133* (1901), 709–12.

- "Sur quelques propriétés du rayonnement des corps radioactifs." *CR,* *134* (1902), 208–11.
- "Sur le rayonnement du polonium et du radium." *CR, 136* (1903), 431–4.
- "Sur une propriété des rayons α du radium." *CR, 136* (1903), 1517–22.
- "Über die von der Strahlung radioactiver Körper hervorgerufene sekundäre Strahlung." *PZ, 5* (1904), 561–3.
- "Sur la radio-activité de la matière." *PRI, 17* (1906), 85–94.
Benoist, L., and D. Hurmuzescu. "Nouvelles propriétés des rayons x." *CR, 122* (1896), 235–6.
Birge, R. T. "Probable values of the general physical constants." *PR Supplement, 1* (1929), 1–73.
Black D. H. "The β ray spectrum of mesothorium 2." *PRS, 106A* (1924), 632–40.
- "β ray spectra of thorium disintegration products." *PRS, 109A* (1925), 166–76.
Blondlot, R. "Sur la vitesse de propagation des rayons x." *CR, 135* (1902), 666–70, and related studies, *ibid.,* 721–4, 763–6, 1293–5.
Bohr, N. "On the constitution of atoms and molecules [Part I]." *PM, 26* (1913), 1–25.
- "L'application de la theorie des quanta aux problèmes atomiques." In Solvay III (1923), 228–47.
- "Linienspektren und Atombau." *AP, 71* (1923), 228–88.
- "Über die Anwendung der Quantentheorie auf den Atombau. I. Die Grundpostulate der Quantentheorie." *ZP, 13* (1923), 117–65. Translated as "On the application of the quantum theory to atomic structure. Part I, The fundamental postulates of the quantum theory." *PCPS Supplement* (1924).
- *Collected works,* vol. 2. Ulrich Hoyer, ed. Amsterdam, 1981.
Bohr, N., H. A. Kramers, and J. C. Slater. "The quantum theory of radiation." *PM, 47* (1924), 785–802.
Boltzmann, L. "Röntgen's neue Strahlen." *Elektro-Techniker, 14* (1896), 385–9. Reported in *Electrician, 36* (1896), 449. Reprinted in Boltzmann, *Populäre Schriften,* pp. 188–97. Leipzig, 1896.
- "Über ein Medium, dessen mechanische Eigenschaften auf die von Maxwell für den Elektromagnetismus aufgestellen Gleichungen führen." *AP, 48* (1893), 78–99.
Born, M. "Zur Wellenmechanik der Stossvorgänge." Göttingen *Nachrichten* (1926), 146–60.
- "Zur Quantentheorie der Stossvorgänge." *ZP, 37* (1926), 863–7; *38* (1926), 803–27.
Bose, S. N. "Planck's Gesetz und Lichtquantenhypothese." *ZP, 26* (1924), 178–81.

Bothe, W., and H. Geiger. "Ein Weg zur experimentellen Nachprüfung der Theorie von Bohr, Kramers, und Slater." *ZP, 26* (1924), 44.

Bragg, W. H. "The 'elastic medium' method of treating electrostatic theorems." AusAAS *Report, 3* (1891), 57–71.

– "Presidential address to Section A." AusAAS *Report, 4* (1892), 31–47.

– "The energy of the electromagnetic field." AusAAS *Report, 6* (1895), 223–31.

– "On some recent advances in the theory of the ionization of gases." AusAAS *Report, 10* (1904), 47–77.

– "On the ionization of various gases by the α particles of radium." *PM, 11* (1906), 617–32.

– "A comparison of some forms of electric radiation." *TPRRSSA, 31* (1907), 79–93.

– "The nature of Röntgen rays." *TPRRSSA, 31* (1907), 94–8.

– "On the properties and natures of various electric radiations." *PM, 14* (1907), 429–49

– "The nature of γ and x rays." *Nature, 77* (1908), 270–1, 560; *Nature, 78* (1908), 271.

– "The consequences of the corpuscular hypothesis of the γ and x rays, and the range of β rays." *PM, 20* (1910), 385–416. German translation in *JRE, 7* (1910), 348–86.

– "The mode of ionization by x rays." *PM, 22* (1911), 222–3.

– "Corpuscular radiation." BAAS *Report,* (1911), 340–41.

– "On the direct or indirect nature of the ionization by x rays." *PM, 23* (1912), 647–50.

– "Radiations old and new." BAAS *Report,* (1912), 750–3; *Nature, 90* (1913), 529–32, 557–60.

– *Studies in radioactivity.* London, 1912.

– "X rays and crystals." *Nature, 90* (1912), 219, 360–1.

– "The reflection of x rays by crystals." *PRS, 89A* (1913), 246–8.

– "Aether waves and electrons." *Nature, 107* (1921), 374.

– "Craftsmanship and science." BAAS *Report, 98* (1928), 1–20.

Bragg, W. H., and W. L. Bragg. "The reflection of x rays by crystals." *PRS, 88A* (1913), 428–38.

– *X rays and crystal structure.* London, 1915.

Bragg, W. H., and J. L. Glasson. "On a want of symmetry shown by secondary x rays." *TPRRSSA, 32* (1908), 301–10; *PM, 17* (1909), 855–64.

Bragg, W. H., and J. P. V. Madsen. "An experimental investigation of the nature of the γ rays – No. 1." *TPRRSSA, 32* (1908), 1–10; *PM, 15* (1908), 663–75.

– "An experimental investigation of the nature of the γ rays – No. 2." *TPRRSSA, 32* (1908), 35–54; *PM, 16* (1908), 918–939.

Bragg, W. H., and S. E. Peirce. "The absorption coefficients of x rays."
   PM, 28 (1914), 626–30.
Bragg, W. H., and H. L. Porter. "Energy transformations of x rays." *PRS,
   85A* (1911), 349–65.
Bragg, W. L. "The diffraction of short electromagnetic waves by a
   crystal." *PCPS, 17* (1912), 43–57.
– *The start of x ray analysis. Chemistry background book.* London, 1967.
Breit, G. "The propagation of a fan-shaped group of waves in a dispersing
   medium." *PM, 44* (1922), 1149–52.
Bridgman, P. W. "Biographical memoir of William Duane, 1872–1935."
   NAS *Biographical memoirs, 18* (1937), 23–41.
Brillouin, L. "L'agitation moléculaire et les lois du rayonnement ther-
   mique." *JP, 2* (1921), 142–55.
Brillouin, M. "Actions méchaniques à hérédité discontinue par propagation: Essai de théorie dynamique de l' atom à quanta." *CR, 168*
   (1919), 1318–20.
de Broglie, L. "Sur la dégradation du quantum dans les transformations
   successives des radiations de haute fréquence." *CR, 173* (1921),
   1160–2.
– "Sur la theorie de l'absorption des rayons x par la matière et le principe
   de correspondance." *CR, 173* (1921), 1456–8.
– "Considérations théoriques sur l'absorption des rayons x par la matière." SFP *PVRC, 1921* (1922), 15–16.
– "Rayons x et equilibre thermodynamique." *JP, 3* (1922), 33–45.
– "Rayonnement noir et quanta de lumière." *JP, 3* (1922), 422–8.
– "Sur les interférences et la theorie des quanta de lumière." *CR, 175*
   (1922), 811–13.
– "Ondes et quanta." *CR, 177* (1923), 507–10.
– "Quanta de lumière, diffraction et interferences." *CR, 177* (1923),
   548–50.
– "Les quanta, le théorie cinetique des gaz et le principe de Fermat." *CR,
   177* (1923), 630–2.
– "A tentative theory of light quanta." *PM, 47* (1924), 446–58.
– *Thèses. Recherches sur la théorie des quanta.* Paris, 1924. Republished
   in *Annales de physique, 3* (1925), 22–128, with a useful historical
   introduction by de Broglie, pp. 22–30. German translation, *Unter-
   suchungen zur Quantentheorie.* Leipzig, 1927. The French text was
   republished by Masson. Paris, 1963.
– "Allocution prononcée au jubilé de M. Maurice de Broglie le 13 Juin
   1946." *Savants et decouverts,* pp. 298–305. Paris, 1951.
– "Vue d'ensemble sur mes travaux scientifiques." In *Louis de Broglie:
   Physicien et penseur,* A. Georges, ed., pp. 457–86. Paris, 1953.
– *Le dualisme des ondes et les corpuscules dans l'oeuvre de Albert
   Einstein.* Paris, 1955.

– "Une rencontre avec Einstein au conseil Solvay en 1927." In *Nouvelles perspectives en microphysique*, pp. 232–7. Paris, 1956.
– "The beginnings of wave mechanics." In *Wave mechanics: The first fifty years*, W. C. Price, S. S. Chissick, and T. Ravensdale, eds., pp. 12–18. New York, 1973.
– "Avant propos." *Recherches d'un demi-siecle*, pp. 7–16. Paris, 1976.
de Broglie, L., and A Dauvillier. "Sur le système spectral des rayons Röntgen." *CR, 175* (1922), 685–8.
– "Sur les analogies de structure entre les séries optiques et les séries de Röntgen." *CR, 175* (1922), 755–7.
de Broglie, M. *Thèses. Recherches sur les centres électrisés de faible mobilité dans les gaz.* Paris, 1908.
– "Sur un nouveau procédé permettant d'obtenir la photographie des spectres de raies des rayons de Röntgen." *CR, 157* (1913), 924–6.
– "Sur la spectroscopie des rayons secondaires émis hors des tube à rayons de Röntgen, et les spectres d'absorption," *CR, 158* (1914), 1493–5.
– "La nature des rayons de "Röntgen . . ." *Revue scientifique, 53* (1915), 577–582.
– "Sur une système de bandes d'absorption correspondant aux rayons L des spectres de rayons x des éléments . . ." *CR, 163* (1916), 352–5.
– "La portée des nouvelles découvertes dans la région des rayons de très haute fréquence." *Scientia, 27* (1920), 102–11.
– "Les écrans renforçateurs et le spectre des rayons x." SFP *PVRC, 1920* (1921), 32–3.
– "Sur les spectres corpusculaires des éléments." *CR, 172* (1921), 274–5.
– "Remarques sur l'équation photo-électrique d'Einstein." SFP *PVRC, 1921* (1922), 10, 13–14.
– "Sur les spectres corpusculaires des élements." *CR, 172* (1921), 527–9.
– "Sur les spectres corpusculaires. Lois de l'émission photo-électrique pour les hautes fréquences." *CR, 172* (1921), 806–7.
– "Les phénomènes photo-électriques correspondant aux rayons x et les spectres corpusculaires des éléments." SFP *PVRC, 1921* (1922), 48–9.
– "Les phénomènes photo-électriques pour les rayons x et les spectres corpusculaires des éléments." *JP, 2* (1921), 265–87.
– "Sur les spectres corpusculaires et leur utilization pour l'étude des spectres de rayons x." *CR, 173* (1921), 1157–60.
– *Les rayons x.* Paris, 1922.
– "La relation $hv = \epsilon$ dans les phénomènes photoélectriques." In Solvay III, pp. 80–100.
– "Le prince Louis de Broglie enfant." *Grandes souvenirs, belles actualités: Le recueil du jeunes, 2* (July 1948), 4–6.

- "La jeunesse et les orientations intellectuelles de Louis de Broglie." In *Louis de Broglie: Physicien et penseur,* A. Georges, ed., pp. 423–9. Paris, 1953.

de Broglie, M., and L. de Broglie. "Sur le modèle d'atom de Bohr et les spectres corpusculaires." *CR, 172* (1921), 746–8.

- "Sur les spectres corusculaires des éléments." *CR, 173* (1921), 527–9.

- "Remarques sur les spectres corpuscularies et l'effet photo-électrique." *CR, 175* (1922), 1139–41.

de Broglie, M., and J. Cabrera. "Sur les rayons gamma de la famille du radium et du thorium étudiés par leur effet photo-électrique." *CR, 176* (1923), 295–6.

Brunhes, B. "Une méthode de mesure de la vitesse des rayons Röntgen." *CR, 130* (1900), 127–30.

Buchwald, E. "Zur Berechnung der γ Strahlschwankungen." *AP, 39* (1912), 41–52.

Buisson, H., and C. Fabry. "Sur la mesure des intensités des diverses radiations d'un rayonnement complexe." *CR, 152* (1911), 1838–41.

Bumstead, H. A. "The heating effects produced by Röntgen rays in different metals, and their relation to the question of change in the atom." *PM, 11* (1906), 292–317.

Burbidge, P. W. "The fluctuation in the ionization due to γ rays." *PRS, 89A* (1913), 45–57.

Campbell, N. *Modern electrical theory.* Cambridge, 1907.

- "The study of discontinuous phenomena." *PCPS, 15* (1909), 117–36.

- "Discontinuities in light emission." *PCPS, 15* (1909), 310–28; *15* (1910), 513–25.

- "The aether." *PM, 19* (1910), 181–91. German translation in *JRE, 7* (1910), 15–28.

- "Über Schweidlersche Schwankungen." *PZ, 11* (1910), 826–33.

- "Nachschrift." *PZ, 13* (1912), 81–3.

Carvallo, E. "Sur la nature de la lumière blanche." *CR, 130* (1900), 79–82.

- "Sur la nature de la lumière blanche et des rayons x." *CR, 130* (1900), 130–2.

Cassidy, D. "Heisenberg's first core model of the atom: The formation of a professional style." *HSPS, 10* (1979), 187–224.

Chadwick, J. "Intensitätsverteilung im magnetischen Spektrum der β Strahlen von Radium B + C." *VDpG, 16* (1914), 383–91.

- "The charge on the atomic nucleus and the law of force." *PM, 40* (1920), 734–46.

Chwolson. See Khvol'son.

Compton, A. H. "The size and shape of the electron." WAS *Journal, 8* (1918), 2–11; *PR, 14* (1919), 20–43.

- "The degradation of gamma ray energy." *PM, 41* (1921), 749–67.

- "The absorption of γ rays by magnetized iron." *PR, 17* (1921), 38–41.
- "Secondary high frequency radiation." *PR, 18* (1921), 96.
- "The spectrum of secondary x rays." *PR, 19* (1922), 267–8.
- "Secondary radiations produced by x-rays, and some of their applications to physical problems." NRC *Bulletin, 4* (1922), 1–56.
- "A quantum theory of the scattering of x rays by light elements." *PR, 21* (1923), 483–502.
- "The scattering of x rays." *JFI, 198* (1924), 57–72.

Compton, A. H., and A. W. Simon. "Measurements of β-rays associated with scattered x-rays." *PR, 25* (1925), 306–13.
- "Directed quanta of scattered x-rays." *PR, 26* (1925), 289–99.

Compton, K. T., and O. Richardson. "The photoelectric effect II." *PM, 26* (1913), 549–67.

Cooksey, C. D. "On the corpuscular rays produced in different metals by Röntgen rays." *AJS, 24* (1907), 285–304.
- "The nature of γ and x rays." *Nature, 77* (1908), 509–10.
- "On the asymmetry in the distribution of secondary cathode rays produced by the x rays; and its dependence on the penetrating power of the exciting rays." *PM, 24* (1912), 37–45.

Coolidge, W. "A powerful Röntgen ray tube with a pure electron discharge." *PR, 2* (1913), 409–30.

Des Coudres, T. "Über Kathodenstrahlen unter dem Einflusse magnetischer Schwingungen." *VDNA, 14* (1895), 86–8.
- "Elektrodynamisches über Kathodenstrahlen." *VDNA, 16* (1897), 157–62.

Crookes, W. "Bakerian lecture on the illumination of lines of molecular pressure and the trajectories of molecules." *PTRS, 170:1* (1879), 135–64.
- *On radiant matter.* London, 1879.
- "Radio-activity of uranium." *PRS, 66* (1900), 409–23.

Curie, M. *Thèses. Recherches sur les substances radioactives.* Paris, 1903. Reprinted in the Mallinkrodt classics of radiology. St. Louis, 1966.

Curie, P. "Action du champ magnétique sur les rayons de Becquerel. Rayons déviés et rayons non déviés." *CR, 130* (1900), 73–6.

Curie, P., and M. Curie. "Sur une substance nouvelle radio-active, contenue dans la pechblende. *CR, 127* (1898), 175–8.
- "Sur la radioactivité provoquée par les rayons de Becquerel." *CR, 129* (1898), 714–16.
- "Les nouvelles substances radioactives et les rayons qu'elles émittent." In *Congres international de physique,* C. Guillaume and L. Poincaré, eds., vol. 3, pp. 79–114. Paris, 1900.

Curie, P., and G. Sagnac. "Électrisation negative des rayons secondaires produits au moyen des rayons Röntgen." *CR, 130* (1900), 1013–15.

Danysz, J. "Sur les rayons β de la famille du radium." *CR, 153* (1911), 339–41, 1066–8; *Le Radium, 9* (1912), 1–5.

- "Sur les rayons β des radiums B, C, D." *Le Radium, 10* (1913), 4–6.

Danysz, J., and W. Duane. "Sur les charges électroniques transportées par les rayons α et β." *CR, 155* (1912), 500–3.

Dauvillier, A. *Thèses. Recherches spectrométriques sur les rayons x.* Paris, 1920.

- *La technique des rayons x.* Paris, 1924.

Dauvillier, A., and L. de Broglie. "Sur la structure éléctronique des atomes lourds." *CR, 172* (1921), 1650–3.

- "Sur la distribution des électrons dans les atomes lourdes." *CR, 173* (1921), 137–9.

Davisson, C. "Note on radiation due to impact of β particles upon solid matter." *PR, 28* (1909), 469–70.

Davisson, C., and L. H. Germer. "The scattering of electrons by a single crystal of nickel." *Nature, 119* (1927), 558–60.

Davisson, C., and C. H. Kunsman. "The scattering of low speed electrons by platinum and magnesium." *PR, 22* (1923), 242–58.

Debye, P. "Zustandsgleichung und Quantenhypothese mit einem Anhang über Wärmeleitung." In *Vorträge über die kinetische Theorie der Materie und der Elektrizität. Mathematische Vorlesungen an der Universität Göttingen,* D. Hilbert, ed., vol. 6, pp. 17–60. Leipzig, 1914. An abstract appeared in *PZ, 14* (1913), 259–260.

- "Über den Einfluss der Wärmebewegung auf die Interferenzerscheinungen bei Röntgenstrahlen." *VDpG,15* (1913), 678–89.

- "Über die Intensitätsverteilung in den mit Röntgenstrahlen erzeugten Interferenzbildern." *VDpG, 15* (1913), 738–52.

- "Spektrale Zerlegung der Röntgenstrahlen mittels Reflexion und Wärmebewegung." *VDpG, 15* (1913), 857–75.

- "Zerstreuung von Röntgenstrahlen und Quantentheorie." *PZ, 24* (1923), 161–6.

Debye, P., and A. Sommerfeld. "Theorie des lichtelektrischen Effektes vom Standpunkt des Wirkungquantums." *AP, 41* (1913), 873–930.

Donath, B. *Die Einrichtung zur Erzeugung der Roentgenstrahlen und ihr Gebrauch.* Berlin, 1899.

Dorn, E. "Über die erwärmende Wirkung der Röntgenstrahlen." *AP, 63* (1897), 160–76.

- "Versuche über Sekundärstrahlen." *Archives Néerlandaises* (2), 5 (1900), 595–608. Issued separately as *Recueil de travaux offerts par les auteurs à H. A. Lorentz.* The Hague, 1900.

- "Versuche über Sekundärstrahlen und Radiumstrahlen." Halle *Abhandlungen, 22* (1901), 39–43.

- "Elektrostatische Ablenkung der Radiumstrahlen." Halle *Abhandlungen, 22* (1901), 47–50.
- "Eine merkwürdige Beobachtung mit Radium." *PZ, 4* (1903), 507–8.
Drude, P. "Zur Geschichte der elektromagnetischen Dispersionsgleichungen." *AP, 1* (1900), 437–40.
Duane, W. "Relation between the wavelength and absorption of x rays." Title only in *PR, 4* (1914), 544.
- "On [the] relation between the wavelengths of x rays and the voltages required to produce them." *PR, 6* (1915), 166–76.
- "Planck's radiation formula deduced from hypotheses suggested by x ray phenomena." *PR, 7* (1916), 143–7.
- "Radiation and matter." *Science, 46* (1917), 347–9.
- "The transfer in quanta of radiation momentum to matter." NAS *Proceedings, 9* (1923), 158–64.
Duane, W., and K. F. Hu. "On the critical absorption and characteristic emission x-ray frequencies." *PR, 11* (1918), 489–91.
Duane, W., and F. L. Hunt. "On x-ray wave-lengths." *PR, 6* (1915), page numbers repeat in this volume; first 166–71.
Eckart, C. "Operator calculus and the solutions of the equations of quantum dynamics." *PR, 28* (1926), 711–26.
Edison, T. "Are Röntgen ray phenomena due to sound waves?" *Electrical engineer, 21* (1896), 353–4.
Einstein, A. "Über einen Erzeugung und Verwandlung des Lichtes betreffenden heuristischen Gesichtspunkt." *AP, 17* (1905), 132–48.
- "Zur Theorie der Lichterzeugung und Lichtabsorption." *AP, 20* (1906), 199–206.
- "Zum gegenwärtigen Stand des Strahlungsproblems." *PZ, 10* (1909), 185–93. See also the discussions between Einstein and Ritz, *ibid.*, 224–5, 323–4.
- "Über die Entwicklung unserer Anschauungen über das Wesen und die Konstitution der Strahlung." *PZ, 10* (1909), 817–25.
- Contribution to the discussion following Sommerfeld's presentation at the 1912 Naturforscherversammlung. *PZ, 12* (1911), 1068–9.
- "Zur Quantentheorie der Strahlung." Zürich *Mitteilungen, 18* (1916), 47–62. Reprinted in *PZ, 18* (1917), 121–8.
- "Über ein den Elementarprozess der Lichtemission betreffendes Experiment." Berlin *Sb* (1921), 882–3.
- "Zur Theorie der Lichtfortpflanzung in dispergierenden Medien." Berlin *Sb* (1922), 18–22.
- "Bietet die Feldtheorie Möglichkeiten für die Lösung des Quantenproblems?" Berlin *Sb* (1923), 359–64.
- "Quantentheorie des einatomigen idealen Gases." Berlin *Sb* (1924), 261–7; (1925), 3–14.

Einstein, A., and J. Laub. "Über die Elektromagnetische Grundglei-
chungen für bewegte Körper." *AP, 26* (1908), 532–40.

– "Über die im elektromagnetischen Felde auf ruhende Körper aus-
geübten ponderomotorischen Kräfte." *AP, 26* (1908), 541–50.

Eisberg, R. M. *Fundamentals of modern physics.* New York, 1961.

Ellis, C. D. "The magnetic spectrum of the β rays excited by γ rays." *PRS,
99A* (1921), 261–71.

– "β ray spectra and their meaning." *PRS, 101A* (1922), 1–17.

– "Über die Deutung der β-Strahlspektren radioaktiver Substanzen."
ZP, 10 (1922), 303–7.

– "The high energy groups in the magnetic spectrum of the radium C β
rays." *PCPS, 22* (1924), 369–78.

Ellis, C. D., and H. W. B. Skinner. "The interpretation of β ray spectra."
PRS, 105A (1924), 185–98.

Elsasser, W. "Bemerkungen zur Quantenmechanik freier Elektronen."
Die Naturwissenschaften, 13 (1925), 711.

Elster, J., and H. Geitel. "Der photoelektrische Effekt am Kalium bei sehr
geringen Lichtstärken." *PZ, 13* (1912), 468–76.

Emden, R. "Über Lichtquanten." *PZ, 22* (1921), 513–17.

Epstein, P. S. "Versuch einer Anwendung der Quantenlehre auf die
Theorie des lichtelektrischen Effektes und der β Strahlung radioaktiver Substanzen." *PZ, 17* (1916), 313–16; *AP, 50* (1916), 815–40.

Eve, A. S. "Röntgen rays and the γ rays from radium." *Nature, 69* (1904),
436.

– "A comparison of the ionization produced in gases by penetrating
Röntgen and radium rays." *PM, 8* (1904), 610–18.

– "On the secondary radiation caused by the β and γ rays of radium."
PM, 8 (1904), 669–85.

– "On the secondary radiation due to the γ rays of radium." *Nature, 70*
(1904), 454.

– "The absorption of the γ rays of radioactive substances." *PM, 11*
(1906), 586–95.

Faxén, H., and J. P. Holtsmark. "Beitrag zur Theorie des Durchganges
langsamer Elektronen durch Gase." *ZP, 45* (1927), 307–24.

Florance, D. C. H. "Primary and secondary γ rays." *PM, 20* (1910),
921–38.

Fomm, L. "Die Wellenlänge der Röntgen-Strahlen." München *Sb, 26*
(1896), 283–6.

Forman, P. "The doublet riddle and atomic physics *circa* 1924." *Isis, 59*
(1968), 156–74.

– "The discovery of the diffraction of x-rays by crystals: A critique of the
myths." *AHES, 6* (1969), 38–71.

- "Alfred Landè and the anomalous Zeeman effect." *HSPS, 2* (1970), 153-261.
- "Weimar culture, causality, and quantum theory, 1918-1927: Adaptation by German physicists and mathematicians to a hostile intellectual environment." *HSPS, 3* (1971), 1-115.
- "Bragg, William Henry." *DSB, 2* (1973), 397-400.
- "Duane, William." *DSB, 4* (1971), 194-197.
Forman P., J. L. Heilbron, and S. Weart. "Physics *circa* 1900." *HSPS, 5* (1975), 1-185.
Franck, J., and G. Hertz. "Über Zusammenstösse zwischen Elektronen und den Molekülen des Quecksilberdampfes und die Ionisierungsspannung derselben." *VDpG, 14* (1912), 457-67.
- "Über die Erregung der Quecksilberresonanzlinie 253,6µµ durch Elektronenstösse." *VDpG, 14* (1912), 512-17.
Franck, J., and R. Pohl. "Zur Frage nach der Geschwindigkeit der Röntgenstrahlen." *VDpG, 10* (1908), 117-36, 489-94.
- "Bemerkung zu den Versuchen des Hrn. Marx über die Geschwindigkeit der Röntgenstrahlen." *AP, 34* (1911), 936-40.
Frank, P. *Einstein, his life and times.* New York, 1953.
Friedman, R. "Nobel physics prize in perspective." *Nature, 292* (1981), 793-8.
Friedrich, W. "Röntgenstrahlinterferenzen." *PZ, 14* (1913), 1079-84.
Friedrich, W., P. Knipping, and M. Laue. "Interferenz-Erscheinungen bei Röntgenstrahlen." München *Sb, 42* (1912), 303-22. Reprinted with additional notes in *AP, 41* (1913), 971-88.
Frilley, M. *Thèses. Spectrographie par diffraction cristalline des rayons γ et la famille de radium.* Paris, 1928.
Füchtbauer, C. "Über die Geschwindigkeit der von Kanalstrahlen und von Kathodenstrahlen beim Auftreffen auf Metalle erzeugten negativen Strahlen." *PZ, 7* (1906), 748-50.
Gehrenbeck, R. K. *C. J. Davisson, L. H. Germer, and the discovery of electron-diffraction.* Unpublished Ph.D dissertation, University of Minnesota, 1973.
Geiger, H. "The irregularities in the radiation from radioactive bodies." *PM, 15* (1908), 539-47.
Geiger, H., and W. Bothe. "Über das Wesen des Comptoneffektes, ein experimenteller Beitrag zur Theorie der Strahlung." *ZP, 32* (1925), 639-63.
Georges, André, ed. *Louis de Broglie: Physician et penseur.* Paris, 1953.
Giesel, F. O. "Über die Ablenkbarkeit der Becquerelstrahlen im magnetischen Felde," *AP, 69* (1899), 834-7.
Gifford, J. W. "Are Röntgen rays polarized?" *Nature, 54* (1896), 172.

322   Bibliography

Gocht, H. *Die Röntgen-Literatur,* 5 vols. Stuttgart, 1911–25.
Godlewski, T. "On the absorption of the β and γ rays of actinium." *PM,* *10* (1905), 375–9.
Goldstein, E. "Über die Entladung der Elektrizität in verdünnten Gasen." Berlin *Mb* (1888), 82–124.
Gouy, L. G. "Sur le mouvement lumineux." *JP, 5* (1886), 354–62.
– "Sur le réfraction et la diffraction des rayons x." *CR, 123* (1896), 43–4.
– "Sur la constitution de la lumière blanche." *CR, 130* (1900), 241–4.
– "Sur le mouvement lumineux et les formules de Fourier." *CR, 130* (1900), 560–2.
Gray, J. A. "Secondary γ rays produced by β rays." *PRS, 85A* (1911), 131–9.
– "The nature of the γ rays excited by β rays." *PRS, 86A* (1912), 513–29.
– "The similarity in nature of x and primary γ rays." *PRS, 87A* (1912), 489–501.
– "The scattering and absorption of the γ rays of radium." *PM, 26* (1913), 611–23.
Green, G. "On the laws of the reflection and refraction of light at the common surface of two non-crystallized media." *TCPS, 7* (1839), 2–24.
– "On the propagation of light in crystallized media." *TCPS, 7* (1839), 121–40.
Guillaume, C. E. *Les rayons x.* Paris, 1896.
Guillaume, C. E., and L. Poincaré, eds. *Rapports présentés au congrès international de physique.* 3 vols. Paris, 1900.
Haga, H., and C. H. Wind. "De buiging van x-stralen." Amsterdam *Verslag, 7* (1899), 387–8.
– "De buiging der Röntgen-stralen." Amsterdam *Verslag, 7* (1899), 500–7. An English translation appeared in Amsterdam *Proceedings, 1* (1899), 321, 420–6. A German version appeared as "Die Beugung der Röntgenstrahlen." *AP, 68* (1899), 884–95.
– "Die Beugung der Röntgenstrahlen." *AP, 10* (1903), 305–12.
Hanle, P. "Erwin Schrödinger's reaction to Louis de Broglie's thesis on the quantum theory." *Isis, 68* (1977), 606–9.
Heaviside, O. "On the electromagnetic effects due to the motion of electrification through a dielectric." *PM, 27* (1889), 324–39.
Heilbron, J. L. *A history of the problem of atomic structure from the discovery of the electron to the beginning of quantum mechanics.* Unpublished Ph.D. dissertation, University of California, Berkeley, 1964.
– "The Kossel–Sommerfeld theory and the ring atom." *Isis, 58* (1967), 451–85.

- "The scattering of α and β particles and Rutherford's atom." *AHES, 4* (1968), 247–307.
- *H. G. J. Moseley: The life and letters of an English physicist, 1887–1915.* Berkeley, 1974.
- "Lectures on the history of atomic physics." In *History of twentieth century physics,* C. Weiner, ed., pp. 40–108. New York, 1977.
- "Physics at McGill in Rutherford's time." In *Rutherford and physics at the turn of the century,* M. Bunge and W. Shea, eds., pp. 42–73. New York, 1979.

Heilbron, J. L., and T. S. Kuhn. "The genesis of the Bohr atom." *HSPS, 1* (1969), 211–90.

Heilbron, J. L., and B. R. Wheaton. *Literature on the history of physics in the 20th century.* Berkeley papers in history of science, vol 5. Berkeley, 1981.

Heisenberg, W. "Über quantentheoretische Umdeutung kinematischer und mechanischer Beziehungen." *ZP, 33* (1925), 879–93.
- "Über den anschaulichen Inhalt der quantentheoretischen Kinematik und Mechanik." *ZP, 43* (1927), 172–98.

Heisenberg, W., and H. A. Kramers. "Über die Streuung von Strahlen durch Atome." *ZP, 31* (1925), 681–708.

Helmholtz, H. von. *Vorlesungen über die elektromagnetische Theorie des Lichts.* Hamburg, 1897.

Hendry, J. "The development of attitudes toward the wave-particle duality of light and quantum theory, 1900–1920." *Annals of Science, 37* (1980), 59–79.

Hermann, A. "Albert Einstein und Johannes Stark: Briefwechsel und Verhältnis der beiden Nobelpreisträger." *Sudhoffs Archiv, 50* (1966), 267–85.
- "Die frühe Diskussion zwischen Stark und Sommerfeld über die Quantenhypothese." *Centaurus, 12* (1968), 38–59.
- *Frühgeschichte der Quantentheorie 1889–1913.* Mosbach, 1969.

Hermann, K. "Zur Theorie des lichtelektrischen Effektes." *VDpG, 14* (1912), 936–45.

Hertz, H. "Versuche über die Glimmentladung." *AP, 19* (1883), 782–816.
- "Über die Ausbreitungsgeschwindigkeit der elektrodynamischen Wirkung." *AP, 34* (1888), 551–69. A first announcement appeared in Berlin *Sb* (1888), 197–210.
- "Über Strahlen elektrischer Kraft," *AP, 36* (1889), 769–83.
- "Über die Beziehungen zwischen Licht und Elektrizität." In GDNA *Tageblatt,* pp. 144–9. Heidelberg, 1890.
- *Die Principien der Mechanik, in neuem Zusammenhange.* Leipzig, 1894.

- *Gesammelte Werke*, vol. 3, *Schriften vermischten Inhalts*, P. Lenard, ed. Leipzig, 1895.

Herweg, J. "Über die Beugungserscheinungen der Röntgenstrahlen am Gips." *PZ, 14* (1913), 417–20.

Hirosige, T. "Origins of Lorentz' theory of electrons and the concept of the electromagnetic field." *HSPS, 1* (1969), 151–209.

Home, R. W. "W. H. Bragg and J. P. V. Madsen: Collaboration and correspondence, 1905–1911." *Historical records of Australian science, 5:2* (1981), 1–29.

Hu, K. F. "Some preliminary results in a determination of the maximum emission velocity of the photoelectrons from metals at x-ray frequencies." *PR, 11* (1918), 505–7.

Hughes, A. L. "On the velocities of the electrons produced by ultra-violet light." *PCPS, 16* (1911), 167–74.

- "On the emission velocities of photo-electrons." *PTRS, 212A* (1913), 209–26.

Hull, A., and M. Rice. "The law of absorption of x rays at high frequencies." *PR, 8* (1916), 326–8.

Hund, F., "Theoretische Betrachtungen über die Ablenkung von freien langsamen Elektronen in Atomen." *ZP, 13* (1923), 241–63.

Iklé, M. "Literatur der Radioaktivität vor dem Jahre 1904." *JRE, 1* (1904), 413–42.

Innes, P. D. "On the velocity of the cathode particles emitted by various metals under the influence of Röntgen rays, and its bearing on the theory of atomic disintegration." *PRS, 79A* (1907), 442–62.

Ioffe, A. F. [Joffé]. "Eine Bemerkung zu der Arbeit von E. Ladenburg: 'Über Anfangsgeschwindigkeit und Menge der photoelektrischen Elektronen usw.'." *AP, 24* (1907), 939–40.

- "Beobachtungen über den photoelektrischen Elementareffekt." München *Sb, 43* (1913), 19–37.

Jammer, M. *The conceptual development of quantum mechanics.* New York, 1966.

Jaumann, G. "Longitudinales Licht." Wien *Sb, 104:IIa* (1895), 747–92; revised in *AP, 57* (1895), 147–84.

- "Über die Interferenz und die elektrostatische Ablenkung der Kathodenstrahlen." Wien *Sb, 106:IIa* (1897), 533–50.

Jeans, J. H. "Discussion of radiation." BAAS *Report*, (1913), 380.

- *Report on radiation and the quantum theory.* London, 1914.

Jenkin, J. G., R. C. G. Leckey, and J. Liesegang. "The development of x-ray photoelectron spectroscopy: 1900–1960." *Journal of electron spectroscopy and related phenomena, 12* (1977), 1–35.

Joffé. See Ioffe.

Kahan, T. "Un document historique de l'académie des sciences de Berlin sur l'activité scientifique d' Albert Einstein." *Archives internationales d'historie des sciences, 15* (1962), 337–42.

Kargon, R. H. "Model and analogy in Victorian science: Maxwell's critique of the French physicists." *JHI, 30* (1969), 423–36.

Karnozhitzkiy, A. N., and B. Golitsyn. "Researches concerning the properties of x-rays." *Nature, 53* (1896), 528; described by Stokes, "On the nature of the Röntgen rays." Manchester *LP, 41* (1897), no. 15, p. 2.

Kaufmann, W. "Die magnetische und elektrische Ablenkbarkeit der Becquerelstrahlen und die scheinbare Masse der Elektronen." Göttingen *Nachrichten,* (1901), 143–55.

Kaufmann, W., and E. Aschkinass. "Über die Deflexion der Kathodenstrahlen." *AP, 62* (1897), 588–95.

Kaye, G. W. C. "The emission and transmission of Röntgen rays." *PTRS, 209A* (1908), 123–51.

– "The emission of Röntgen rays from thin metallic sheets." *PCPS, 15* (1909), 269–72.

Kelvin. See Thomson, William.

Khvol'son, O. D. [Chwolson]. *Lehrbuch der Physik,* vol. 2, H. Pflaum, tr. Braunschweig, 1904.

Kirchhoff, G. "Zur Theorie der Lichtstrahlen." Berlin *Mb* (1882), 641–69.

– *Vorlesungen über mathematische Physik,* vol. 2, *Mathematische Optik,* K. Hensel, ed. Leipzig, 1891.

Kleeman, R. D. "On the recombination of ions made by α, β, and x rays." *PM, 12* (1906), 273–97.

– "On the ionisation of various gases by α-, β-, and γ-rays." *PRS, 79A* (1907), 220–33.

– "On the secondary cathode rays emitted by substances when exposed to the γ rays." *PM, 14* (1907), 618–44.

– "On the different kinds of γ rays of radium, and the secondary γ rays which they produce." *PM, 15* (1908), 638–63.

– "A difference in the photoelectric effect caused by incident and divergent light." *Nature, 83* (1910), 339.

– "On the direction of motion of an electron ejected from an atom by ultra-violet light." *PRS, 84A* (1910), 92–9.

Klein, M. J. "Max Planck and the beginnings of the quantum theory." *AHES, 1* (1962), 459–79.

– "Einstein's first paper on quanta." *The natural philosopher, 2* (1963), 59–86.

- "Einstein and the wave particle duality." *The natural philosopher, 3* (1964), 3–49.
- "Einstein, specific heats and the early quantum theory." *Science, 148* (1965), 173–80.
- "Thermodynamics in Einstein's thought. *Science, 157* (1967), 509–16.
- "The first phase of the Bohr–Einstein dialogue." *HSPS, 2* (1970), 1–39.
- "Mechanical explanations at the end of the nineteenth century." *Centaurus, 17* (1972), 58–82.

Kleinert, A., and C. Schönbeck. "Lenard und Einstein. Ihr Briefwechsel und ihr Verhältnis vor der Nauheimer Diskussion von 1920." *Gesnerus, 35* (1978), 318–33.

Klickstein, H. *Wilhelm Conrad Röntgen. On a new kind of rays.* Mallinkrodt classics of radiology, vol. 1. St. Louis, 1966.

Koch, P. P. "Über die Messung der Schwärzung photographischer Platten in sehr schmalen Bereichen. Mit Anwendung auf die Messung der Schwärzungsverteilung in einigen mit Röntgenstrahlen aufgenommenen Spaltphotogrammen von Walter und Pohl." *AP, 38* (1912),507–22.

Kohlrausch, K. W. F. "Über Schwankungen der radioaktiven Umwandlung." Wien *Sb, 115:IIa* (1906), 673–82.

Kossel, W. "Über die sekundäre Kathodenstrahlung in Gasen in der Nähe des Optimums der Primärgeschwindigkeit." *AP, 37* (1912), 393–424.
- "Bemerkung zur Absorption homogener Röntgenstrahlen." *VDpG, 16* (1914), 898–909, 953–63.

Kovarik, A. "On the high frequency rays in the γ ray spectrum of radium." *PR, 19* (1922), 433.

Kramers, H. A. "The law of dispersion and Bohr's theory of spectra." *Nature, 113* (1924), 673–4.

Kronig, R. "The turning point." In *Theoretical physics in the twentieth century: A memorial volume to Wolfgang Pauli,* M. Fierz and V. Weisskopf, eds., pp. 5–39. New York, 1960.

Kubli, F. "Louis de Broglie und die Entdeckung der Materiewellen." *AHES, 7* (1970), 26–68.

Kuhn, T. S. "The quantum theory of specific heats: A problem in professional recognition." XIV congrès internationale d'histoire des sciences, Tokyo, *Proceedings, 1* (1974), 183–94.
- *Black-body theory and the quantum discontinuity, 1894–1912.* New York, 1978.

Kuhn, T. S., J. L. Heilbron, P. Forman, and L. Allen. *Sources for history of quantum physics: An inventory and report.* Philadelphia, 1967.

Kunz, J. *Theoretische Physik auf mechanischer Grundlage.* Stuttgart, 1907.

Laby, T. H. "A string electrometer." *PCPS, 15* (1909), 106–13.

Laby, T. H., and P. Burbidge. "The nature of γ rays." *Nature, 87* (1911), 144.

– "The observation by means of a string electrometer of fluctuations in the ionization produced by γ rays." *PRS, 86A* (1912), 333–48.

Ladenburg, E. "Über Anfangsgeschwindigkeit und Menge der photoelektrischen Elektronen in ihrem Zusammenhange mit der Wellenlänge des auslösenden Lichtes." *VDpG, 9* (1907), 504–14.

Ladenburg, R. "Die neuren Forschungen über die durch Licht- und Röntgenstrahlen hervorgerufene Emission negativer Elektronen." *JRE, 6* (1909), 425–84.

– "Die quantentheoretische Deutung der Zahl der Dispersionselektrons." *ZP, 4* (1921), 451–68.

Landé, A. "Schwierigkeiten in der Quantentheorie des Atombaues, besonders magnetischer Art." *PZ, 24* (1923), 441–4.

Langevin, P., and M. de Broglie, eds. *La théorie du rayonnement et les quanta* (Solvay I). Paris, 1912.

Larmor, J. "On the theory of the magnetic influence on spectra; and on the radiation from moving ions." *PM, 44* (1897), 503–12.

– *Aether and matter.* Cambridge, 1900.

– "On the constitution of natural radiation." *PM, 10* (1905), 574–84.

Laub, J. J. "Über sekundäre Kathodenstrahlen." *AP, 23* (1907), 285–300.

– "Zur Optik der bewegten Körper." *AP, 23* (1907), 738–44.

– "Zur Optik der bewegten Körper." *AP, 25* (1908), 175–84.

– "Über die durch Röntgenstrahlen erzeugten sekundären Kathodenstrahlen." *AP, 26* (1908), 712–26.

Laue, M. "Eine quantitative Prüfung der Theorie für die Interferenz-Erscheinungen bei Röntgenstrahlen." München *Sb, 42* (1912), 363–73.

– "Kritische Bemerkungen zu den Deutungen der Photogramme von Friedrich und Knipping." *PZ, 14* (1913), 421–3.

– "Röntgenstrahlinterferenzen," *PZ, 14* (1913), 1075–9.

– "Zusätze (März 1913)." *AP, 41* (1913), 989–1002.

– "Ein Versagen der klassischen Optik." *VDpG, 19* (1917), 19–21.

– "Antrittsrede." Berlin *Sb* (1921), 479–81.

Ledoux-Lebard, R., and A. Dauvillier. *La physique des rayons x.* Paris, 1921.

Lehmann, O. *Die elektrischen Lichterscheinungen oder Entladungen.* Halle, 1898.

Lenard, P. "Über Kathodenstrahlen in Gasen von atmosphärischem Druck und in äussersten Vakuum." *AP, 51* (1894), 225–68.

– "Über die elektrostatischen Eigenschaften der Kathodenstrahlen." *AP, 64* (1898), 279–89.

– "Erzeugung von Kathodenstrahlen durch ultraviolettes Licht." *AP, 2* (1900), 359–75.

– "Über die lichtelektrische Wirkung." *AP, 8* (1902), 149–98.

– "Über Äther und Materie." Heidelburg *Sb, 1* (1910), Abh. 16, separately paginated.

– "Über Elektrizitätsleitung durch freie Elektronen und Träger, II." *AP, 41* (1913), 53–98.

Lenard, P., and C. Ramsauer. "Über die Wirkungen sehr kurzwelligen ultravioletten Lichtes auf Gase und über eine sehr reiche Quelle dieses Lichtes." With various sub-titles in the Heidelberg *Sb, 1* (1910), Abh. 28, 31, 32; *2* (1911), Abh. 16, 24, all separately paginated.

– *Ibid.,* part 2 (Abh. 31), subtitled "Wenig absorbierbares und doch auf Luft wirkendes Ultraviolett."

– *Ibid.,* part 5 (Abh. 24), subtitled "Wirkung des stark absorbierbaren Ultraviolett und Zusammenfassung."

Lodge, O. "On the rays of Lenard and Röntgen." *Electrician, 36* (1896), 438–40.

– "On the present hypotheses concerning the nature of Röntgen's rays." *Electrician, 36* (1896), 471–3.

– "The surviving hypothesis concerning the x-rays." *Electrician, 37* (1896), 370–3.

Lorentz, H. A. "Vereenvoudigde theorie der electrische en optische verschijnselen in lichamen die zich bewegen." Amsterdam *Verslag, 7* (1899), 507–22. Reprinted in *Collected papers,* P. Zeeman and A. Fokker, eds., vol. 5, pp. 139–55, as "Théorie simpliêe des phénomènes électriques et optiques dans des corps en mouvement." The Hague, 1937.

– Untitled Nobel lecture in *Les prix Nobel en 1902,* separately paginated. Stockholm, 1905.

– *Lehrbuch der Physik zum Gebrauche bei akademischen Vorlesungen,* 2 vols. G. Siebert, tr. Leipzig, 1907.

– "[Het licht en de bouw der materie]." NNGC *Handelingen, 11* (1907), 6–21. *Collected papers,* P. Zeeman and A. Fokker, eds., vol. 9, pp. 167–81. The Hague, 1939. German translation in *PZ, 8* (1907), 542–9.

– "De hypothese der lichtquanta." NNGC *Handelingen, 12* (1909), 129–39; *Collected papers,* P. Zeeman and A. Fokker, eds., vol. 7, pp.

374–84. The Hague, 1934. German translation in *PZ, 11* (1910), 349–54.
- "Alte und Neue Fragen der Physik." *PZ, 11* (1910), 1234–57; *Collected papers*, P. Zeeman and A. Fokker, eds., vol. 7, pp. 205–57, The Hague, 1934.
- "Nature of light." In *Encyclopaedia Britannica*, 11th ed., vol. 16, pp. 617–23. London, 1911.
- "Over den aard der Röntgenstralen." Amsterdam *Verslag, 21* (1913), 911–23; *Collected papers*, P. Zeeman and A. Fokker, eds., vol. 3, pp. 281–94. The Hague, 1936.
- "The radiation of light." *Nature, 114* (1924), 608–11.
- *Problems in modern physics*. New York, 1927.
McClelland, J. A. "The penetrating radium rays." RDS *Sci trans, 8* (1904), 99–108; *PM, 8* (1904), 67–77.
- "On secondary radiation." RDS *Sci trans, 8* (1904), 169–82; *PM, 9* (1905), 230–43.
- "On secondary radiation (Part II); and atomic structure." RDS *Sci trans, 9* (1905), 1–8; and three succeeding works, *ibid.*, 9–26, 27–36, and 37–50.
McClung, R. K. "A preliminary account of an investigation of the effect of temperature on the ionization produced in gases by the action of Röntgen rays." *PCPS, 12* (1903), 191–98.
McCormmach, R. "J. J. Thomson and the structure of light." *BJHS, 3* (1967), 362–87.
- "H. A. Lorentz and the electromagnetic view of nature." *Isis, 61* (1970), 459–97.
- "Einstein, Lorentz, and the electron theory." *HSPS, 2* (1970), 41–87; and "Editor's foreword," *ibid.*, ix–xx.
- "Hertz, Heinrich Rudolf." *DSB, 6* (1972), 340–50.
- *Night thoughts of a classical physicist*. Cambridge, Mass., 1981.
Mach, E. "Durchsicht-Stereoskopbilder mit X-Strahlen." *Zeitschrift für Elektrotechnik, 14* (1896), 259–61.
MacKenzie, A. S. "Secondary radiations from a plate exposed to rays from radium." *PM, 14* (1907), 176–87.
MacKinnon, E. "De Broglie's thesis: A critical retrospective." *AJP, 44* (1976), 1047–55.
Madsen, J. P. V. "Secondary γ radiation." *TPRRSSA, 32* (1908), 163–92; *PM, 17* (1909), 423–48.
Malley, M. *From hyperphosphorescence to nuclear decay: A history of the early years of radioactivity 1896–1914*. Unpublished Ph.D. dissertation, University of California, Berkeley, 1976.
Marx, E. "Die Geschwindigkeit der Röntgenstrahlen." *VDpG, 7* (1905),

302–21; *PZ, 6* (1905), 768–77. An expanded report is in *AP, 20* (1906), 677–722.

- *Grenzen in der Natur und in der Wahrnehmung vom Standpunkte der Elektronentheorie und des elektromagnetischen Weltbildes.* Leipzig, 1908.
- "Theorie der Vorgänge im Nullapparat zur Geschwindigkeitsmessung der Röntgenstrahlen." *VDpG, 10* (1908), 137–56.
- "Zur Frage der Geschwindigkeit der Röntgenstrahlen, Erwiderung auf die gleichlautende Arbeit der Herren Franck und Pohl." *VDpG, 10* (1908), 157–201.
- "Experimentelles Verhalten und Theorie des Apparates zur Geschwindigkeitsmessung der Röntgenstrahlen." *AP, 28* (1909), 37–56.
- "Über den Einfluss der Röntgenstrahlen auf das Einsetzen der Glimmentladung." *AP, 28* (1909), 153–74.
- "Zweite Durchführung der Geschwindigkeitsmessung der Röntgenstrahlen." *AP, 33* (1910), 1305–91.
- "Sind meine Versuche über die Geschwindigkeit der Röntgenstrahlen durch Interferenz elektrischer Luftwellen erklärbar?" *AP, 35* (1911), 397–400.
- "Die Theorie der Akkumulation der Energie bei intermittierender Belichtung und die Grundlage des Gesetzes der schwarzen Strahlung." *AP, 41* (1913), 161–90.
Marx, E., and K. Lichtenecker. "Experimentelle Untersuchung des Einflusses der Unterteilung der Belichtungszeit auf die Elektronenabgabe in Elster und Geitelschen Kaliumhydrürzellen bei sehr schwacher Lichtenergie." *AP, 41* (1913), 124–60.
Mauguin, C. "La thèse de doctorat de Louis de Broglie." In *Louis de Broglie: Physicien et penseur,* A. Georges, ed., pp. 430–6. Paris, 1953.
Maxwell, J. C. "On physical lines of force." *PM, 21* (1861), 161–75, 338–48; *23* (1862), 12–24, 85–95. Reprinted in Maxwell, *Scientific papers,* W. D. Niven, ed. vol. 1, pp. 451–513. Cambridge, 1890.
Mayer, H. F. "Über das Verhalten von Molekülen gegenüber freien langsamen Elektronen." *AP, 64* (1921), 451–80.
Meitner, L. "Über die Entstehung der β-Strahl-Spektren radioaktiver Substanzen." *ZP, 9* (1922), 131–44.
- "Über den Zusammenhang zwischen β- und γ-Strahlen." *ZP, 9* (1922), 145–52.
Mendelssohn, K. *The world of Walther Nernst.* London, 1973.
Meyer, E. "Bericht über die Untersuchungen der zeitlichen Schwankungen der radioaktiven Strahlung." *JRE, 5* (1908), 423–50; *6* (1909), 242–5.

- "Über Stromschwankungen bei Stossionisation." *VDpG, 12* (1910), 253-74.
- "Über die Struktur der γ Strahlen." Berlin *Sb, 32* (1910), 624-62; *JRE, 7* (1910), 279-95.
- "Über Schweidlersche Schwankungen." *PZ, 13* (1912), 73-81.
- "Über die Struktur der γ Strahlen. II." *AP, 37* (1912), 700-20.
- "Zur Diskussion über die Struktur der γ Strahlen." *PZ, 13* (1912), 253-4.
Meyer, E., and W. Gerlach, "Sur l'émission photo-électrique d'électrons par des particules métalliques ultra-microscopiques," *ASPN Genève, 35* (1913), 398-400.
- "Sur l'effet photoélectrique des particules ultra-microscopiques aux basses pressions." *ASPN Genève, 37* (1914), 253-45.
Meyer, E., and E. Regener. "Über Schwankungen der radioactiven Strahlung und eine Methode zur Bestimmung des elektrischen Elementarquantums." *VDpG, 10* (1908), 1-13; *AP, 25* (1908), 757-74.
Meyer, S., and E. von Schweidler. "Über das Verhalten von Radium und Polonium im magnetischen Felde." *PZ, 1* (1899), 90-1, 113-14.
Michelson, A. A. "A theory of the 'x-rays'." *AJS, 1* (1896), 312-14; *Nature, 54* (1896), 66-7.
Miller, A. I. *Albert Einstein's special theory of relativity: Emergence (1905) and early interpretations (1905-1911).* Reading, Mass., 1981.
Millikan, R. A. "Some new values of the positive potentials assumed by metals in a high vacuum under the influence of ultra-violet light." *PR, 30* (1910), 287-8.
- "The effect of the character of the source upon the velocities of emission of electrons liberated by ultra-violet light." *PR, 35* (1912), 74-6.
- "Über hohe Anfangsgeschwindigkeiten durch ultraviolettes Licht ausgelöster Elektronen." *VDpG, 10* (1912), 712-26.
- "On the cause of the apparent differences between spark and arc sources in the imparting of initial speeds to photo-electrons." *PR, 1* (1913), 73-5.
- "A direct determination of 'h'." *PR, 4* (1914), 73-5.
- "New tests of Einstein's photo-electric equation." *PR, 6* (1915), 55.
- "A direct photoelectric determination of Planck's 'h'." *PR, 7* (1916), 355-88.
- *The electron.* Chicago, 1917.
- "The electron and the light-quant from the experimental point of view." Separately paginated in *Les prix Nobel en 1923*. Stockholm, 1924.
- *The autobiography of Robert A. Millikan.* New York, 1950.
Millikan, R. A., and G. Winchester. "Upon the discharge of electrons

from ordinary metals under the influence of ultra-violet light." *PR, 24* (1907), 116-18.

- "The absence of photoelectric fatigue in a very high vacuum." *PR, 29* (1909), 85.

Millikan, R. A., and J. Wright. "The effect of prolonged illumination on photo-electric discharge in a high vacuum." *PR, 34* (1912), 68-70.

Moseley, H. "The number of β particles emitted in the transformation of radium." *PRS, 87A* (1912), 230-55.

Moseley, H., and C. G. Darwin. "The reflection of the x rays." *Nature, 90* (1913), 594.

- "The reflection of the x rays." *PM, 26* (1913), 210-32.

Moseley, H., and W. Makower. "γ radiation from radium B." *PM, 23* (1912), 203-10.

Mott, N. F. "The quantum theory of electronic scattering by helium." *PCPS, 25* (1929), 304-9.

Nernst, W. *Theoretische Chemie vom Standpunkte der Avogadroschen Regel und der Thermodynamik,* 5th ed. Stuttgart, 1907.

Nisio, S. "Sommerfeld's theory of the photoelectric effect." XIV congrès internationale d'histoire des sciences, Tokyo, *Proceedings, 2* (1975), 302-4.

Les prix Nobel en 1902. Stockholm, 1905.

Les prix Nobel en 1921-1922. Stockholm, 1923.

Les prix Nobel en 1929. Stockholm, 1930.

Les prix Nobel en 1937. Stockholm, 1938.

Nye, M. "N-rays: An episode in the history and psychology of science." *HSPS, 11* (1980), 125-56.

Oseen, C. W. "Die Einsteinsche Nadelstichstrahlung und die Max-wellsche Gleichungen." *AP, 69* (1922), 202-4.

Owen, E. A. "The passage of homogeneous Röntgen rays through gases." *PRS, 86A* (1912), 426-39.

Pais, A. "How Einstein got the Nobel Prize." *American scientist, 70* (1982), 358-65.

Parkinson, E. M. "Stokes, George Gabriel." *DSB, 13* (1976), 74-9.

Paschen, L. C. H. F. "Über die durchdringenden Strahlen des Radiums." *AP, 14* (1904), 164-71.

- "Über die Kathodenstrahlen des Radiums." *AP, 14* (1904), 389-405.

- "Über eine von den Kathodenstrahlen des Radiums in Metallen erzeugte Sekundärstrahlung." *PZ, 5* (1904), 502-4.

- "Über die γ Strahlen des Radiums." *PZ, 5* (1904), 563-8.

- "Über die Wärmeentwickelung des Radiums in einer Bleihülle." *PZ, 6* (1905), 97.

Phillips, C. E. S., ed. *Bibliography of x-ray literature and research.* London, 1897.

Planck, M. "Über die Natur des weissen Lichtes." *AP, 7* (1902), 390–400.
- Remarks following Einstein's talk to the GDNA. *PZ, 10* (1909), 826.
- "Zur Theorie der Wärmestrahlung." *AP, 31* (1910), 758–68.
- "Eine neue Strahlungshypothese." *VDpG, 13* (1911), 138–48.
- "La loi du rayonnement noir et l'hypothèse des quantités élémentaires d'action." In *Solvay I* (1912), 93–114.
- "Über die Begründung des Gesetzes der schwarzen Strahlung." *AP, 37* (1912), 642–56.
- "Über das Gleichgewicht zwischen Oszillatoren, freien Elektronen und strahlender Wärme." Berlin *Sb* (1913), 350–63.
- *Das Wesen des Lichts.* Berlin, 1920. Originally in *Die Naturwissenschaften, 7* (1919), 903–9.
- "The physical reality of light quanta." *JFI, 204* (1927), 13–18.
Plücker, J. "Über die Einwirkung des Magneten auf die elektrischen Entladungen in verdünnten Gasen." *AP, 103* (1858), 88–106.
J. C. Poggendorff's biographisch-literarisches Handwörterbuch zur Geschichte der exacten Naturwissenschaften. Vol. 4, Leipzig 1904. Vol. 5, Leipzig, 1926. Vol. 6, Leipzig, 1936–40.
Pohl, R. *Die Physik der Röntgenstrahlen.* Braunschweig, 1912.
Pohl, R., and P. Pringsheim. "Die lichtelektrische Empfindlichkeit der Alkalimetalle als Funktion der Wellenlänge." *VDpG, 12* (1910), 215–28, 249–60.
- "The normal and the selective photoelectric effect." *PM, 21* (1910), 155–61.
- "Weitere Versuche über den selektiven lichtelektrischen Effekt." *VDpG, 12* (1910), 682–96.
- "Der selektive lichtelektrische Effekt an K-Hg-Legierungen." *VDpG, 12* (1910), 697–710.
- "Zur Frage hoher Geschwindigkeiten lichtelektrischer Elektronen." *VDpG, 14* (1912), 974–82.
- "Über die langwellige Grenze des normalen Photoeffektes." *VDpG, 15* (1913), 637–44.
Poincaré, H. *Théorie mathématique de la lumière*, vol. 1, . *Leçons sur la théorie mathématique de la lumière.* Paris, 1889.
- "Les rayons cathodiques et les rayons Röntgen." *RGS, 7* (1896), 52–9.
Poynting, J. H. "On the transfer of energy in the electromagnetic field." *PTRS, 175* (1884), 343–61.
Precht, J. "Untersuchungen über Kathoden- und Röntgenstrahlen." *AP, 61* (1897), 350–62.
- "Die Wärmeabgabe des Radiums." *VDpG, 6* (1904), 101–3.
Przibram, K., ed. *Briefe zur Wellenmechanik.* Vienna, 1963.

Pyenson, L. "Einstein's early scientific collaborators." *HSPS, 7* (1976), 83–123.

Quincke, G. Response to Hertz. GDNA *Tageblatt,* p. 149. Heidelberg, 1890.

Raman, V. V., and P. Forman. "Why was it Schrödinger who developed de Broglie's ideas?" *HSPS, 1* (1969), 291–314.

Ramsauer, C. "Über die Wirkung sehr kurzwelligen ultravioletten Lichtes auf Gase." *PZ, 12* (1911), 997–8.

-   "Über eine direkte magnetische Methode zur Bestimmung der lichtelektrischen Geschwindigkeitsverteilung." Heidelberg *Sb, 5A* (1914), Abh. 19, separately paginated; *AP, 45* (1914), 961–1002.

-   "Über die lichtelektrische Geschwindigkeitsverteilung und ihre Abhängigkeit von der Wellenlänge." Heidelberg *Sb, 5A* (1914), Abh. 20, separately paginated; *AP, 45* (1914), 1121–59.

-   "Über den Wirkungsquerschnitt der Gasmoleküle gegenüber langsamen Elektronen." *AP, 64* (1921), 513–40.

-   "Über den Wirkungsquerschnitt der Gasmoleküle gegenüber langsamen Elektronen." *PZ, 21* (1920), 576–8.

-   "Über den Wirkungsquerschnitt der Edelgase gegenüber langsamen Elektronen." *PZ, 22* (1921), 613–15.

Rayleigh. See Strutt.

Richardson, H. "Analysis of the γ rays from uranium products." *PM, 27* (1914), 252–6.

Richardson, O. W. "Some applications of the electron theory of matter." *PM, 23* (1912), 594–627.

-   "The theory of photoelectric action." *PM, 24* (1912), 570–4.

-   "The laws of photoelectric action and the unitary theory of light (Lichtquanten Theorie)." *Science, 36* (1912), 57–8.

-   "The application of statistical principles to photoelectric effects and some allied phenomena." *PR, 34* (1912), 146–9.

-   "The asymmetric emission of secondary rays." *PM, 25* (1913), 144–50.

-   "The theory of photoelectric and photochemical action." *PM, 27* (1914), 476–88.

-   "The complete photoelectric emission." *PM, 31* (1916), 149–55.

-   "The photoelectric action of x rays." *PRS, 94A* (1918), 269–80.

Richardson, O. W., and K. T. Compton. "The photoelectric effect." *PM, 24* (1912), 575–94.

Richardson, O. W., and F. J. Rogers, "The photoelectric effect III." *PM, 29* (1915), 618–23.

Righi, A. "Sur les changements de longueur d'onde obtenus par la rotation d'un polariseur, et sur le phénomène des battements produits avec les vibrations lumineuses." *JP, 2* (1883), 437–46.

-   "Sulla propagazione dell' elettricità nei gas attraversati dai raggi de Röntgen." Bologna *Memorie, 6* (1896), 231–301.

- "Sull' elettrizzazione prodotta dai raggi del radio." Lincei *Atti, 14* (1905), 556-9.
Robinson, H., and W. F. Rawlinson. "The magnetic spectrum of the β rays excited in metals by soft x rays." *PM, 28* (1914), 277-81.
Röntgen, W. C. "Über eine neue Art von Strahlen (vorläufige Mittheilung)." Würzburg *Sb* (1895), 132-41; *AP, 64* (1898), 1-11. (Röntgen I).
- "Über eine neue Art von Strahlen II. Mittheilung." Würzburg *Sb* (1896), 11-16; *AP, 64* (1898), 12-17. (Röntgen II).
- "Weitere Beobachtungen über die Eigenschaften der X-Strahlen." Berlin *Sb* (1897), 576-92; *AP, 64* (1898), 18-37. (Röntgen III).
Rubens, H., and E. Ladenburg. "Über die lichtelektrische Erscheinung an dünnen Goldblättchen." *VDpG, 9* (1907), 749-52.
Runge, C., and J. Precht. "Die Wärmeabgabe des Radiums." Berlin *Sb* (1903), 783-6.
Russell, A. S. "The effect of temperature upon radioactive disintegration." *PRS, 86A* (1911), 240-53.
Russell, A. S., and F. Soddy. "The γ rays of thorium and actinium." *PM, 21* (1911), 130-54.
Russo, A. "Fundamental research at Bell Laboratories: The discovery of electron diffraction." *HSPS, 12* (1981), 117-60.
Rutherford, E. "Uranium radiation and the electrical conduction produced by it." *PM, 47* (1899), 109-63.
- "Penetrating rays from radio-active substances." *Nature, 66* (1902), 318-19. Translated, with omissions, in *PZ, 3* (1902), 517-20.
- "The magnetic and electric deviation of the easily absorbed rays from radium." *PM, 5* (1903), 177-87. Translated in *PZ, 4* (1903), 235-40.
- "Radio-active processes." *Nature, 68* (1903), 163.
- "Does the radio-activity of radium depend on its concentration?" *Nature, 69* (1904), 222.
- "The heating effect of the radium emanation." AusAAS *Report* (1904), 86-91.
- "Nature of the γ rays from radium." *Nature, 69* (1904), 436-7.
- *Radioactivity.* Cambridge, 1904. (2nd ed., Cambridge, 1905.)
- "Some properties of the α rays from radium," *PM, 10* (1905), 163-76.
- "The origin of β and γ rays from radioactive substances." *PM, 24* (1912), 453-62.
- *Radioactive substances and their radiations.* Cambridge, 1913.
- "The connexion between the β and γ ray spectra." *PM, 28* (1914), 305-19.
- "The structure of the atom." *PM, 27* (1914), 488-98.
- "Penetrating power of the x radiation from a Coolidge tube." *PM, 34* (1917), 153-62.
- Comment on M. de Broglie's report. In *Solvay III* (1923), 107-9.

Rutherford, E., and E. N. Andrade. "The wavelength of the soft γ rays from radium B." *PM, 27* (1914), 854–68. A brief first announcement appeared in *Nature, 92* (1913), 267.

– "The spectrum of the penetrating γ rays from radium B and radium C." *PM, 28* (1914), 263–73.

Rutherford, E., and H. T. Barnes. "Heating effect of the γ rays from radium." *PM, 9* (1905), 621–28.

– "The charge and nature of the α particle." *PRS, 81A* (1908), 162–73.

Rutherford, E., and H. Geiger. "An electrical method of counting the number of α particles from radio-active substances." *PRS, 81A* (1908), 141–61.

Rutherford, E., and R. K. McClung. "Energy of Röntgen and Becquerel rays, and the energy required to produce an ion in gases," *PTRS, 196A* (1901), 25–59; German translation in *PZ, 2* (1900), 53–5.

Rutherford, E., and H. Richardson. "The analysis of the gamma rays from radium B and radium C." *PM, 25* (1913), 722–34.

– "Analysis of the γ rays from radium D and radium E." *PM, 26* (1913), 324–32.

– "Analysis of the γ rays of the thorium and actinium products." *PM, 26* (1913), 937–44.

Rutherford, E., and H. Robinson. "The analysis of the β rays from radium B and radium C." *PM, 26* (1913), 717–29.

Rutherford, E., and W. A. Wooster. "The natural x ray spectrum of radium B. *PCPS, 22* (1925), 834–837.

Rutherford, E., J. Barnes, and H. Robinson. "Maximum frequency of the x rays from a Coolidge tube for different voltages." *PM, 30* (1915), 339–60.

Rutherford, E., J. Chadwick, and C. D. Ellis. *Radiations from radioactive substances.* Cambridge, 1930.

Rutherford, E., H. Robinson, and W. F. Rawlinson. "Spectrum of the β rays excited by γ rays." *PM, 28* (1914), 281–86.

Sadler, C. A. "Transformations of x rays." *PM, 18* (1909), 106–32.

– "Homogeneous corpuscular radiation." *PM, 19* (1910), 337–56.

Sagnac, G. "Sur la transformation des rayons x par les métaux." *CR, 125* (1897), 942–4.

– "Transformation des rayons x par transmission." *CR, 126* (1898), 467–70.

– "Émission de rayons secondaires par l'air sous l'influence des rayons x." *CR, 126* (1898), 521–3.

– *Thèses. De l'optique des rayons de Röntgen et des rayons secondaires qui en dérivent.* Paris, 1900.

Schlegel, R. "Louis de Broglie's thesis." *AJP, 45* (1977), 871–2.

Schmidt, H. W. "Über Reflexion und Absorption von β Strahlen." *AP, 23* (1907), 671–97.

- "Einige Versuche mit β Strahlen von Radium E." *PZ, 8* (1907), 361–73.
- "Über die Strahlung des Uranium x." *PZ, 10* (1909), 6–16.
Schrödinger, E. "Dopplerprinzip und Bohrsche Frequenzbedingung." *PZ, 23* (1922), 301–3.
- "Über eine bemerkenswerte Eigenschaft der Quantenbahnen eines einzelnen Elektrons." *ZP, 12* (1922), 13–23.
- "Zur Einsteinschen Gastheorie." *PZ, 27* (1926), 95–101.
- "Quantisierung als Eigenwertproblem (erste Mitteilung)." *AP, 79* (1926), 361–76.
- "Über das Verhältnis der Heisenberg–Born–Jordanschen Quantenmechanik zu der meinen." *AP, 79* (1926), 734–56.
Schuster, A. "The discharge of electricity through gases." *PRS, 57* (1890), 526–59.
- "On interference phenomena." *PM, 37* (1894), 509–45.
- "On the new kind of radiation." *British Medical Journal* (18 Jan. 1896), 172–3.
- "On Röntgen's rays." *Nature, 53* (1896), 268.
Schweidler, E. von "Über Schwankungen der radioaktiven Umwandlung." *Premier congrès international pour l'étude de la radiologie et de l' ionization, Liége,* paginated section no. 3. Brussels, 1906.
- "Zur experimentellen Entscheidung der Frage nach der Natur der γ Strahlen." *PZ, 11* (1910), 225–7, 614–19.
Seelig, C. *Albert Einstein, eine dokumentarische Biographie.* Zurich, 1954.
- *Albert Einstein, Leben und Werk eines Genies unserer Zeit.* Zurich, 1960.
Seitz, W. "Bemerkung zu der B. Walterschen Mitteilung über das Röntgensche Absorptionsgesetz." *AP, 26* (1909), 448.
- "Geschwindigkeit von Elektronen, welche durch weiche Röntgenstrahlen erzeugt werden." *PZ, 11* (1910), 705–8.
Serwer, D. *The rise of radiation protection: Science, medicine, and technology in society, 1896–1935.* Unpublished Ph.D. dissertation, Princeton University, 1976.
- "*Unmechanischer Zwang:* Pauli, Heisenberg, and the rejection of the mechanical atom." *HSPS, 8* (1977), 189–256.
Siegbahn, M. "Über den Zusammenhang zwischen Absorption und Wellenlänge bei Röntgenstrahlen." *PZ, 15* (1914), 753–6.
Simons, L. "The beta ray emission from thin films of the elements exposed to Röntgen rays." *PM, 41* (1921), 120–40.
Slater, J. C. "The development of quantum mechanics in the period 1924–1926." In *Wave mechanics: The first fifty years,* W. C. Price, S. S. Chissick, and T. Ravensdale, eds., pp. 19–25. New York, 1973.

Soddy, F., W. M. Soddy, and A. S. Russell. "The question of the homogeneity of γ rays." *PM, 19* (1910), 725–57.

Solvay I. *La théorie du rayonnement et les quanta,* P. Langevin and M. de Broglie, eds. Paris, 1912.

Solvay III. *Atomes et electrons.* Paris, 1923.

Sommerfeld, A. "Zur mathematischen Theorie der Beugungserscheinungen." Göttingen *Nachrichten* (1894), 338–42.

- "Diffractionsprobleme in exacter Behandlung." Deutsche Mathematiker-Vereinigung *Jahresbericht, 4* (1895), 172–4.

- "Mathematische Theorie der Diffraction." *Mathematische Annalen, 47* (1896), 317–74.

- "Theoretisches über die Beugung der Röntgenstrahlen." *PZ, 1* (1899), 105–11; *PZ, 2* (1900), 55–60

- "Die Beugung der Röntgenstrahlen unter der Annahme von Aetherstössen." *VDNA* (1900), part 2, first half, separately paginated.

- "Theoretisches über die Beugung der Röntgenstrahlen." *Zeitschrift für Mathematik und Physik, 46* (1901), 11–97.

- "Über die Verteilung der Intensität bei der Emission von Röntgenstrahlen." *PZ, 10* (1909), 969–76.

- "Über die Verteilung der Intensität bei der Emission von Röntgenstrahlen." *PZ, 11* (1910), 99–101.

- "Das Plancksche Wirkungsquantum und seine allgemeine Bedeutung für die Molekularphysik." *PZ, 12* (1911), 1057–68.

- "Über die Struktur der γ Strahlen." München *Sb, 41* (1911), 1–60.

- "Application de la théorie de l'élément d'action aux phénomènes moléculaires non périodiques." In *Solvay I* (1912), 313–72.

- "Über die Beugung der Röntgenstrahlen." *AP, 38* (1912), 473–506.

- "Unsere gegenwärtige Anschauungen über Röntgenstrahlung." *Die Naturwissenschaften, 1* (1913), 705–13.

- "Über das Spektrum der Röntgenstrahlen." *AP, 46* (1915), 721–48.

- "Die neueren Fortschritte in der Physik der Röntgenstrahlen." *Münchener medizinische Wochenschrift, 62* (1915), 1424–30.

- "Die neueren Fortschritte in der Physik der Röntgenstrahlen." *Die Naturwissenschaften, 4* (1916), 1–8, 13–18.

- "Die Wellenlänge der Röntgenstrahlen als Härtemass." *Münchener medizinische Wochenschrift, 63* (1916), 458–60.

- *Atombau und Spektrallinien.* Braunschweig, 1919.

- "Das Spektrum der Röntgenstrahlung als Beispiel für die Methodik der alten und neuen Mechanik." *Scientia, 51* (1932), 41–50.

- "Autobiographische Skizze." In *Geist und Gestalt: Biographische Beitrage zur Geschichte der Bayerischen Akademie der Wissenschaften,* vol. 2, pp. 100–109. Munich, 1959.

Sommerfeld, A., and G. Wentzel. "Über reguläre und irreguläre Dubletts." *ZP, 7* (1922), 86–92.

Sommerfeld, A., and E. Wiechert. "Harmonischer Analysator."
Deutsche Mathematiker-Vereinigung *Katalog mathematischer Modelle, Apparate und Instrumente,* pp. 214–221. Munich, 1892.

Stark, J. "Elektricität in Gasen." In *Handbuch der Physik,* 2nd ed., A. Winkelmann, ed., vol. 4, pp. 454–653. Leipzig, 1905.

– "Über Absorption und Fluoreszenz im Bandenspektrum und über ultraviolette Fluoreszenz des Benzols." *PZ, 8* (1907), 81–5.

– "Beziehung des Doppler-Effektes bei Kanalstrahlen zur Planckschen Strahlungstheorie." *PZ, 8* (1907), 913–19.

– "Elementarquantum der Energie, Modell der negativen und der positiven Elektrizität." *PZ, 8* (1907), 881–4.

– "Zur Energetik und Chemie der Bandenspektra." *PZ, 9* (1908), 85–94, 356–8.

– "Neue Beobachtungen an Kanalstrahlen in Beziehung zur Lichtquantenhypothese." *PZ, 9* (1908), 767–73.

– "Die Valenzlehre auf atomistisch elektrischer Basis." *JRE, 5* (1908), 124–53.

– "The wave-length of Röntgen rays." *Nature, 77* (1908), 320.

– "Über die Ionisierung von Gasen durch Licht." *PZ, 10* (1909), 614–23.

– "Über Röntgenstrahlen und die atomistische Konstitution der Strahlung," *PZ, 10* (1909), 579–86.

– "Zur experimentellen Entscheidung zwischen Ätherwellen- und Lichtquantenhypothese. I. Röntgenstrahlen." *PZ, 10* (1909), 902–13.

– *Die Prinzipien der Atomdynamik,* vol. 1, *Die elektrischen Quanten.* Leipzig, 1910.

– "Zur experimentelle Entscheidung zwischen der Lichtquantenhypothese und der Ätherimpulstheorie der Röntgenstrahlen." *PZ, 11* (1910), 24–31.

– "Folgerungen über die Natur des Lichtes aus Beobachtungen über die Interferenz." *JRE, 7* (1910), 386–404.

– "Weitere Beobachtungen über die dissymmetrische Emission von Röntgenstrahlen." *PZ, 11* (1910), 107–12.

– "Zur experimentellen Entscheidung zwischen Lichtquantenhypothese und Ätherwellentheorie. II. Sichtbares und ultraviolettes spektrum." *PZ, 11* (1910), 179–87.

– Comment following Sommerfeld's presentation to the GDNA. *PZ, 12* (1911), 1068–9.

– *Die Prinzipien der Atomdynamik,* vol. 2, *Die elementare Strahlung.* Leipzig, 1911.

– "Zur Diskussion über die Struktur der γ Strahlen." *PZ, 13* (1912), 161–2.

- "Bemerkung über Zerstreuung und Absorption von β Strahlen und Röntgenstrahlen." *PZ, 13* (1912), 973–6.
- "Zur Diskussion des Verhaltens der Röntgenstrahlung in Kristallen." *PZ, 14* (1913), 319–21.
- "Schwierigkeiten für die Lichtquantenhypothese im Falle der Emission von Serienlinien." *VDpG, 16* (1914), 304–6.
Stark, J., and W. Steubing. "Fluoreszenz und lichtelektrische Empfindlichkeit organischer Substanzen." *PZ, 9* (1908), 481–95.
- "Weitere Beobachtungen über die Fluoreszenz organischer Substanzen." *PZ, 9* (1908), 661–9.
Stark, J., and G. Wendt. "Pflanzt sich der Stoss von Kanalstrahlen in einem festen Körper fort?" *AP, 38* (1912), 941–57.
Starke, H. "Die magnetische und elektrische Ablenkbarkeit reflektierter und von dünnen Metallblättchen hindurchgelassener Kathodenstrahlen." *VDpG, 5* (1903), 14–22.
- "Recherches sur les rayons secondaires du radium." *Le Radium, 5* (1908), 35–41.
Stephenson, R. J. "The scientific career of Charles Glover Barkla," *AJP, 35* (1967), 140–52.
Stewart, O. "Becquerel rays, a resumé." *PR, 11* (1900), 155–75.
Stokes, G. G. "On the dynamical theory of diffraction." *TCPS, 9* (1849), 1–62. Reprinted in *Mathematical and physical papers,* J. Larmor, ed., vol. 2, pp. 243–328. Cambridge, 1883.
- *On the nature of light.* London, 1884.
- *On light.* London, 1887.
- "On the nature of Röntgen rays." *PCPS, 9* (1896), 215–16.
- "On the Röntgen rays." *Nature, 54* (1896), 427–30.
- "An annual address (chiefly on the subject of the Röntgen rays)." VI *Journal of Transactions, 30* (1898), 13–25.
- "On the nature of the Röntgen rays." Manchester *MP, 41* (1897), no. 15, 29 pp. Reprinted in *Mathematical and physical papers,* J. Larmor, ed., vol. 5, pp. 256–77. Cambridge, 1905.
- *Memoir and scientific correspondence of the late Sir George Gabriel Stokes,* J. Larmor, ed., vol. 2. Cambridge, 1907.
Stoletov, A. G. "Aktinoelektrische Untersuchungen." *Physikalische Revue, 1* (1892), 723–80.
Stoney, G. J. "Evidence that Röntgen rays are ordinary light." *PM, 45* (1898), 532–6; *PM, 46* (1898), 253–4.
Strutt, J. W. [third Lord Rayleigh]. "Wave theory." *Encyclopaedia Britannica,* 9th ed., vol. 24, pp. 421–59. London, 1888.
- "On the character of the complete radiation at a given temperature." *PM, 27* (1889), 460–9.
- "Röntgen rays and ordinary light." *Nature, 57* (1898), 607.

Strutt, R. J. [fourth Lord Rayleigh]. "On the behavior of the Becquerel and Röntgen rays in a magnetic field." *PRS, 66* (1900), 75–9.
- "On the conductivity of gases under the Becquerel rays." *PTRS, 196A* (1901), 507–27.
- "On the intensely penetrating rays of radium." *PRS, 72* (1903), 208–10.
- *The Becquerel rays and the properties of radium.* London, 1904.
- *The life of Sir J. J. Thomson.* Cambridge, 1942.
Stuewer, R. "Non-Einsteinian interpretations of the photoelectric effect." *Minnesota studies in philosophy of science,* R. Stuewer, ed., vol. 5, pp. 246–63. Minneapolis, 1970.
- "William H. Bragg's corpuscular theory of x rays and γ rays." *BJHS, 5* (1971), 258–81.
- *The Compton effect: Turning point in physics.* New York, 1975.
Stuhlmann, O., Jr. "A difference in the photoelectric effect caused by incident and divergent light." *Nature, 83* (1910), 311.
- "A difference in the photoelectric effect caused by incident and emergent light." *PM, 20* (1910), 331–9; *22* (1911), 854–64.
Sudre, R. "L'oeuvre scientifique de Maurice de Broglie." *Revue des deux mondes* (July–August 1960), 577–82.
Swenson, L. "Richardson, Owen Willans," *DSB, 11* (1975), 419–23.
Taylor, G. I. "Interference fringes with feeble light." *PCPS, 15* (1909), 114–15.
Thibaud, J. *Thèses. La spectroscopie des rayons γ. Spectres β secondaires et diffraction cristalline.* Paris, 1925.
Thomson, G. P. "Experiments on the diffraction of cathode rays." *PRS, 117A* (1928), 600–9.
Thomson, J. J. *Notes on recent researches in electricity and magnetism.* Oxford, 1893.
- "Longitudinal electric waves and Röntgen's x-rays." *PCPS, 9* (1896), 49–61.
- "On the discharge of electricity produced by the Röntgen rays, and the effects produced by these rays on dielectrics through which they pass." *PRS, 59* (1896), 274–6.
- "Presidential address to Section A," BAAS *Report* (1896), 699–706.
- "The Röntgen rays." *Nature, 53* (1896), 391–2.
- "The Röntgen rays." *Nature, 54* (1896), 302–6. The Rede Lecture delivered June 10, 1896.
- "Cathode rays." *PM, 44* (1897), 293–316.
- "On the charge of electricity carried by the ions produced by Röntgen rays." *PM, 46* (1898), 528–45.
- "On the diffuse reflection of Röntgen rays." *PCPS, 9* (1898), 393–7.
- "On the connection between the chemical composition of a gas and the

ionization produced in it by Röntgen rays." *PCPS, 10* (1898), 10–14.
- *The discharge of electricity through gases.* New York, 1898.
- "A theory of the connection between cathode and Röntgen rays." *PM, 45* (1898), 172–83.
- *Conduction of electricity through gases.* Cambridge, 1903.
- "The magnetic properties of systems of corpuscles describing circular orbits." *PM, 6* (1903), 673–93.
- "Radium." *Nature, 67* (1903), 601–2.
- *Electricity and matter.* New Haven, 1904.
- "On the positive electrification of α rays, and the emission of slowly moving cathode rays by radio-active substances." *PCPS, 13* (1904), 49–54.
- "On the emission of negative corpuscles by the alkali metals." *PM, 10* (1905), 584–90.
- "Do the γ rays carry a charge of negative electricity?" *PCPS, 13* (1905), 121–3.
- *Conduction of electricity through gases,* 2nd ed. Cambridge, 1906.
- "On the number of corpuscles in an atom." *PM, 11* (1906), 769–81.
- "On secondary Röntgen radiation." *PCPS, 13* (1906), 322–4; *14* (1906), 109–14.
- *Corpuscular theory of matter.* London, 1907.
- "On the ionization of gases by ultra-violet light and on the evidence as to the structure of light afforded by its electrical effects." *PCPS, 14* (1907), 417–24.
- "On the light thrown by recent investigations on electricity on the relation between matter and ether." Victoria University of Manchester *Lectures, 8* (1908), separately paginated.
- "The nature of the γ rays." *PCPS, 14* (1908), 540.
- "On the velocity of secondary cathode rays from gases." *PCPS, 14* (1908), 541–5.
- "On the theory of radiation." *PM, 20* (1910), 238–47.
- "On a theory of the structure of the electric field and its application to Röntgen radiation and to light." *PM, 19* (1910), 301–13. Hints of Thomson's radical proposal were given in BAAS *Report* (1909), 3–29; and in *Nature, 81* (1909), 248–57.
- "Röntgen rays," *Encyclopaedia Britannica,* 11th ed., vol. 23. London, 1911.
- "On the structure of the atom." *PM, 26* (1913), 792–9.
- *The structure of light.* Cambridge, 1925.
- *Recollections and reflections.* London, 1936.
Thomson, J. J., and E. Rutherford. "On the passage of electricity through gases exposed to Röntgen rays." *PM, 42* (1896), 392–407.

Thomson, W. [Lord Kelvin]. *Baltimore lectures on molecular dynamics and the wave theory of light.* London, 1904. Notes by Hathaway distributed by Johns Hopkins University, 1884.

- "On the generation of longitudinal waves in ether." *PRS, 59* (1896), 270–3.
- "Note on Lord Blythswood's paper." *PRS, 59* (1896), 332–3.
- "Continuity in undulatory theory of condensational–rarefactional waves in gases, liquids, and solids, of distortional waves in solids, of electric waves in all substances capable of transmitting them, and of radiant heat, visible light, and ultra-violet light." *PM, 46* (1898), 494–500; BAAS *Report* (1898), 783–7.
- "On the reflection and refraction of solitary plane waves at a plane interface between two isotropic elastic mediums – fluid, solid, or ether." *PM, 47* (1899), 179–91.
- "On the application of force within a limited space, required to produce spherical solitary waves, or trains of periodic waves, of both species, equivoluminal and irrotational in an elastic solid." *PM, 47* (1899), 480–93; *48* (1899), 227–36, 388–93.
- "Contribution by Lord Kelvin to the discussion on the nature of the emanations from radium . . ." BAAS *Report* (1903), 535–7; *PM, 7* (1904), 220–2.
Topper, D. "Commitment to mechanism: J. J. Thomson, the early years." *AHES, 7* (1971) 393–410.
Trenn, T. "Rutherford and the alpha–beta–gamma classification of radioactive rays." *Isis, 67* (1976), 61–75.
Trillat, J. J. "Maurice de Broglie." Societé Française de minéralogie et de cristallographie *Bulletin, 83* (1960), 239–41.
La Varende. *Les Broglie.* Paris, 1950.
Villard, P. *Les rayons cathodiques.* Paris, 1900. Originally presented in *Rapports presentes au congrès international de physique,* C. Guillaume and L. Poincaré, eds., vol. 3, pp. 115–37. Paris, 1900.
- "Sur le rayonnement du radium." *CR, 130* (1900), 1178–9.
- "Sur la réflexion et la réfraction des rayons cathodiques et des rayons déviables du radium." *CR, 130* (1900), 1010–12.
- "La formation des rayons cathodiques." In *Ions, électrons, corpuscles,* H. Abraham and P. Langevin, eds., pp. 1013–28. Paris, 1905.
Volterra, V. *Leçons sur les équations intégrales et les équations intégro-différentielles.* Paris, 1913.
- "Sur la théorie mathematique des phénomènes héréditaires." *Journal de mathématiques pures et appliquées, 7* (1928), 249–98.
van der Waals, J. D., Jr. "Zur Frage der Wellenlänge der Röntgenstrahlen." *AP, 22* (1907), 603–5; *23* (1907), 395–6.

344   Bibliography

Wagner, E. "Spektraluntersuchungen an Röntgenstrahlen nach Versuchen gemeinschaft mit Joh. Brentano." München *Sb, 44* (1914), 329–38.
- "Spectraluntersuchungen an Röntgenstrahlen I." *AP, 46* (1915), 868–92
Walter, B. "Die Nature der Röntgenstrahlen." *AP, 66* (1898), 74–81.
- "Über die Haga und Wind'schen Beugungsversuche mit Röntgenstrahlen." *VDNA* (1901), part 2, p. 43; *PZ, 3* (1902), 137–40.
- "Über die Geschwindigkeit der Röntgenstrahlen." *Fortschritte auf dem Gebiete der Röntgenstrahlen, 9* (1906), 223–5.
Walter, B., and R. Pohl. "Zur Frage der Beugung der Röntgenstrahlen." *AP, 25* (1908), 715–24.
- "Weitere Versuche über die Beugung der Röntgenstrahlen." *AP, 29* (1909), 331–54.
Webster, D. L. "The x-ray spectrum of tungsten at a constant potential." *PR, 6* (1915), 56.
- "Experiments on the emission quanta of characteristic x rays." *PR, 7* (1916), 599–613.
Weidner, R., and R. Sells. *Elementary modern physics.* Boston, 1960.
Weill-Brunschvicg, A. R., and J. L. Heilbron. "Broglie, Louis-César-Victor-Maurice de." *DSB, 2* (1973), 487–9.
Wentzel, G. "Die unperiodischen Vorgänge in der Wellenmechanik." *PZ, 29* (1928), 321–37.
Wessels, L. "Schrödinger's route to wave mechanics." *SHPS, 10* (1979), 311–40.
Wheaton, B. R. *The photoelectric effect and the origin of the quantum theory of free radiation.* Unpublished M.A. thesis, University of California, Berkeley, 1971.
- *On the nature of x- and gamma-rays: Attitudes toward localization of energy in the "new radiations," 1896–1922.* Unpublished Ph.D. dissertation, Princeton University, 1978.
- "Philipp Lenard and the photoelectric effect, 1889–1911." *HSPS, 9* (1978), 299–323.
- "Impulse x-rays and radiant intensity: The double edge of analogy." *HSPS, 11* (1981), 367–90.
Wheaton, B. R., and J. L. Heilbron. *An inventory of published letters to and from physicists, 1900–1950.* Berkeley papers in history of science, vol. 6. Berkeley, 1982.
Whiddington, R. "Preliminary note on the properties of easily absorbed Röntgen radiation." *PCPS, 15* (1910), 574–5.
- "The production and properties of soft Röntgen radiation." *PRS, 85A* (1911), 99–118.

- "The production of characteristic Röntgen radiations." *PCPS, 16* (1911), 150-4; *PRS, 85A* (1911), 323-32. A first announcement appeared in *Nature, 88* (1911), 143.
- "The transmission of cathode rays through matter." *PRS, 86A* (1912), 360-70.
- "The velocity of the secondary cathode particles ejected by the characteristic Röntgen rays." *PRS, 86A* (1912), 370-8.
- "X-ray electrons." *PM, 43* (1922), 1116-26.

Wiechert, E. "Bedeutung des Weltäthers." Königsberg *Schriften, 35* (1894), 4-11.
- "Über die Grundlagen der Electrodynamik." *AP, 59* (1896), 283-323.
- "Die Theorie der Elektrodynamik und die Röntgen'sche Entdeckung." Königsberg *Schriften, 37* (1896), 1-48.
- "Experimentelle Untersuchungen über die Geschwindigkeit und die magnetische Ablenkbarkeit der Kathodenstrahlen." *AP, 69* (1899), 739-66.

Wien, W. "Temperatur und Entropie der Strahlung." *AP, 52* (1894), 132-65.
- "Über die Energievertheilung im Emissionsspektrum eines schwarzen Körpers." *AP, 58* (1896), 662-9.
- "Über die Möglichkeit einer elektromagnetischen Begründung der Mechanik." *Archives Néerlandaises* (2), *5* (1900), 96-107.
- "Untersuchungen über die elektrische Entladung in verdünnten Gasen." *AP, 5* (1901), 421-35; *AP, 8* (1902), 244-66.
- "Über die Natur der positiven Elektronen." *AP, 9* (1902), 660-4.
- "Über die Selbstelektrisierung des Radiums und die Intensität der von ihm ausgesandten Strahlen." *PZ, 4* (1903), 624-6.
- "Über die Energie der Röntgenstrahlen." *PZ, 5* (1904), 128-30.
- "Über die Theorie der Röntgenstrahlen." *JRE, 1* (1904), 215-20.
- "Über die Energie der Kathodenstrahlen im Verhältnis zur Energie der Röntgen- und Sekundärstrahlen." *Festschrift Adolph Wüllner gewidmet zum siebzigsten Geburtstag,"* pp. 1-14. Leipzig, 1905.
- "Über die Energie der Kathodenstrahlen im Verhältnis zur Energie der Röntgen- und Sekundärstrahlen." *AP, 18* (1905), 991-1007.
- "Über eine Berechnung der Wellenlänge der Röntgenstrahlen aus dem Planckschen Energie Element." Göttingen *Nachrichten* (1907), 598-601.
- "Über die Berechnung der Impulsbreite der Röntgenstrahlen aus ihrer Energie." *AP, 22* (1907), 793-7.
- "Über die absolute von positiven Ionen ausgestrahlte Energie und die Entropie der Spektrallinien." *AP, 23* (1907), 415-38.
- "Gesetze und Theorien der Strahlung." Würzburg *Sb* (1907), 103-12.

– *Vorlesungen über neuere Probleme der theoretischen Physik.* Leipzig, 1913; WAS *Journal, 3* (1913), 273–84.

– "Zur Theorie der Strahlung." *AP, 46* (1915), 749–52.

Wigger, O. "Zur Characteristik der α- und γ-Strahlen." *JRE, 2* (1905), 391–433.

Wilson, C. T. R. "On a method of making visible the paths of ionizing particles through a gas." *PRS, 85A* (1911), 285–8.

Wilson, W. "On the absorption of homogeneous β rays by matter, and on the variation of the absorption of the rays with velocity." *PRS, 82A* (1909), 612–28.

– "The decrease in velocity of the β-particles on passing through matter." *PRS, 84A* (1910), 141–50.

Wind, C. H. "Über die Deutung der Beugungserscheinungen bei Röntgenstrahlen." *AP, 68* (1899), 896–901.

– "Over helderheidsmaxima en -minima als gevolg van een gezichtsbedrog." Amsterdam *Verslag, 7* (1899), 12–19.

– "Zur Anwendung der Fourierschen Reihenentwickelung in der Optik." *PZ, 2* (1900), 189–96.

– "Zur Beugung der Röntgenstrahlen." *PZ, 2* (1900), 297–8.

Winkelmann, A., ed. *Handbuch der Physik,* 2nd. ed. Leipzig, 1905.

Wright, J. R. "The positive potential of aluminum as a function of the wavelength of the incident light." *PR, 33* (1911), 43–52.

Wüllner, A. *Lehrbuch der Experimentalphysik,* vol. 3, *Die Lehre vom Magnetismus und von der Elektricität,* 5th ed. Leipzig, 1897.

Zehnder, L. *W. C. Röntgen, Briefe an L. Zehnder.* Zurich, 1935.

Zwicky, F. "Das Verhalten von langsamen Elektronen in Edelgasen." *PZ, 24* (1923), 171–83.

INDEX

All names mentioned in the text and substantive footnotes, as well as selected subjects, are noted here. Important contributions are listed by topic; significant discussions are indicated by boldface numbers. References to authors cited in the bibliography are in italic type. Discussions that extend well beyond the page cited are marked here by "ff", but no attempt has been made to index subsequent page numbers for topics that extend simply to the next page.

350   *Index*

352   *Index*

Thomson, William (Lord Kelvin) (1824–1907), 18, 19, 21, 29, 57, *343*
"tiger and shark," 306
Topper, David, *343*
Townsend, John Sealy (1868–1957), 85
Trenn, Thaddeus, *343*
triggering hypothesis, **74ff**, 85, 136, 171, 179
Trillat, Jean (1899–), 274, *343*
Trowbridge, John (1843–1923), 250
tubes of electric force, 78, 140

ultraviolet, 9
unit hypothesis, 138ff, 188

Varend, La, *343*
Victoria Institute, 21
Villard, Paul Ulrich (1860–1934), **53ff**, 246, *343*
Volterra, Vito (1860–1940), 287, *343*

Waals, J. D. van der, Jr. (1873–1971), 207, *343*
Wagner, Ernst (1876–1928), **226**, 254, 265, *343*
Walter, Bernhard Ludwig (1861–1950), 48, 151, 201, *344*
Warburg, Emil (1846–1931), 194
wave packets, 293
wave theory of light, 9ff, 189ff
Weart, Spencer R., *321*
Webster, David Locke (1888–1976), 250, 251, *344*
Weidner, Richard, *344*
Weill-Brunschvicg, A. R., *344*
Wendt, Gerald (1891–1973), *340*
Wentzel, Gregor (1898–1978), *338, 344*
Wessels, Linda A., *344*
Wheaton, Bruce R., *323, 344*
Whiddington, Richard D. (1885–1970), **242ff**, 256, *344*
Whiddington's law, 256
Whiddington's rule, 243
white light, 40ff
white x-rays, 214
Wiechert, Emil (1861–1928), 8, **30**, 33, *338, 345*
Wien, Wilhelm (1864–1928), 2, 105, 160, 184, 194, 207, 234, *345;* on blackbody law, 106; on electromagnetic world view, 8, 110; and Lorentz, 110; on paradox of quality, 112; and Stark, 116; on triggering hypothesis, 75, 113; on x-rays: azimuthal variation, 111, frequency, 114, and light, 113, quantum theory, 114
Wigger, Otto, 66,*346*
Wilberforce, Lionel Robert, 45
Wilson, Charles Thomas Rees (1869–1959), 165, 190, *346*
Wilson, Harold Albert (1874–1964), 51
Wilson, William (1887–1948), 159, 245, *346*
Winchester, George, *331*
Wind, Cornelis H. (1867–1911), 38, 201, *322, 346;* diffraction of x-rays, 31ff
Winkelman, Adolph (1848–1910), *346*
Wolfke, Mieczysław (1883–1947), 193
Wood, Robert Williams (1868–1955), 282
Wooster, W. A., *336*
Wright, J. R., 239, *331, 346*
Wüllner, Friedrich Adolph (1835–1908), 43, 110, *346*

x-rays: absorption coefficient, 17, 79; absorption edges, 226, 254; *Bremsstrahlung,* 127, 202, 216, 219; confused with γ-rays, 220ff; discovered, 11; energy, 72, 110; impulse hypothesis, 15ff, 20ff; intensity 79, 294; interference, 200ff; interpretations, 16ff; photoeffect, 266ff, 276; quality, 78; radiative nature, 17; reflection, 44, 73, 210; scattered, 95, 283ff; tertiary, 45; velocity, 46ff; white, 214

Zeeman effect, 30, 297
Zehnder, Ludwig A. (1854–1949), 17, *346*
Zwicky, Fritz (1898–1974), *346*

Printed in the United States
By Bookmasters